ELECTRICAL MACHINE ANALYSIS USING FINITE ELEMENTS

POWER ELECTRONICS AND APPLICATIONS SERIES

Muhammad H. Rashid, Series Editor
University of West Florida

PUBLISHED TITLES

Complex Behavior of Switching Power Converters
Chi Kong Tse

DSP-Based Electromechanical Motion Control
Hamid A.Toliyat and Steven Campbell

Advanced DC/DC Converters
Fang Lin Luo and Hong Ye

Renewable Energy Systems: Design and Analysis with Induction Generators
M. Godoy Simões and Felix A. Farret

Uninterruptible Power Supplies and Active Filters
Ali Emadi, Abdolhosein Nasiri, and Stoyan B. Bekiarov

Modern Electric, Hybrid Electric, and Fuel Cell Vehicles: Fundamentals, Theory, and Design
Mehrdad Eshani, Yimin Gao, Sebastien E. Gay, and Ali Emadi

Electric Energy: An Introduction
Mohamed El-Sharkawi

Electrical Machine Analysis Using Finite Elements
Nicola Bianchi

ELECTRICAL MACHINE ANALYSIS USING FINITE ELEMENTS

NICOLA BIANCHI

Boca Raton London New York Singapore

A CRC title, part of the Taylor & Francis imprint, a member of the Taylor & Francis Group, the academic division of T&F Informa plc.

Published in 2005 by
CRC Press
Taylor & Francis Group
6000 Broken Sound Parkway NW, Suite 300
Boca Raton, FL 33487-2742

International Standard Book Number-10: 0-8493-3399-7 (Hardcover)
International Standard Book Number-13: 978-0-8493-3399-6 (Hardcover)
Library of Congress Card Number 2005041853

Library of Congress Cataloging-in-Publication Data

Bianchi, Nicola.
Electrical machine analysis using finite elements / Nicola Bianchi.
p. cm. -- (Power electronics and applications series ; 7)
Includes bibliographical references and index.
ISBN 0-8493-3399-7 (alk. paper)
1. Electric machinery--Mathematical models. 2. Finite element method. I. Title. II. Series.

TK2189.B48 2005
621.31′042--dc22 2005041853

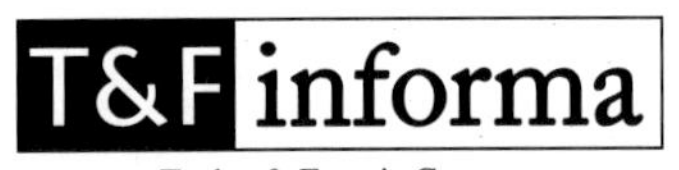

Taylor & Francis Group
is the Academic Division of T&F Informa plc.

Visit the Taylor & Francis Web site at
http://www.taylorandfrancis.com

and the CRC Press Web site at
http://www.crcpress.com

Preface

For a long time the course *Design Methodology of Electrical Machines* has been a teaching course for electrical engineering aimed at electrical applications. This is due both to the subject that is dealt with and to the organization of the course by Prof. Luciano Merigliano, based on many years' experience with electromechanical design.

The laboratory activities of the course started in 1996, directed by myself along with Prof. Silverio Bolognani, who was the teacher in charge in those years, and to whom the credit for course innovation has to be given.

This book reports some of the laboratory exercitations of the course *Design Methodology of Electrical Machines*. The aim is to examine the key concepts of the finite element method, as well as to focus attention on the applications of such a method to electrical machine analysis.

Not long after the course began, the students expressed the need to have a text illustrating the analysis of the electric and magnetic devices by means of the finite element method. In addition, the electromechanical industries, which require accurate analyses, also had need for such a text. This was a result of the popularity of software programs for the finite element analysis of electromagnetic problems, some of which are even available on the Web. Among them, some well-known packages are ANSYS, CADEMA, FEMM, FLUX, MAXWELL, MEGA, QFIELD, and VECTOR-FIELD.

A first draft of this text was published in Italian in 2001, for the electrical engineering students at the University of Padova. It was also adopted at other Italian universities and by some industries. The present text is essentially a translation of the Italian version of the same text, with a few revisions.

The book is organized to meet the goal described above. Several examples of finite element analysis of the electrical machines are pointed out. Some repetitions in the various chapters are intentional, for the sake of a more fluent treatment. Some theoretical concepts, numerical techniques, and simulation methods are reported gradually in the various chapters, in an effort to make for an easier read and to avoid many boring introductory chapters.

To the aim of facilitating understanding, some electrotechnical principles and some electrical machines concepts are sometimes reviewed. Moreover, simple numerical algorithms, written using MATLAB or BASIC software, are also reported. They are examples of processing the field analysis results and of the automatic computation procedures.

Let me thank my two generous teachers, Prof. L. Merigliano and Prof. S. Bolognani, for the unbroken respect and the constant teaching, and not only in the technical field. Let me also thank my friends and colleagues Prof. F. Dughiero and Dr. A. Tortella, whose help was essential for the fulfillment of this text. Last, but not least, I would like to express my thanks to my students, whose lively curiosity and diligent work contributed significantly to the improvement of the text.

Nicola Bianchi

The Author

Nicola Bianchi, Ph.D., received the Laurea degree in electrical engineering from the University of Padova in 1991, obtaining full marks and praise. His Laurea thesis dealt with the electromechanical analysis and design of a synchronous reluctance motor. The thesis was rewarded by the AEI (Italian Electrical Association) with the "Stefano e Flora Bodoni" award for new engineers in 1991. He received his Ph.D. degree in 1995, presenting a thesis titled "PM Motors for AC Drives: Electromagnetic Analysis and Innovative Design Procedures" (in Italian).

Since 1998, he has worked at the Electric Drives Laboratory at the Department of Electric Engineering of the University of Padova, as an assistant professor. His research activity is in the field of analysis of the performance, the determination of the parameters, and the development of design procedures for electrical motors, considering conventional and innovative machines, design by using optimization techniques, together with finite element analysis.

Since 1998, he has been an IEEE Member and Electric Machines Committee member. He is a technical reviewer of technical transactions, journals, and conferences. In addition, he has served as chairman at various conferences in the same field.

Since 2002 he has been recognized as an associate professor in the scientific area "Electrical Machines, Drives and Converters."

He contributes to several research projects between the Electric Drives Laboratory and local companies, concerning analysis and design of electrical motors and drives.

Dr. Bianchi is author and coauthor of several papers on the subject of electrical machines and drives. The paper "Design Criteria of Embedded PM Motors for Squarewave Current Control" (with S. Bolognani), presented at the International Conference *OPTIM 2000* (sponsored by IEEE IAS), Brasov Poiana, Romania, was given the First Award in the section "Electrical Machines and Drives."

Dedication

To Giacomo, Chiara, the little angel, and their wonderful mother

"...a man traveling to a far country called his own servants and delivered his goods to them. And to one he gave five talents, to another two, and to another one, to each according to his own ability..."

Mt. 25, 14–15

Contents

List of Symbols

A	Tm	magnetic vector potential
A_m	m^2	surface of the m-th finite element
a	—	transformation ratio
B	T	magnetic flux density
B_{dM} B_{qM}	T	d-axis and q-axis fundamental harmonic flux density
B_g	T	air-gap flux density
B_r	T	radial component of the flux density
B_{res}	T	residual flux density of the permanent magnet
B_ϑ	T	azimuthal component of the flux density
C	F	capacitance
D	C/m^2	electric displacement
D	m	diameter, air-gap diameter
D_e	m	outer diameter
$\mathcal{D}$		integration domain (surface or volume)
d	m	diameter
d_{Cu}	m	wire diameter
E	V/m	specific electric force
E, e	V	EMF
E_1 E_2	V	EMF induced in primary or secondary winding
E_{rms}	V	RMS voltage
E_c	V/m	Coulombian electric field
E_i	V/m	induced electric field
E_m	V/m	motional specific electric force
F	N	force
F_t F_n	N	tangential and normal force (to a surface or to a line)
$\mathcal{F}$		functional
f		forcing functional
f	Hz	frequency
f_r	Hz	rotor frequency
f_a f_b f_c	A	magnetic voltage of the a, b, c phases
f_d f_q	A	d-axis and q-axis magnetic voltages
g	m	air-gap thickness
H	A/m	magnetic field strength
H_c	A/m	coercitive force of the permanent magnet
H_{knee}	A/m	magnetic field strength at the knee of the BH curve

H_n H_t	A/m	normal and tangential component of the magnetic field strength
h_w	m	winding height
h_m	m	permanent magnet height
h_t	m	tooth (or slot) height
I, i	A	electric current intensity
i_1 i_2	A	primary and secondary currents
I_{1o}	A	primary current at no-load
I_c	A	current flowing in each coil
I_e	A	excitation current
I_{rms}	A	RMS current
I_L	A	line current
I_M	A	maximum value of the current
I_m	A	magnetizing current
I_o	A	phase current at no-load
I_q	A	equivalent current within the q-th slot
i_a i_b i_c	A	current of the a, b, and c phase
i_d i_q	A	d-axis and q-axis current
J	kgm^2	inertia
J	A/m^2	electric current density
J_{sd}	A/m	direct axis linear current density
J_{sq}	A/m	quadrature axis linear current density
k_B	Nms	coefficient of viscous friction
k_f	—	lamination packing factor
k_{jq}	—	fill factor of the q-th slot by the j-th phase
k_w	—	winding factor
$\mathcal{L}$	—	generic differential operator (e.g., laplacian operator)
L	m	axial length of a structure with planar symmetry
L	H	magnetic inductance
L_1 L_2	H	primary and secondary self-inductance
L_{1m} L_{2m}	H	primary and secondary magnetizing inductances
$L_{1\sigma}$ $L_{2\sigma}$	H	primary and secondary leakage inductances
L_a M_{ba}	H	self- and mutual inductances of each phase
L_d L_q	H	d-axis and q-axis synchronous inductances
L_{app}	H	apparent inductance
L_{dif}	H	differential inductance
L_{Fe}	m	actual iron length of machine with planar symmetry
L_M	H	magnetizing inductances of an induction motor
L_σ	H	leakage inductance
l	m	line
l_g	m	line passing in the middle of the air-gap

l_{cv}	m	ventilation channel length
l_m	m	permanent magnet thickness
l_{ave}	m	average length of a turn
M	A/m	magnetization of a permanent magnet
M	H	mutual inductance
M	—	number of finite elements in the domain
N	—	number of nodes of the finite elements
N	A/m	electric potential vector
N	—	number of series conductors per phase
N_1 N_2	—	number of turns of primary and secondary winding
N_c	—	number of sampled points
N_{cv}	—	number of channels for cooling
N_e	—	number of turns of the excitation winding
N_{sp}	—	number of turns
n	—	normal unity vector (to a line or to a surface)
n_b	rpm	base speed
n_{fw}	rpm	flux-weakening speed (in the constant power region)
n_{pm}	—	number of parallel paths
n_q	—	number of conductors within a slot
P	—	dependence on the point P, e.g., on the coordinate (x,y,z)
p	—	number of pole pairs
p_{Fe}	W	iron losses
p_J	W	Joule power losses
p_n	N/m^2	magnetic pressure
p_s	W/kg	specific iron losses
Q	—	number of total slots
Q_r Q_s	—	number of rotor and stator slots
q	C	electric charge
q_s	—	number of stator slots per pole per phase
r, ϑ, z	—	cylindrical coordinates
R, r	—	residue
R	Ω	electric resistance
R_r	Ω	rotor resistance
R_g	H^{-1}	magnetic reluctance corresponding to the air-gap
S	m^2	surface
S_{Cu+} S_{Cu-}	m^2	equivalent surface of the conductors
S_{Fe}	m^2	actual iron surface
S_n	VA	nominal VA power
S_q	m^2	surface of the slot
s	—	slip
[SS]	—	stiffness matrix
[T]	—	known terms vector matrix

T	Nm	electromagnetic torque
T_b	Nm	base torque
T_{cog}	Nm	cogging torque
T_{fw}	Nm	flux-weakening torque (in the constant power region)
$T_{abc/dq}$	—	transformation matrix abc/dq
$\mathbf{t}$	—	tangent unity vector to a line or to a surface
t	s	time
t^*	s	reference instant time
U_0	V	line voltage at no-load
$U_{\nu cos}$ $U_{\nu sin}$	V	voltage harmonic component of order ν
$\mathbf{u}$	—	coordinate unity vector
V, v	V	electric voltage
V	V	electric potential
V_{dc}	V	direct-current source voltage
v_1 v_2	V	primary and secondary voltage
v_d v_q	V	d-axis and q-axis electric voltages
v_m	m/s	speed of the medium on which the charge is bound
v_N	V	star-center potential
v_q	m/s	speed of the electric charge
v_ρ	m/s	speed of the electric charge density
x, y, z	—	Cartesian coordinates
y_q	—	slot pitch (in number of slots)
w_i	—	i-th weight function
w_m	J/m³	magnetic energy density
w_t	m	tooth width
W_m	J	magnetic energy
W'_m	J	magnetic coenergy
W_{el}	J	electric work
W_{mech}	J	mechanic work

Greek symbols

α_c	rad	electric slot angle
β_r	rad	electric shortened angle
β_r β_s	rad	rotor and stator width angle
δ	rad	load angle
Δi	A	current variation
$\Delta\vartheta$	red	mechanical angle variation
ε	F/m	electric permittivity
ϕ, φ, ψ	—	unknown potential to be determined
ϕ^*	—	function that approaches the unknown potential
Φ	Wb	magnetic flux
Φ_o	Wb	magnetic flux at no-load

φ	rad	power angle
φ	rad	initial voltage phase
Φ_j	—	j-th coefficient to approach the potential ϕ
$[\phi]$	—	matrix vector of the unknown coefficients Φi
Γ	—	boundary of the domain (line or surface)
γ_{Fe}	kg/m^3	specific iron weight
Λ, λ	Wb, Vs	flux linkage
Λ_d Λ_q	Vs	d-axis and q-axis flux linkages
Λ_{rms}	Vs	RMS flux linkage
Λ_j	Vs	flux linkage of the j-th phase (j = a, b, c)
λ_1 λ_2	Vs	primary and secondary flux linkages
$\lambda_{1\sigma}$ $\lambda_{2\sigma}$	Vs	primary and secondary leakage flux linkages
λ_{pm}	Vs	flux linkage due to the permanent magnet
λ_r λ_s	rad	rotor and stator pole pitch angle
μ	H/m	magnetic permeability
μ_r	—	relative magnetic permeability
h	—	efficiency
ν_j	—	j-th interpolanting function
ρ	C/m^3	volume density of electric charge
ρ_s	C/m^2	surface density of electric charge
ρ	Ωm	electric resistivity, $\rho = \sigma^{-1}$
ϑ	rad	angular coordinate (electric angle)
ϑ_m	rad	angular coordinate (mechanic angle)
ϑ_{off}	rad	switch-off angle
ϑ_{on}	rad	switch-on angle
ϑ_r	rad	angular coordinate referred to the rotor polar axis
ϑ_s	rad	angular coordinate referred to the stator a-phase axis
ξ	—	saliency ratio, $\xi = L_q/L_d$
ω	rad/s	electric frequency
ω_m	rad/s	mechanic angular speed
ω_r	rad/s	rotor electric angular speed $\omega_r = 2\pi f_r$
σ	S/m	electric conductivity
τ	m^3	volume
τ_{Cu}	m^3	conductive material volume
τ_m	Nm	instantaneous electromagnetic torque
τ_L	Nm	load torque
Ψ	A	magnetic voltage

Further Symbols

Im	imaginary part
Re	real part
~	conjugate complex

1

Outline of Electromagnetic Fields

In this chapter, some basic concepts of vector analysis and electromagnetic fields are outlined. These concepts will be used in the remainder of the book. The aim is to furnish a concise treatment and a practical formulary, but keep in mind that it is not an exhaustive one. Any reader interested in mastering this subject should look it up in the books listed in the References.

1.1 VECTOR ANALYSIS

In a three-dimensional space, let $V = V(P)$ be a scalar quantity that defines a scalar field, and $\mathbf{A} = \mathbf{A}(P)$ be a vector quantity that defines a vector field. The dependence on the point P is expressed by using a reference system with curvilinear orthogonal coordinates: they are Cartesian (x,y,z), cylindrical (r,ϑ,z), or spherical (r,ϑ,φ) coordinates.

In general, each of these quantities can be a function of the time, that is, $V = V(P,t)$ and $\mathbf{A} = \mathbf{A}(P,t)$. This dependence is considered in Section 1.1.8. However, in order to simplify the text, the explicit notation pointing out the dependence of the various quantities on the point P and the time t is omitted.

1.1.1 Operations Among Vectors

Let **A** and **B** be two vectors, with $|\mathbf{A}|$ and $|\mathbf{B}|$ their magnitude respectively and γ the angle between their oriented directions. The *scalar product* is the real number given by

$$\mathbf{A} \cdot \mathbf{B} = |\mathbf{A}|\, |\mathbf{B}| \cos\gamma \tag{1.1}$$

The *vector product* is the vector given by

$$\mathbf{A} \times \mathbf{B} = (|\mathbf{A}|\, |\mathbf{B}| \sin\gamma)\, \mathbf{n}_{AB} \tag{1.2}$$

where $\mathbf{n}_{AB}$ is the unity vector normal to the plane defined by **A** and **B**. The vector (1.2) results normal to both vector **A** and vector **B**, and its direction

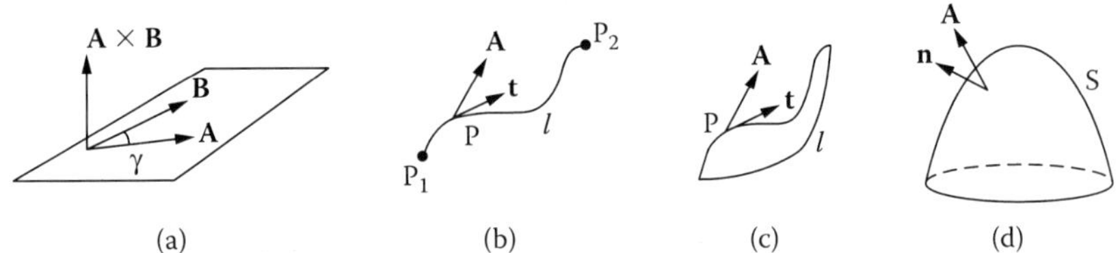

FIGURE 1.1
Operations among the vectors.

corresponds to the advancement of a right screw, rotating so as to place the vector **A** on the vector **B**, as shown in Figure 1.1(a).

Using the Cartesian coordinates, the two vectors are $\mathbf{A} = (A_x, A_y, A_z) = A_x\mathbf{u}_x + A_y\mathbf{u}_y + A_z\mathbf{u}_z$ and $\mathbf{B} = (B_x, B_y, B_z) = B_x\mathbf{u}_x + B_y\mathbf{u}_y + B_z\mathbf{u}_z$, where $\mathbf{u}_x$, $\mathbf{u}_y$, and $\mathbf{u}_z$ are the coordinate unity vectors. The scalar and the vector products become

$$\mathbf{A}\cdot\mathbf{B} = A_xB_x + A_yB_y + A_zB_z \tag{1.3}$$

$$\begin{aligned}\mathbf{A}\times\mathbf{B} &= \begin{vmatrix} \mathbf{u}_x & \mathbf{u}_y & \mathbf{u}_z \\ A_x & A_y & A_z \\ B_x & B_y & B_z \end{vmatrix} \\ &= \left(A_yB_z - A_zB_y\right)\mathbf{u}_x + \left(A_zB_x - A_xB_z\right)\mathbf{u}_y + \left(A_xB_y - A_yB_x\right)\mathbf{u}_z\end{aligned} \tag{1.4}$$

The following properties exist:

$$\begin{aligned} &\mathbf{A}\cdot\mathbf{A} = \left|\mathbf{A}\right|^2 && \mathbf{A}\cdot(\mathbf{B}+\mathbf{C}) = \mathbf{A}\cdot\mathbf{B} + \mathbf{A}\cdot\mathbf{C} \\ &\mathbf{A}\cdot\mathbf{B} = \mathbf{B}\cdot\mathbf{A} && \mathbf{A}\times(\mathbf{B}+\mathbf{C}) = \mathbf{A}\times\mathbf{B} + \mathbf{A}\times\mathbf{C} \\ &\mathbf{A}\times\mathbf{B} = -\mathbf{B}\times\mathbf{A} && \mathbf{A}\times(\mathbf{B}\times\mathbf{C}) = (\mathbf{A}\cdot\mathbf{C})\mathbf{B} - (\mathbf{A}\cdot\mathbf{B})\mathbf{C} \end{aligned} \tag{1.5}$$

1.1.2 Linear Integral and Flux of a Vector Field

The *linear integral* of a vector field **A** along a curve l, from point P_1 to point P_2 and oriented in each point by the unity vector **t**, tangent to the line as shown in Figure 1.1(b), is

$$\Gamma_{12} = \int_{P_1}^{P_2} \mathbf{A}\cdot\mathbf{t}\, \mathrm{d}l \tag{1.6}$$

The *loop integral* of the vector field corresponds to the linear integral along a closed curve, shown in Figure 1.1(c), which is

$$\Gamma = \oint_l \mathbf{A} \cdot \mathbf{t}\, dl \tag{1.7}$$

The *flux* of the vector field **A** through the surface S, oriented by the normal unity vector **n** in each point, as shown in Figure 1.1(d), is given by

$$\Phi = \int_S \mathbf{A} \cdot \mathbf{n}\; dS \tag{1.8}$$

In the case in which the surface is closed, and **n** is assumed conventionally to have external direction, Φ defines the flux going out through the surface S itself.

1.1.3 Differential Operators

The *gradient* of a scalar field V is a vector function defined in each point P by a direction corresponding to the maximum directional derivative of V, defined by the unity vector $\mathbf{u}_{max}$, and by a magnitude equal to this derivative, given by

$$\mathrm{grad}V = \max\left[\frac{\partial V}{\partial l}\right] \mathbf{u}_{max} \tag{1.9}$$

Along any direction defined by the unity vector **u**, the scalar product of gradV by **u** gives the derivative of V along the direction of **u**, which is

$$\mathrm{grad}V \cdot \mathbf{u} = \left.\frac{\partial V}{\partial l}\right|_{\mathbf{u}} \tag{1.10}$$

Finally, if the field V is not continuous through a surface Σ, characterized by the normal unity vector $\mathbf{n}_\Sigma$, and it assumes the values V_1 and V_2 on the two sides of Σ, then a *surface gradient* can be defined as

$$\mathrm{grad}_\Sigma V = \left(V_2 - V_1\right)\, \mathbf{n}_\Sigma \tag{1.11}$$

The *divergence* of a vector field **A** is a scalar function defined in each point P by the ratio between the flux of **A** going out through a closed surface S_c and the volume τ contained by S_c, when the volume tends to the point P itself, which is

$$\mathrm{div}\mathbf{A} = \lim_{\tau \to 0} \frac{\oint_{S_c} \mathbf{A} \cdot \mathbf{n}\; dS}{\tau} \tag{1.12}$$

The infinitesimal flux $d\Phi$ of the vector **A** going out through a closed surface, that contains the infinitesimal volume $d\tau$, is given by $d\Phi = \text{div}\mathbf{A}d\tau$. Then the divergence $\text{div}\mathbf{A} > 0$ means a point where the lines of the field **A** arise, while $\text{div}\mathbf{A} < 0$ means a point where the lines of the field **A** terminate.

Finally, if the field **A** is not continuous through a surface Σ, with normal unity vector $\mathbf{n}_\Sigma$, and assumes the values $\mathbf{A}_1$ and $\mathbf{A}_2$ on the two sides of Σ, then a *surface divergence* can be defined as

$$\text{div}_\Sigma \mathbf{A} = \left(\mathbf{A}_2 - \mathbf{A}_1\right) \cdot \mathbf{n}_\Sigma \tag{1.13}$$

The *curl* of a vector field **A** is a vector function defined in each point P by the ratio between the integral of **A** along a closed loop l and the surface S bordered by l, when S tends to the point P itself, which is

$$\text{curl}\mathbf{A} = \lim_{S\to 0} \frac{\oint_l \mathbf{A} \cdot \mathbf{t}\, dl}{S} \tag{1.14}$$

The infinitesimal loop integral $d\Gamma$ of the vector **A** along a closed loop l that borders the infinitesimal surface dS, oriented by **n**, is given by $\text{curl}\mathbf{A}\cdot\mathbf{n}dS$. Then the curl represents the vortexes around which the lines of the vector **A** go round.

If the field **A** is not continuous through a surface Σ, with normal unity vector $\mathbf{n}_\Sigma$, and assumes the values $\mathbf{A}_1$ and $\mathbf{A}_2$ on the two sides of Σ, then a *surface curl* is defined as

$$\text{curl}_\Sigma \mathbf{A} = -\left(\mathbf{A}_2 - \mathbf{A}_1\right) \times \mathbf{n}_\Sigma \tag{1.15}$$

The *laplacian of a scalar field* V is given by the divergence of the vector gradient of the scalar function V, as

$$\nabla^2 V = \text{div grad} V \tag{1.16}$$

The laplacian of a vector field **A** is given by

$$\nabla^2 \mathbf{A} = \text{grad div}\mathbf{A} - \text{curl curl}\mathbf{A} \tag{1.17}$$

Referring to the differential operators defined above, some properties are reported in the following. U and V are scalar fields, **A** and **B** are vector fields, k and h are generic constants.

1.1.4 Integral Identities

The *gradient theorem* states that the linear integral from point P_1 and point P_2 of a vector gradient gradV depends only on the value of V in the points P_1 and P_2:

$$\Gamma_{12} = \int_{P_1}^{P_2} \text{gradV} \cdot \mathbf{t}\; dl = V_2 - V_1 \tag{1.18}$$

that, in the case of loop integral, is equal to zero. In addition, it is

$$\int_\tau \text{gradV}\; d\tau = \oint_S V\mathbf{n}\; dS \tag{1.19}$$

The *divergence theorem* (or Gauss's theorem) states that the volume integral of a divergence div**A** is equal to the integral of the vector **A** through the closed surface that contains the volume, which is the flux of **A** going out through the closed surface:

$$\Phi = \oint_{S_c} \mathbf{A} \cdot \mathbf{n}\; dS = \int_\tau (\text{div}\mathbf{A})\; d\tau \tag{1.20}$$

The *curl theorem* (or the Stokes theorem) states that the integral through the surface S of a vector curl**A** is equal to the linear integral of the vector **A** along a closed line l bordering the surface S itself. The unity vector **t** tangential to the line l has to be chosen in accordance with the unity vector **n** normal to the surface S (with the rule of the right screw). Finally, it is

$$\oint_l \mathbf{A} \cdot \mathbf{t}\; dl = \int_S \text{curl}\mathbf{A} \cdot \mathbf{n}\; dS \tag{1.21}$$

First Green's scalar theorem states that

$$\int_\tau U\; \text{div}(h\; \text{gradV}) + h\; \text{gradU} \cdot \text{gradV}\; d\tau = \oint_{S_c} hU \frac{\partial V}{\partial n}\; dS \tag{1.22}$$

Second Green's scalar theorem states that

$$\int_\tau U\; \text{div}(a\; \text{gradV}) - V\text{div}(a\; \text{gradU})\; d\tau = \oint_{S_c} A\left(U\frac{\partial V}{\delta n} - V\frac{\partial U}{\delta n}\right) dS \tag{1.23}$$

First Green's vector theorem states that

$$\int_{\tau} \text{k curl}\mathbf{A}\cdot\text{curl}\mathbf{B} - \mathbf{A}\cdot\left[\text{curl(k curl}\mathbf{B})\right] d\tau = \oint_{S_c} \text{k}(\mathbf{A}\times\text{curl}\mathbf{B})\cdot\mathbf{n}\ dS \quad (1.24)$$

Second Green's vector theorem states that

$$\begin{aligned}&\int_{\tau} \mathbf{B}\cdot\left[\text{curl(k curl}\mathbf{A})\right] - \mathbf{A}\cdot\left[\text{curl(k curl}\mathbf{B})\right] d\tau \\ &= \oint_{S_c} \text{k}\left(\mathbf{A}\times\text{curl}\mathbf{B} - \mathbf{B}\times\text{curl}\mathbf{A}\right)\cdot\mathbf{n}\ dS\end{aligned} \quad (1.25)$$

1.1.5 Differential Identities

Gradient:

$$\text{grad}\left(\text{VU}\right) = \text{V(gradU)} + \text{U(gradV)} \quad (1.26)$$

$$\text{grad}\left(\text{kU} + \text{hV}\right) = \text{k gradU} + \text{h gradV} \quad (1.27)$$

$$\text{gradV}\left(\text{U}\right) = \frac{\partial V}{\partial U}\text{gradU} \quad (1.28)$$

Divergence:

$$\text{div}\left(\text{U}\mathbf{A}\right) = \text{U(div}\mathbf{A}) + \text{gradU}\cdot\mathbf{A} \quad (1.29)$$

$$\text{div}\left(\text{k}\mathbf{A} + \text{h}\mathbf{B}\right) = \text{k div}\left(\mathbf{A}\right) + \text{h div}\left(\mathbf{B}\right) \quad (1.30)$$

$$\begin{aligned}&\text{div}\left(\mathbf{A}\times\mathbf{B}\right) = -\mathbf{A}\cdot(\text{curl}\mathbf{B}) + (\text{curl}\mathbf{A})\cdot\mathbf{B} \\ &\text{div}\left(\text{curl}\mathbf{A}\right) = 0\end{aligned} \quad (1.31)$$

Curl:

$$\text{curl}\left(\text{U}\mathbf{A}\right) = \text{U}\cdot(\text{curl}\mathbf{A}) + (\text{gradU})\times\mathbf{A} \quad (1.32)$$

$$\text{curl}\left(\text{k}\mathbf{A} + \text{h}\mathbf{B}\right) = \text{k curl}\left(\mathbf{A}\right) + \text{h curl}\left(\mathbf{B}\right) \quad (1.33)$$

$$\text{curl gradU} = 0 \quad (1.34)$$

Laplacian:

$$\nabla^2(UV) = U \cdot \nabla^2 V + V\nabla^2 U + 2\text{grad}U \cdot \text{grad}V \tag{1.35}$$

$$\nabla^2(kU + hV) = k\ \nabla^2 U + h\ \nabla^2 V \tag{1.36}$$

1.1.6 Expression of the Differential Operators within Different Coordinate Systems

1.1.6.1 Cartesian Coordinates

Using Cartesian coordinates, the point P is given by P(x, y, z) and the coordinate unity vectors are $\mathbf{u}_x$, $\mathbf{u}_y$ and $\mathbf{u}_z$ as shown in Figure 1.2(a). The differential operators are as follows:

$$\text{grad}V = \frac{\partial V}{\partial x}\mathbf{u}_x + \frac{\partial V}{\partial y}\mathbf{u}_y + \frac{\partial V}{\partial z}\mathbf{u}_z \tag{1.37}$$

$$\text{div}\mathbf{A} = \frac{\partial A_x}{\partial x} + \frac{\partial A_y}{\partial y} + \frac{\partial A_z}{\partial z} \tag{1.38}$$

$$\text{curl}\mathbf{A} = \begin{vmatrix} \mathbf{u}_x & \mathbf{u}_y & \mathbf{u}_z \\ \dfrac{\partial}{\partial x} & \dfrac{\partial}{\partial y} & \dfrac{\partial}{\partial z} \\ A_x & A_y & A_z \end{vmatrix} = \left(\frac{\partial A_z}{\partial y} - \frac{\partial A_y}{\partial z}\right)\mathbf{u}_x + \left(\frac{\partial A_x}{\partial z} - \frac{\partial A_z}{\partial x}\right)\mathbf{u}_y + \left(\frac{\partial A_y}{\partial x} - \frac{\partial A_x}{\partial y}\right)\mathbf{u}_z \tag{1.39}$$

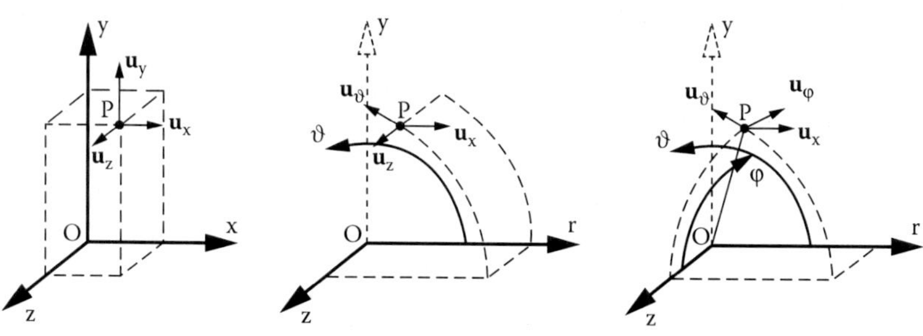

FIGURE 1.2
Systems of orthogonal coordinates: Cartesian (a), cylindrical (b), and spherical (c) coordinates.

$$\nabla^2 U = \text{div gradU} = \frac{\partial^2 U}{\partial x^2} + \frac{\partial^2 U}{\partial y^2} + \frac{\partial^2 U}{\partial z^2} \tag{1.40}$$

$$\begin{aligned}\nabla^2 \mathbf{A} &= \text{grad}\left(\text{div}\mathbf{A}\right) - \text{curl}\left(\text{curl}\mathbf{A}\right) \\ &= \nabla^2 A_x \mathbf{u}_x + \nabla^2 A_y \mathbf{u}_y + \nabla^2 A_z \mathbf{u}_z \\ &= \left(\frac{\partial^2 A_x}{\partial x^2} + \frac{\partial^2 A_x}{\partial y^2} + \frac{\partial^2 A_x}{\partial z^2}\right)\mathbf{u}_x + \left(\frac{\partial^2 A_y}{\partial x^2} + \frac{\partial^2 A_y}{\partial y^2} + \frac{\partial^2 A_y}{\partial z^2}\right)\mathbf{u}_y + \\ &\quad \left(\frac{\partial^2 A_z}{\partial x^2} + \frac{\partial^2 A_z}{\partial y^2} + \frac{\partial^2 A_z}{\partial z^2}\right)\mathbf{u}_z\end{aligned} \tag{1.41}$$

1.1.6.2 Cylindrical Coordinates

Using cylindrical coordinates, the point P is given by P(r, ϑ, z) and the coordinate unity vectors are $\mathbf{u}_r$, $\mathbf{u}_\vartheta$ and $\mathbf{u}_z$, as shown in Figure 1.2(b). The differential operators are

$$\text{gradV} = \frac{\partial V}{\partial r}\mathbf{u}_r + \frac{1}{r}\frac{\partial V}{\partial \theta}\mathbf{u}_\theta + \frac{\partial V}{\partial z}\mathbf{u}_z \tag{1.42}$$

$$\text{div}\mathbf{A} = \frac{1}{r}\frac{\partial\left(rA_r\right)}{\partial r} + \frac{1}{r}\frac{\partial A_\theta}{\partial \theta} + \frac{\partial A_z}{\partial z} \tag{1.43}$$

$$\text{curl}\mathbf{A} = \left(\frac{1}{r}\frac{\partial A_z}{\partial \theta} - \frac{\partial A_\theta}{\partial z}\right)\mathbf{u}_r + \left(\frac{\partial A_r}{\partial z} - \frac{\partial A_z}{\partial r}\right)\mathbf{u}_\theta + \frac{1}{r}\left(\frac{\partial\left(rA_\theta\right)}{\partial r} - \frac{\partial A_r}{\partial \theta}\right)\mathbf{u}_z \tag{1.44}$$

$$\begin{aligned}\nabla^2 V &= \frac{1}{r}\frac{\partial}{\partial r}\left(r\frac{\partial V}{\partial r}\right) + \frac{1}{r^2}\frac{\partial^2 V}{\partial \theta^2} + \frac{\partial^2 V}{\partial z^2} \\ &= \frac{1}{r}\frac{\partial V}{\partial r} + \frac{\partial^2 V}{\partial r^2} + \frac{1}{r^2}\frac{\partial^2 V}{\partial \theta^2} + \frac{\partial^2 V}{\partial z^2}\end{aligned} \tag{1.45}$$

1.1.6.3 Spherical Coordinates

Using spherical coordinates, the point P is given by P(r, ϑ, φ) and the coordinate unity vectors are $\mathbf{u}_r$, $\mathbf{u}_\vartheta$ and $\mathbf{u}_\varphi$, as shown in Figure 1.2(c). The differential operators are

$$\text{gradV} = \frac{\partial V}{\partial r}\mathbf{u}_r + \frac{1}{r}\frac{\partial V}{\partial \theta}\mathbf{u}_\theta + \frac{1}{r\,\sin\theta}\frac{\partial V}{\partial \varphi}\mathbf{u}_\varphi \tag{1.46}$$

$$\text{div}\mathbf{A} = \frac{1}{r^2}\frac{\partial\left(r^2 A_r\right)}{\partial r} + \frac{1}{r\ \sin\theta}\frac{\partial\left(\sin\theta\ A_\theta\right)}{\partial\theta} + \frac{1}{r\ \sin\theta}\frac{\partial A_\varphi}{\partial\varphi} \tag{1.47}$$

$$\begin{aligned}\text{curl}\mathbf{A} = &\frac{1}{r\ \sin\theta}\left(\frac{\partial\left(\sin\theta\ A_\varphi\right)}{\partial\theta} - \frac{\partial A_\theta}{\partial\varphi}\right)\mathbf{u}_r + \\ &\frac{1}{r\ \sin\theta}\left(\frac{\partial A_r}{\partial\varphi} - \frac{\partial\left(\sin\theta\ rA_\varphi\right)}{\partial r}\right)\mathbf{u}_\theta + \frac{1}{r}\left(\frac{\partial\left(rA_\theta\right)}{\partial r} - \frac{\partial A_r}{\partial\theta}\right)\mathbf{u}_\varphi\end{aligned} \tag{1.48}$$

$$\nabla^2 V = \frac{1}{r^2}\frac{\partial}{\partial r}\left(r^2\frac{\partial V}{\partial r}\right) + \frac{1}{r^2\sin\theta}\frac{\partial}{\partial\theta}\left(\sin\theta\frac{\partial V}{\partial\theta}\right) + \frac{1}{\left(r\ \sin\theta\right)^2}\frac{\partial^2 V}{\partial\varphi^2} \tag{1.49}$$

1.1.7 Conservative, Irrotational, Solenoidal, and Harmonic Fields

A vector field **B** is *conservative* if it exhibits a null loop integral along each closed line in the domain of definition. As a consequence, the linear integral of **B** along a generic open line results dependent only on the end points of the line and not on its path. It is possible to combine a *scalar potential* Φ with the conservative field **B**. The potential Φ is defined so that $\mathbf{B} = -(\text{grad}\Phi + C)$, where C is an arbitrary constant. The surfaces where Φ assumes an equal value are equipotential surfaces and are always normal to the lines of the field **B**.

A vector field **B** is *irrotational* if its curl is null in all the domain of definition. A conservative field is of course an irrotational field. Conversely an irrotational field is a conservative field only if it is defined in a simply connected domain (a domain is simply connected when each closed line along which the integral is computed may be reduced to a point, remaining always within the domain).

A vector field **B** is *solenoidal* if its divergence is null in the whole domain of definition. As a consequence, the flux of **B** along any closed surface results in null. It is possible to express a solenoidal field **B** by means of a curl of a vector potential **A**. This vector potential **A** is defined so that $\mathbf{B} = \text{curl}(\mathbf{A} + \mathbf{C})$, where **C** is an arbitrary irrotational vector field. It follows that the loop integral of the vector potential along the closed line l is equal to the flux of the solenoidal field through any surfaces bordered by l.

Let **B** be a vector field with divergence and curl given by $\text{div}\mathbf{B} = f$ and $\text{curl}\mathbf{B} = \mathbf{g}$, respectively. It is convenient to split the vector field **B** as the sum of an irrotational field $\mathbf{B}_{irr}$ and a solenoidal field $\mathbf{B}_{sol}$. Such a splitting up is called the resolution of Clebsh–Helmholtz. The field $\mathbf{B}_{irr}$ has to exhibit a divergence equal to f, and the field $\mathbf{B}_{sol}$ has to exhibit a curl equal to **g**. The solution is unique if the field is normal at the infinite (a field whose magnitude tends to $1/r^2$ when the distance r from the origin tends to infinite). Thus, it is

$$\mathbf{B} = \mathbf{B}_{irr} + \mathbf{B}_{sol} \quad \text{with} \quad \begin{cases} \text{div}\mathbf{B} = \text{div}\mathbf{B}_{irr} \\ \text{curl}\mathbf{B} = \text{curl}\mathbf{B}_{sol} \end{cases} \tag{1.50}$$

A scalar field is *harmonic* if it satisfies the Laplace equation:

$$\nabla^2\phi = \text{div}\left(\text{grad}\phi\right) = 0 \tag{1.51}$$

Then, a vector field **B** that is obtained as gradient of a harmonic scalar field Φ, i.e., **B** = gradΦ, is irrotational, since curl(gradΦ) = 0, and solenoidal.

A vector field **A** is *harmonic* if it is irrotational and solenoidal. Hence, it results

$$\nabla^2\mathbf{A} = 0 \tag{1.52}$$

Then, the vector field **B** that is obtained as curl of a harmonic vector field **A**, i.e., **B** = curl**A**, is solenoidal because div(curl**A**) = 0, and irrotational.

1.1.8 Time Dependence

Up to now, the scalar fields and vector fields have been considered as a function exclusively of the points of the space. However, they may be also a function of the time. Of course, the space variables and the time variable t are considered separately. Then, it is obtained

$$\begin{aligned} &\text{grad}\frac{\partial V}{\partial t} = \frac{\partial}{\partial t}\text{grad}V \qquad && \text{div}\frac{\partial \mathbf{A}}{\partial t} = \frac{\partial}{\partial t}\text{div}\mathbf{A} \\ &\text{curl}\frac{\partial \mathbf{A}}{\partial t} = \frac{\partial}{\partial t}\text{curl}\mathbf{A} && \nabla^2\frac{\partial \mathbf{A}}{\partial t} = \frac{\partial}{\partial t}\nabla^2\mathbf{A} \end{aligned} \tag{1.53}$$

It is usual to consider that the dependence on the time is sinusoidal, characterized by the pulsation ω. In this case the complex notation can be used: each quantity is represented by means of a complex phasor. For instance, referring to the component A_x(P,t) along the x–axis of the vector field **A**(P,t), it is possible to introduce the correspondence:

$$\begin{aligned} A_x = A_{xM}\cos(\omega t + \alpha_x) \leftrightarrow \dot{A}_x &= A_{xM}\, e^{j\alpha_x} \\ &= A_{xr} + jA_{xi} \\ &= A_{xM}\cos\alpha_x + jA_{xM}\sin\alpha_x \end{aligned} \tag{1.54}$$

and analogously for the other components. The vector field **A**, whose components are described by the symbolic function (1.54), is indicated as

$$\dot{\mathbf{A}} = \dot{A}_x \mathbf{u}_x + \dot{A}_y \mathbf{u}_y + \dot{A}_z \mathbf{u}_z \tag{1.55}$$

1.2 Electromagnetic Fields

1.2.1 Electric Charge and Electric Charge Density

The electric charges in one region of the space may be distinguished as free charges and charges of polarization. The former can carry out macroscopic displacement, while the latter depend inevitably on the molecular structure of the media. The phenomena of attraction and repulsion among the various bodies, on which the electric charge are posed, are generated by the total charge, the sum of the free charges and of those of polarization. Only the free charges are dealt with hereafter, while the charges of polarization are considered by defining conveniently the characteristics of the media.

In any point of the space, where the infinitesimal volume $\Delta\tau$ contains the (free) charge Δq, it is possible to define a volume density of (free) charge ρ as

$$\rho = \lim_{\Delta\tau \to 0} \frac{\Delta q}{\Delta\tau} \tag{1.56}$$

that are measured in (C/m^3) and generally variable with the time.

If the element on which the charge is distributed is characterized by a dimension negligible with respect to the other two dimensions, it is possible to define a surface density of charge ρ_s, measured in (C/m^2), as

$$\rho_s = \lim_{\Delta S \to 0} \frac{\Delta q}{\Delta S} \tag{1.57}$$

1.2.2 Electric Displacement Field

The spatial description of the effects due to the free electric charges that are in restricted regions is represented by the vector field **D**, called the electric displacement field. The measure unit of its magnitude is (C/m^2). The fundamental property of the field **D**, in differential form and in integral form respectively, is

$$\mathrm{div}\mathbf{D} = \rho \tag{1.58}$$

$$\oint_{S_c} \mathbf{D} \cdot \mathbf{n} \, dS = q \tag{1.59}$$

The second equation is the well-known Gauss's law. It indicates that the flux of the vector **D** going out through the closed surface S_c, oriented by the unity vector **n** normal to the surface and with external direction, is equal to the free charge q, which is contained by the surface S_c itself.

1.2.3 Current Density Field

The movement of the electric charges is described by means of the electric current density. Referring to a volume charge density ρ moving at a velocity $\mathbf{v}_\rho$, the electric current density vector **J** can be defined as

$$\mathbf{J} = \rho \, \mathbf{v}_\rho \tag{1.60}$$

whose reference positive direction is that of the positive charges. Its magnitude is measured in (A/m^2). The vector **J** defines a vector field, called the current field. Let S be an open surface, with normal unity vector **n**, then the current intensity i measured in (A) is given by

$$i = \int_S \mathbf{J} \cdot \mathbf{n} \, dS \tag{1.61}$$

The continuity equation of the current field in differential form is

$$\text{div}\mathbf{J} = -\frac{\partial \rho}{\partial t} \tag{1.62}$$

In integral form, letting S_c a closed surface, with normal unity vector **n**, the current intensity leaving the surface S_c in the time interval Δt corresponds to the variation of the electric charge Δq in the volume enclosed by S_c, which is

$$i_{out} = \oint_{S_c} \mathbf{J} \cdot \mathbf{n} \, dS = -\frac{\Delta q}{\Delta t} \tag{1.63}$$

It is possible to define the total current density vector as

$$\mathbf{J}_{tot} = \mathbf{J} + \frac{\partial \mathbf{D}}{\partial t} \tag{1.64}$$

which is the sum of the current density vector **J** and the displacement current density vector $\partial\mathbf{D}/\partial t$. By property (1.58) about **D** and the continuity Equation (1.62), the current field $\mathbf{J}_{tot}$ is solenoidal.

1.2.4 Magnetic Flux Density Field

The movement of the electric charges causes effects in the points of the surrounding space, so that it is possible to define the magnetic flux density field **B**. Its magnitude is measured in (T). One can observe that a test charge δq moving within the magnetic flux density field **B** at a velocity $\mathbf{v}_q$ experiences a force given by $\delta F = \delta q\, \mathbf{v}_q \times \mathbf{B}$. This force is called Lorentz's force and will be described hereafter.

The fundamental property of the field **B** is that it is a solenoidal field, i.e., the flux of **B** through any closed surface S_c is null. Thus, it is

$$\mathrm{div}\mathbf{B} = 0 \tag{1.65}$$

$$\oint_{S_c} \mathbf{B} \cdot \mathbf{n}\, dS = 0 \tag{1.66}$$

in differential and integral forms, respectively.

1.2.5 Vector Magnetic Potential Field

Since the magnetic flux density field **B** is solenoidal in the whole space, it is suitable to define a vector magnetic potential field **A**, whose magnitude is measured in (Tm), such as

$$\mathbf{B} = \mathrm{curl}\mathbf{A} \tag{1.67}$$

This relationship defines the field **A** apart from a generic irrotational field. The divergence of **A** can be defined in an arbitrary way; the positions that are commonly adopted are as follows.

Coulomb's position, mainly used in stationary or quasi-stationary magnetic fields and also used in the book:

$$\mathrm{div}\mathbf{A} = 0 \tag{1.68}$$

Lorentz's position, which is well suited in the study of rapidly time-variable electromagnetic fields:

$$\mathrm{div}\mathbf{A} = -\mu\varepsilon \frac{\partial V}{\partial t} \tag{1.69}$$

1.2.6 Magnetic Field Strength

Together with the flux density vector **B**, the magnetic field strength vector **H** is introduced. The measure unit of its magnitude is (A/m). The two vector fields are linked by the constitutive law

$$\mathbf{B} = \mu \mathbf{H} \tag{1.70}$$

where μ (H/m) is the magnetic permeability of the medium. Within an uniform medium the two fields **B** and **H** are proportional (i.e., they have same direction and proportional magnitude), while in an anisotropic medium their link exhibits a tensorial nature.

The fundamental property of the field **H** is the definition of its curl (Ampere's law):

$$\text{curl}\mathbf{H} = \mathbf{J} + \frac{\partial \mathbf{D}}{\partial t} \tag{1.71}$$

In stationary or quasi-stationary magnetic condition, i.e., when the displacement current density may be neglected, Equation (1.71) is reduced to

$$\text{curl}\mathbf{H} = \mathbf{J} \tag{1.72}$$

In integral form, the Ampere's law is expressed by equating the line integral of **H** along an oriented closed line l to the current intensity i flowing through the surface enclosed by the line l itself, which is

$$\oint_l \mathbf{H} \cdot \mathbf{t}\, dl = i \tag{1.73}$$

which is called magnetomotive force (MMF). The line integral of **H** along an oriented open line l defines the magnetic voltage between the end points P and Q of the line:

$$\Psi_{PQ} = \int_P^Q \mathbf{H} \cdot \mathbf{t}\, dl \tag{1.74}$$

The magnetic field strength **H** and the magnetic flux density **B** are wholly defined by Equation (1.64) and Equation (1.71) together with the constitutive equation [Equation (1.70)]. From them, the divergence of **H** is

$$\text{div}\mathbf{H} = -\frac{\mathbf{H}}{\mu}\text{grad}\mu \tag{1.75}$$

that is of course null in any homogeneous medium, while the curl of **B** is

$$\text{curl}\mathbf{B} = \mu\mathbf{J} + \text{grad}\mu \times \mathbf{H} \tag{1.76}$$

If the field **H** is irrotational in a simply connected region, a scalar magnetic potential Ψ may be defined in this region, so that

$$\mathbf{H} = -\text{grad}\Psi \tag{1.77}$$

1.2.7 Specific Electric Force, Electric Field

Forces of various nature can exist on the electric charges. Let $\delta\mathbf{F}_k$ be any force on a test positive charge δq, the specific electric force $\mathbf{E}_k$ is defined as

$$\mathbf{E}_k = \lim_{\delta q \to 0} \frac{\delta \mathbf{F}_k}{\delta q} \tag{1.78}$$

whose magnitude is measured in (N/C).

A useful classification of the specific electric forces is given in Table 1.1.

The Coulomb specific electric force $\mathbf{E}_c$ takes into account that the electric charges tend to attract and repel each other. This is a property of the points of the space; thus it defines a vector field. Its fundamental property is to be conservative. Consequently it is irrotational, i.e., $\text{curl}\mathbf{E}_c = 0$. Then, it is possible to introduce a scalar field, the electric potential field V measured in (V), such as

$$\mathbf{E}_c = -\text{grad}V \tag{1.79}$$

Also the specific induced electric force $\mathbf{E}_i$, produced by the rate of change of the magnetic flux density field **B** with the time, is a function of the points of the space, so that it defines a vector field as well. The fundamental property of the induced electric field is the definition of its curl, which is

$$\text{curl}\mathbf{E}_i = -\frac{\partial \mathbf{B}}{\partial t} \tag{1.80}$$

TABLE 1.1

Classification of the Specific Electric Forces

$\mathbf{E}_k$	electromagn.	conservative	$\mathbf{E}_c$	Coulomb	} **E** electric field
		non-conservative	$\mathbf{E}_i$	induced	
			$\mathbf{E}_L$	Lorentz	
			$\mathbf{E}_m$	motional	
	non-electromagn.		$\mathbf{E}_{n.e.}$		

Since there is no constraint on the divergence of $\mathbf{E}_i$, it is generally assumed that it is a solenoidal field, i.e., div$\mathbf{E}_i = 0$. By means of this assumption, together with the Coulomb's position [Equation (1.68)], i.e., div$\mathbf{A} = 0$, the induced electric field is given by

$$\mathbf{E}_i = -\frac{\partial \mathbf{A}}{\partial t} \tag{1.81}$$

This relationship is particularly useful in the computation of the induced currents in conductive media.

The electric field **E**, also called Maxwell's electric field, corresponds to the sum of the Coulomb electric field and the induced electric field, i.e.,

$$\mathbf{E} = \mathbf{E}_c + \mathbf{E}_i = -\text{grad}V - \frac{\partial \mathbf{A}}{\partial t} \tag{1.82}$$

The electric field **E** is defined by its divergence and its curl, that are div$\mathbf{E}$ = div$\mathbf{E}_c$ and curl$\mathbf{E}$ = curl$\mathbf{E}_i$, respectively. In a dielectric medium with electric permittivity ε (in general of tensorial nature), the electric field **E** is linked to the electric displacement field **D** by means of the constitutive relationship

$$\mathbf{D} = \varepsilon \mathbf{E} \tag{1.83}$$

The Lorentz's specific electric force $\mathbf{E}_L$ acts on the electric charges moving in a magnetic flux density field **B** at a velocity $\mathbf{v}_q$ with respect to the adopted reference system. It is

$$\mathbf{E}_L = \mathbf{v}_q \times \mathbf{B} \tag{1.84}$$

The motional specific electric force $\mathbf{E}_m$ acts on the electric charges posed on a conductor moving at a velocity $\mathbf{v}_m$ in a magnetic flux density field **B**. It is

$$\mathbf{E}_m = \mathbf{v}_m \times \mathbf{B} \tag{1.85}$$

Both these two specific forces are not property of the points of the space, but they depend on the velocity of the charges and of the conductor, with respect to the reference system. Thus, they do not define a vector field. However, by means of a change of the reference system it is possible to consider the effects of the motional specific forces as induced specific forces. An example will be shown in the study of the induction motor, in Chapter 13.

The specific electric force of nonelectromagnetic nature, $\mathbf{E}_{ne}$, can be of chemical, piezoelectric, photovoltaic nature, and so on. They are all nonconservative. They do not define vector fields, but they must be considered as specific forces external to the electromagnetic system.

The total specific electric force $\mathbf{E}_t$ is the sum of the field $\mathbf{E}$ and the Lorentz's, motional, and external specific forces, which is

$$\mathbf{E}_t = \mathbf{E} + \mathbf{E}_L + \mathbf{E}_m + \mathbf{E}_{ne} \tag{1.86}$$

In a conductive medium, characterized by the conductivity σ (in general of tensorial nature), the field $\mathbf{E}_t$ is linked to the current density vector field $\mathbf{J}$ by means of the constitutive relationship

$$\mathbf{J} = \sigma \mathbf{E}_t \tag{1.87}$$

The nonconservative specific electric forces $\mathbf{E}_{emf}$ are rotational specific electric forces. They are called specific electromotive forces. Among them, $\mathbf{E}_i$, $\mathbf{E}_L$, and $\mathbf{E}_m$ are of electromagnetic nature.

1.2.8 Electric Voltage and Electromotive Force

The electric voltage v_{AB}, between two points A (+) e B (–), along a line l, oriented by the tangential unity vector $\mathbf{t}$, and fixed in the adopted reference system, is given by the line integral along l from point A to point B of the electric field $\mathbf{E}$. It is

$$v_{AB} = \int_A^B \mathbf{E} \cdot \mathbf{t} \, dl \tag{1.88}$$

If the line l is moving at a velocity $\mathbf{v}_l$ in the adopted reference system, the electric voltage must also take into account the motional specific electric force acting on the charges constrained to stay on the line. The electric voltage becomes

$$v_{AB} = \int_A^B \left(\mathbf{E} + \mathbf{v}_l \times \mathbf{B}\right) \cdot \mathbf{t} \, dl \tag{1.89}$$

Analogously, the electromotive force (EMF) e_{BA}, between two points B (+) e A (–), along a line l is given by the line integral along l from point A to point B of the specific electromotive force $\mathbf{E}_{emf}$. It is

$$e_{BA} = \int_A^B \mathbf{E}_{emf} \cdot \mathbf{t} \, dl \tag{1.90}$$

In analyzing electrical machines, the induced EMF and the motional EMF are of particular interest. The induced EMF is caused by the time-variation of the flux density $\mathbf{B}$. The motional EMF is caused by the movement of the line in a constant field $\mathbf{B}$. The sum of the two EMFs, computed along a closed line l_c, represents the total EMF of electromagnetic nature. It is expressed by means of the Faraday–Neumann law:

$$e_{l_c} = \oint_{l_c} \left(\mathbf{E}_i + \mathbf{E}_m\right) \cdot \mathbf{t}\, dl = -\frac{d}{dt}\int_S \mathbf{B} \cdot \mathbf{n}\, dS = -\frac{d\lambda}{dt} \tag{1.91}$$

where S is the surface enclosed by l_c and λ is the flux of the vector **B** linked by the line l_c. The subdivision of the EMF e_{lc} in its two components, induced and motional EMF, depends by the adopted reference system only.

1.2.9 Poynting's Vector

Poynting's vector is defined as $\mathbf{P} = \mathbf{E} \times \mathbf{H}$ and is a vector normal to the plane defined by the vectors **E** and **H**. Its divergence corresponds to the power density

$$\begin{aligned} \text{div}\mathbf{P} &= \text{div}(\mathbf{E} \times \mathbf{H}) \\ &= \mathbf{H} \cdot (\text{curl}\mathbf{E}) - \mathbf{E} \cdot (\text{curl}\mathbf{H}) \\ &= \mathbf{H} \cdot \left(-\frac{\partial \mathbf{B}}{\partial t}\right) - \mathbf{E} \cdot \left(\mathbf{J} + \frac{\partial \mathbf{D}}{\partial t}\right) \\ &= -\mathbf{H} \cdot \frac{\partial \mathbf{B}}{\partial t} - \mathbf{E} \cdot \frac{\partial \mathbf{D}}{\partial t} - \mathbf{E} \cdot \mathbf{J} \end{aligned} \tag{1.92}$$

Then the integral of the divergence of the vector **P** throughout a volume τ corresponds to the power that goes out the volume through the surface enclosing the volume τ itself:

$$\begin{aligned} p &= \int_\tau \text{div}\mathbf{P}\, d\tau \\ &= -\int_\tau \left(\mathbf{H} \cdot \frac{\partial \mathbf{B}}{\partial t} + \mathbf{E} \cdot \frac{\partial \mathbf{D}}{\partial t}\right) d\tau - \int_\tau (\mathbf{E} \cdot \mathbf{J}) d\tau \end{aligned} \tag{1.93}$$

Assuming the parameters μ (between **H** and **B**) and ε (between **D** and **E**) to be constant, it is possible to write

$$\begin{aligned} \mathbf{H} \cdot \frac{\partial \mathbf{B}}{\partial t} &= \frac{1}{2}\frac{\partial \mathbf{H} \cdot \mathbf{B}}{\partial t} \\ \mathbf{E} \cdot \frac{\partial \mathbf{D}}{\partial t} &= \frac{1}{2}\frac{\partial \mathbf{E} \cdot \mathbf{D}}{\partial t} \end{aligned} \tag{1.94}$$

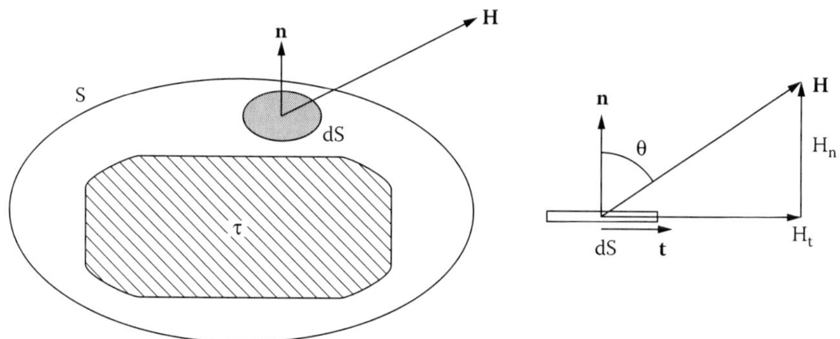

FIGURE 1.3
Computation of the magnetic force by means of the Maxwell's stress tensor.

Using Equation (1.94) and applying the divergence theorem to Equation (1.93), it yields

$$\begin{aligned} p &= \int_\tau \operatorname{div}\mathbf{P}\, d\tau \\ &= \oint_S \mathbf{P}\cdot\mathbf{n}\, dS \\ &= -\frac{\partial}{\partial t}\int_\tau \left(\frac{\mathbf{H}\cdot\mathbf{B}}{2}+\frac{\mathbf{E}\cdot\mathbf{D}}{2}\right) d\tau - \int_\tau \mathbf{E}\cdot\mathbf{J}\, d\tau \end{aligned} \tag{1.95}$$

1.2.10 Maxwell's Stress Tensor

Maxwell's stress tensor allows a rapid computation of the electromagnetic forces acting on an object that is posed within an electromagnetic field. At first, a suitable surface enclosing the object is selected; then the force is computed as the integral over this surface of quantities obtained directly from the potential that describes the field. It could be a scalar potential or a vector potential depending on the study. In the following, let us refer to the magnetic field, since it is more interesting in the applications that are dealt with in this book.

Let a body occupy the volume τ. We suppose to know the value of the vector field **H** over a surface S containing the volume τ and posed in a material characterized by the permeability μ_o, as shown in Figure 1.3(a). In any infinitesimal part dS of the surface S, Maxwell's stress tensor is given by

$$d\mathbf{F} = -\frac{\mu_o}{2}H^2\mathbf{n}dS + \mu_o(\mathbf{H}\cdot\mathbf{n}dS)\mathbf{H} \tag{1.96}$$

where **n** is the unity vector normal to the surface dS. By expressing the vector **H** as the sum of its components tangential and normal to the surface dS, as shown in Figure 1.3(b), as

$$\mathbf{H} = H_t\mathbf{t} + H_n\mathbf{n} \tag{1.97}$$

then the force is given by

$$\begin{aligned} d\mathbf{F} &= -\frac{\mu_o}{2}\left(H_t^2 + H_n^2\right)dS\,\mathbf{t} + \mu_o\left[\left(H_t\mathbf{t} + H_n\mathbf{n}\right)\cdot\mathbf{n}\right]\left(H_t\mathbf{t} + H_n\mathbf{n}\right)dS \\ &= \left(\mu_o H_t H_n\right)dS\mathbf{t} + \frac{\mu_o}{2}\left(H_t^2 - H_n^2\right)dS\mathbf{n} \end{aligned} \tag{1.98}$$

It is possible to identify the two components of the force along the two preferential directions, as

$$\begin{aligned} dF_t &= \left(\mu_o H_t H_n\right)dS \\ dF_n &= \frac{\mu_o}{2}\left(H_n^2 - H_t^2\right)dS \end{aligned} \tag{1.99}$$

The amplitude of the force is computed as

$$\begin{aligned} dF &= \sqrt{\left(\mu_o H_t H_n dS\right)^2 + \left[\frac{\mu_o}{2}\left(H_n^2 - H_t^2\right)dS\right]^2} \\ &\approx \frac{1}{2}\mu_o\left(H_t^2 + H_n^2\right)dS \\ &\approx \frac{1}{2}\mu_o H^2 dS \end{aligned} \tag{1.100}$$

The angle α between the force vector and the unity vector $\mathbf{n}$ normal to the infinitesimal surface dS is computed as

$$\begin{aligned} \tan\alpha &= \frac{H_t H_n}{\frac{1}{2}\left(H_n^2 - H_t^2\right)} \\ &= 2\frac{H_t / H_n}{1 - \left(H_t / H_n\right)^2} \\ &= 2\frac{\tan\theta}{1 - \tan^2\theta} = \tan(2\theta) \end{aligned} \tag{1.101}$$

TABLE 1.2

Particular Cases of the Computation of the Magnetic Force by Means of the Maxwell's Stress Tensor

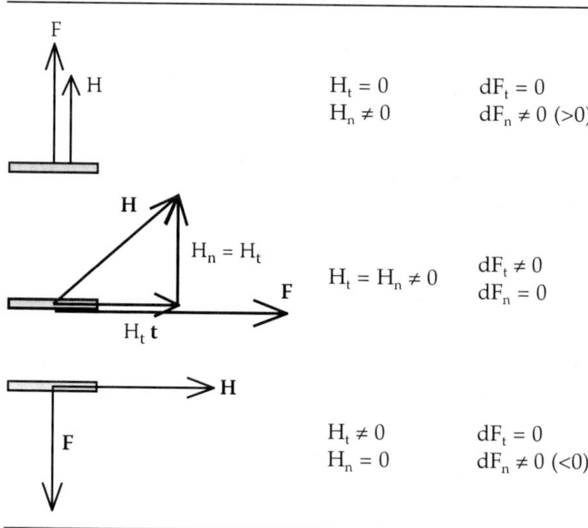

Field components	Force components
$H_t = 0$ $H_n \neq 0$	$dF_t = 0$ $dF_n \neq 0\ (>0)$
$H_t = H_n \neq 0$	$dF_t \neq 0$ $dF_n = 0$
$H_t \neq 0$ $H_n = 0$	$dF_t = 0$ $dF_n \neq 0\ (<0)$

where $\tan\theta = H_t/H_n$, that expresses the angle between the vector **H** and the unity vector **n**. One can notice that the angle α results two times the angle θ, i.e., $\alpha = 2\theta$.

Particular cases are shown in Table 1.2.

1.3 Fundamental Equations Summary

The main vector describing the electric and magnetic fields are

- **B**(P,t) magnetic flux density
- **D**(P,t) electric displacement
- **E**(P,t) Maxwell's electric field, sum of the Coulomb and induced field, $\mathbf{E}(P,t) = \mathbf{E}_c(P,t) + \mathbf{E}_i(P,t)$
- **H**(P,t) magnetic field strength
- **J**(P,t) current density

where P is the point where the vector is considered, while t is the time when the vector is considered. In other words, P indicates the space dependence, while t indicates the time dependence.

1.3.1 Maxwell's Equations

The interactions among the vector fields are described by Maxwell's equations

$$\mathrm{curl}\mathbf{H}(P,t) = \mathbf{J}(P,t) + \frac{\partial \mathbf{D}(P,t)}{\partial t} \tag{1.102}$$

$$\mathrm{curl}\mathbf{E}(P,t) = -\frac{\partial \mathbf{B}(P,t)}{\partial t} \tag{1.103}$$

$$\mathrm{div}\mathbf{B}(P,t) = 0 \tag{1.104}$$

$$\mathrm{div}\mathbf{D}(P,t) = \rho(P,t) \tag{1.105}$$

Of course, in stationary field problems, the time-derivative of the vectors **D** and **B** are null.

1.3.2 Constitutive Relationships and Continuity Equation

The constitutive relationships are

$$\begin{aligned} \mathbf{B}(P,t) &= \mu(P)\,\mathbf{H}(P,t) \\ \mathbf{D}(P,t) &= \varepsilon(P)\,\mathbf{E}(P,t) \\ \mathbf{J}(P,t) &= \sigma(P)\,\mathbf{E}(P,t) \end{aligned} \tag{1.106}$$

and the continuity equation is

$$\mathrm{div}\mathbf{J}(P,t) = -\frac{\partial \rho(P,t)}{\partial t} \tag{1.107}$$

In general the media are not isotropic; then the parameters depending on the material are of tensorial nature:

magnetic permeability electric permittivity electric conductivity

$$\mu = \begin{bmatrix} \mu_x & 0 & 0 \\ 0 & \mu_y & 0 \\ 0 & 0 & \mu_z \end{bmatrix} \quad \varepsilon = \begin{bmatrix} \varepsilon_x & 0 & 0 \\ 0 & \varepsilon_y & 0 \\ 0 & 0 & \varepsilon_z \end{bmatrix} \quad \sigma = \begin{bmatrix} \sigma_x & 0 & 0 \\ 0 & \sigma_y & 0 \\ 0 & 0 & \sigma_z \end{bmatrix} \tag{1.108}$$

In general μ_i, ε_i, and σ_i are not constant along the generic i-th direction. In fact they can be a function of the position (nonhomogeneous media) and/or a function of the magnetic field **H** and the electric field **E** (nonlinear media).

1.3.3 Laplace, Poisson, Helmholtz Equations

Let us refer to a magnetostatic field, described by Equation (1.65), Equation (1.70), and Equation (1.72), where the time-derivative of the vector **D** is null. By assuming the Coulomb's position for the vector magnetic potential **A**, Equation (1.68), the field problem is described by the quasi-harmonic equation

$$\text{curl}\left(\frac{1}{\mu}\text{curl}\mathbf{A}\right) = \mathbf{J} \tag{1.109}$$

If the materials are homogeneous, the permeability μ is constant, then Equation (1.109) is reduced to the Poisson equation, given by

$$\nabla^2\mathbf{A} = -\mu\mathbf{J} \tag{1.110}$$

If the current density field **J** is null in the considered domain, the problem is described by the Laplace equation [Equation (1.52)], which is

$$\nabla^2\mathbf{A} = 0 \tag{1.111}$$

When the magnetic field and the current density field are varying with the time, they are mutually coupled. Let us suppose that such a variation is sinusoidal with the time; then the field quantities can be expressed using the symbolic notation, pointed out in Equation (1.54). A uniform medium is considered, so that a constant electric conductivity σ and a constant magnetic permeability μ are obtained. Finally, the time-derivative of the displacement field **D** could be neglected in comparison with the other fields. Thus the electric field **E** is the unique specific electric force. Hence the current density **J** is obtained as the sum of the source current density $\mathbf{J}_s$ and the induced current density $\mathbf{J}_i$, given by

$$\dot{\mathbf{J}}_i = \sigma\dot{\mathbf{E}}_i = -\sigma\frac{\partial\dot{\mathbf{A}}}{\partial t} = -j\omega\sigma\dot{\mathbf{A}} \tag{1.112}$$

that, substituted in Equation (1.110) gives rise to the Helmholtz equation, combining the space variation of the vector **A** with its time variation, as

$$\nabla^2\dot{\mathbf{A}} - j\omega\sigma\dot{\mathbf{A}} = -\mu\dot{\mathbf{J}}_s \tag{1.113}$$

References

1. J.A. Stratton, *Electromagnetic Theory,* McGraw-Hill, New York, 1941.
2. M. Abramowitz and I.A. Stegun, *Handbook of Mathematical Functions,* Dover Publications, Inc., New York, 1965, pp. 374–376.
3. D.R. Corson and P. Lorrain, *Introduction to Electromagnetic Field and Waves,* W.H. Freeman & Co., San Francisco–London, 1970.
4. J.D. Kraus and K.R. Carver, *Electromagnetics,* McGraw-Hill, New York, 1973.
5. G. Someda, *Elementi di Elettrotecnica Generale,* Patron Editore, Bologna, 1977 (in Italian).
6. P. Lorrain and D.R. Corson, *Electromagnetism. Principles and Applications,* San Francisco, 1978.
7. M.A. Plonus, *Applied Electromagnetics,* McGraw-Hill, New York, 1978.
8. R.E. Collin, *Field Theory of Guided Waves,* IEEE Press, New York, 1991.
9. D.K. Cheng, *Fundamentals of Engineering Electromagnetics,* Addison-Wesley, Reading, MA, 1993.
10. M. Guarnieri and G. Malesani, *Elettromagnetismo Stazionario e Quasi-Stazionario,* Edizioni Progetto, Padova, 1999 (in Italian).
11. J.M. Jin, *Electromagnetic Analysis and Design in Magnetic Resonance Imaging,* CRC Press, Boca Raton, FL, 1998.

2

Basic Principles of Finite Element Methods

Some basic concepts of finite element method are dealt with in this chapter and in the next. In this chapter, the differences of the finite element method compared to the classical methods of field problem analysis are highlighted. The mathematical notions of the method are presented. In the next chapter the construction of the system of equations for the solution is investigated, and the finite element method is applied to some two-dimensional field problems. The concepts here reported are mainly quoted from two excellent books (see References 16 and 17).

These two chapters are useful for acquiring a complete understanding of how the method works. However, they might be passed over by the reader primarily interested in the applications of the finite element method.

2.1 Introduction

The requirement of more and more accuracy during the process of design and analysis of the electrical machines fostered the spreading of numerical models appropriate for computing electric and magnetic fields. These numerical methods are essentially based on the determination of the distribution of the electric and magnetic fields in the structures under study, based on the solution of Maxwell's equations. An analytical solution is barely achieved, because of the complex geometrical machine structures and the nonlinear characteristic of the materials. Then, in most cases only a numerical solution is possible.

The finite element method is a numerical technique that is suitable for this purpose. It allows a field solution to be obtained, even with time-variable fields and with materials that are nonhomogeneous, anisotropic, or nonlinear. Using the finite element method, the whole analysis domain is divided into elementary subdomains, which are called finite elements, and the field equations are applied to each of them.

This method was proposed in the 1940s, but it was firstly applied almost ten years later in aeronautical design and in structural analysis. As years went by, the finite element method was largely adopted in almost all physical and mathematical problems. Today it is the most diffused method for the solution of vector field problems.

The study of the field distributions, and in particular of electromagnetic field problems, exhibits the following advantages. It allows a meticulous local analysis to be carried out, highlighting dangerous field gradient, magnetic field strength, saturation, and so on. It allows a good estimation of the performance of the electromagnetic devices under analysis (especially when the classical methods of analysis give unsatisfactory results). Finally, it permits one to reduce substantially the number of prototypes.

However, the method has some drawbacks, too. Because of its numerical nature, the solution is necessarily approximate. Then, if the method is not correctly applied, it might generate inaccurate results. Finally, since the computed quantities are distributed in the space, the required computation time is generally long.

In order to reduce the computation time, and to improve the analysis at the same time, each periodicity and symmetry (both geometric and electromagnetic symmetry) of the structure is used. The resulting accuracy is influenced by the dimension of the finite elements and by the uniformity of the subdivision. To increase the accuracy, a fine subdivision of the structure is carried out, adopting finite elements of smaller dimension. Nevertheless, an excessive subdivision of the analysis domain causes an aggravation of the computation time.

2.2 Field Problems with Boundary Conditions

Generally, a vector field problem is described by a differential equation, defined in the domain *D*, as

$$L\, \phi(\mathrm{P}, \mathrm{t}) = \mathrm{f}(\mathrm{P}, \mathrm{t}) \tag{2.1}$$

together with the boundary conditions. The latter constrain the fields along the boundary Γ of the domain under analysis. In Equation (2.1) L is a differential operator, ϕ is the unknown function to be determined, and f is the forcing function. Equation (2.1) highlights that both ϕ and f are functions of the position in the space, P(x,y,z), and of the time, t.

2.2.1 Meaning of the Differential Operator *L*

In general, L might be any differential operator. Commonly it represents a linear operation, satisfying the property of additivity and the property of product by a constant.

TEL: 973-328-3200 F

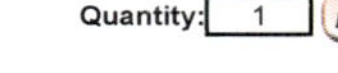

#2350 Tubular Braid (Per MIL Spec QQ-B-575 &/or A-A-59569)

Product Options

Nominal Inside Diameter (rounded): Select Nominal Inside Diameter (rounded)

Package Size (Each): Select Package Size (Each)

Quantity: 1 Add to Cart

Tubular Braid
Per MIL Spec QQ-B-575 &/or A-A-59569

Description: Tinned copper shielded braid manufactured completely round per Mil Spec QQ-B-575 and/or A-A-59569. Maintains its round configuration. Shielding coverage is
Used in military applications where maximum shielding is required against electrostatic interference. Useful against mechanical abrasion and stresses for cables and as groun

This series is RoHS compliant.

DABURN Cat. No.	Nom I.D. Rounded	MIL NO.	CONS.	AWG of Ind. Wires	Approx. AWG	Nom. CMA*	St Pa
2350-1/32	1/32"	QQB575R36T0031	24	36	22	640	50',
2350-1/16	1/16"	QQB575R36T0062	48	36	19	1290	50',
2350-5/64	5/64"	QQB575R36T0078	72	36	17	1900	50',
2350-7/64	7/64"	QQB575R36T0109	96	36	16	2500	50',
2350-1/8	1/8"	QQB575R36T0125	120	36	15	3260	50',
2350-5/32	5/32"	QQB575R36T0156	240	36	12	6200	50',
2350-11/64	11/64"	QQB575R36T0171	168	36	14	4500	50',
2350-3/16	3/16"	**	192	36	13	5100	50',
2350-13/64	13/64"	QQB575R36T0203	312	36	11	8200	50',
2350-1/4	1/4"	QQB575R36T0250	384	36	10	10400	50',

Soft annealed tinned copper braid tightly woven, rolled flat at time of manufacture. It is extremely flexible. Useful as battery grounding, bonding strap for aircraft, vehicles, marine equipment. Eliminates ignition interference.	Woven braid made from soft drawn stainless steel alloy 304. Provides corros and good overall mechanical covering.

2350-3/8	3/8"	QQB575R36T0375	384	36	10	10400	50',
2350-7/16	7/16"	QQB575R30T0437	240	30	6	24120	50',
2350-1/2	1/2"	QQB575R36T0500	528	36	9	13200	50',
2350-9/16	9/16"	**	528	36	9	13200	50',
2351-9/16	9/16"	QQB575R30T0562	480	30	3	48240	50',
2350-21/32	21/32"	QQB575R30T0656	768	30	1	77180	50',
2350-11/16	11/16"	**	720	36	7	19000	50',
2350-25/32	25/32"	QQB575R36T0781	864	36	7	21750	50',
2350-13/16	13/16"	**	864	36	7	21750	50',
2350-7/8	7/8"	QQB575R30T0875	336	30	5	33700	50',
2350-1"	1"	QQB575R30T1000	384	30	4	38600	50',
2350-1 1/8"	1 1/8"	QQB575R30T1125	432	30	4	43330	50',
2350-1 1/4"	1 1/4"	**	480	30	3	48150	50',
2350-1 3/8"	1 3/8"	QQB575R30T1375	528	30	3	53000	50',
2350-1 1/2"	1 1/2"	QQB575R30T1500	576	30	2	57775	50',
2350-2"	2"	QQB575R30T2000	672	30	2	67200	50',

** These sizes are manufactured to standards of QQ-B-575 and/or A-A-59569 although they are not listed in the specification.

Larger spools available - please advise.

* Values based on copper in free air at 65°F and intended as a guide only.
Additional constructions and variations made to order.
Also available in Silver Plated Copper, Bare Copper, Nickel Plated Copper, Stainless Steel, Flattened Tinned Copper, and Oval Tinned Copper.

Related Products

#2355 Tinned Copper Shielding Flat Braid

More Info

#2354 Stainless Steel Tubular Braid (High Tempera

More Info

In the electromagnetic problems, Equation (2.1) is given by the Poisson, Laplace, or Helmholtz equation, in which ϕ is a scalar or a vector field. As an example, in the case of an electrostatic problem, ϕ indicates the scalar electric potential V, and its distribution is described by the Poisson equation. The forcing function is the distribution of the free charge density $f = \rho$. Then, Equation (2.1) is rewritten as

$$-\text{div}\left(\varepsilon \text{ grad V}\right) = \rho \tag{2.2}$$

in which a nonhomogeneous medium is considered, so that the electric permittivity ε could be nonconstant. The differential operator L is then expressed by

$$L = -\text{div}\left(\varepsilon \text{ grad}\right) \tag{2.3}$$

2.2.2 Boundary Conditions

The field problem admits a solution not only if the differential equation that describes its distribution is known in all the points of the domain D, but also if the unknown function ϕ is given on the boundary Γ of the domain D itself. In addition, it can be verified that, once the solution has been found, this solution is unique (this is the unicity theorem).

The conditions that express the behavior of the function ϕ on Γ are called constraint, or boundary, conditions. Among these conditions, one can assign a Dirichlet's condition, which is when a given value of ϕ is assigned on the boundary Γ, or a Neumann's condition, which is when a given value of the derivative of ϕ normal to the boundary Γ is assigned. In addition, periodicity conditions can be assigned, by imposing equal values of ϕ in different parts of the boundary Γ.

If we let Γ_1 be a portion of the boundary Γ, the Dirichlet's condition can be

1. Homogeneous condition (boundary condition of the first type):

$$\phi = 0 \qquad \text{on } \Gamma_1 \tag{2.4}$$

2. Nonhomogeneous condition:

$$\phi = \phi_f \qquad \text{on } \Gamma_1 \tag{2.5}$$

If we let Γ_2 be the remaining portion of the total boundary Γ, the Neumann's conditions can be

1. Homogeneous condition (boundary condition of the second type):

$$\frac{\partial \phi}{\partial n} = 0 \qquad \text{on } \Gamma_2 \tag{2.6}$$

2. Homogeneous condition (boundary condition of the third type):

$$\frac{\partial \phi}{\partial n} + k\phi = 0 \qquad \text{on } \Gamma_2 \tag{2.7}$$

3. Nonhomogeneous condition:

$$\frac{\partial \phi}{\partial n} + k\phi = \phi_g \qquad \text{on } \Gamma_2 \tag{2.8}$$

2.3 Classical Method for the Field Problem Solution

Let the field problem be expressed by Equation (2.1) and by suitable boundary conditions, as given in the previous section. Some methods for solving the field problems are now illustrated. In particular, the classic residual method (or Galerkin's method), the classic variational method (or Rayleigh-Ritz's method), and the finite element method.

All these methods aim to define a function ϕ^* that approximates the unknown function ϕ as closely as possible. Such a function is commonly expressed as a linear combination of basic functions, as

$$\phi^*(P,t) = \sum_{j=1}^{N} \Phi_j \, v_j(P,t) \tag{2.9}$$

where v_j are interpolating functions (that are also called expansion functions or base functions), while Φ_j are unknown coefficients that have to be determined during the computation process. Such a combination has to approximate appropriately the exact solution, satisfying the differential operator [Equation (2.1)] and the boundary conditions at the same time.

The first two methods, the classical residual method and the classical variational method, take into account the whole analysis domain. The functions v_j are defined on the whole domain. Conversely, in the finite element method the whole domain is divided in subdomains; then the function ϕ^* is a combination of functions v_j that are defined in the subdomains. Consequently, since the subdomains are of reduced dimensions, the interpolating functions v_j can be very simple.

Before illustrating the different procedures, let us introduce the inner product between two functions ϕ and φ. Let us refer to the volume τ; then the inner product is defined as

$$\langle \phi, \varphi \rangle = \int_\tau \phi \tilde{\varphi} \, d\tau \tag{2.10}$$

where the symbol ~ indicates the complex conjugate. This inner product is a linear operation, since the properties of additivity and the product by a constant are satisfied:

$$\langle \phi_1 + \phi_2, \varphi \rangle = \langle \phi_1, \varphi \rangle + \langle \phi_2, \varphi \rangle \tag{2.11}$$

$$\langle \alpha\phi, \varphi \rangle = \alpha \langle \phi, \varphi \rangle \tag{2.12}$$

The definition of inner product given by Equation (2.10) could be used to verify the properties of the differential operator L. In particular, it should be verified if the operator L is positive defined:

$$\langle L\phi, \varphi \rangle = \begin{cases} > 0 & \phi \neq 0 \\ = 0 & \phi = 0 \end{cases} \tag{2.13}$$

and if it possible to change the argument of the operator L within the operation of the inner product:

$$\langle L\phi, \varphi \rangle = \langle \phi, L\varphi \rangle \tag{2.14}$$

2.4 The Classical Residual Method (Galerkin's Method)

The classical residual method deals with the differential equation (2.1) directly. It solves the field problem by reducing the residual of the differential equation (2.1). It is based on the following assumption: the function ϕ^* that better approaches the exact solution ϕ corresponds to a residual

$$r = L\phi^* - f \tag{2.15}$$

equal to zero (or at least very low) in the whole analysis domain. Fixing some weight functions w_i, the residual method forces the integral of the residuals, weighed by w_i to be zero over the domain volume τ_D. This is to force the following condition:

$$R_i = \int_{\tau_D} w_i \left(L\,\phi^* - f \right) d\tau = 0 \tag{2.16}$$

There are different weighed residual methods. The best known and most used method is Galerkin's method, where the weight functions w_i are chosen equal to the interpolating function v_i, i.e.,

$$w_i = v_i \qquad i = 1, 2, 3, \ldots, N \tag{2.17}$$

Generally this choice yields to a more accurate solution.

Then, using the approximation of Equation (2.9), Equation (2.16) becomes

$$R_i = \int_{\tau_D} v_i L \left(\sum_{j=1}^{N} \Phi_j \, v_j \right) - v_i f \; d\tau \qquad i = 1, 2, 3, \ldots, N \tag{2.18}$$

This equation yields to a system of equations that can be expressed as

$$[SS][\phi] = [T] \tag{2.19}$$

where $[\phi]$ is the column vector of the unknown coefficients Φ_i. [SS] is a matrix vector that depends on the interpolating functions whose elements are given by

$$s_{ij} = \frac{1}{2} \int_{\tau_D} \left(v_i L v_j + v_j L v_i \right) d\tau \tag{2.20}$$

If the operator L satisfies the property (2.14), the matrix vector [SS] is symmetrical, and its elements are

$$s_{ij} = \int_{\tau_D} v_i L v_j \; d\tau \tag{2.21}$$

In this case, the system of equations that is obtained by Galerkin's method is the same of that obtained by the variational method. At last, [T] in Equation (2.19) is the column vector whose elements depend on the forcing function f. They are given by

$$t_i = \int_{\tau_D} v_i f \; d\tau \tag{2.22}$$

2.5 The Classical Variational Method (Rayleigh-Ritz's Method)

The variational method (also known as Rayleigh-Ritz's method or simply as Ritz's method) solves the field problem by means of an integral approach. Starting from the differential equation (2.1), a suitable functional is built, so that its minimum corresponds to the solution of the field problem, which is when the field equation (2.1) and the boundary conditions have been matched. This functional is just called variational. The process of seeking the solution of the differential problem becomes then a process of seeking the minimum of the functional.

Then, since the minimum of the functional with respect to the function ϕ corresponds to the solution of the Equation (2.1), it can be expressed as

$$F(\phi) = \frac{1}{2}\langle L\phi, \phi\rangle - \frac{1}{2}\langle \phi, f\rangle - \frac{1}{2}\langle f, \phi\rangle \tag{2.23}$$

The function ϕ is substituted by the function ϕ^* given by Equation (2.9), where v_j are again defined in the whole domain *D*. By substituting Equation (2.9) in Equation (2.23), and posing to zero the derivatives of the variational F with respect to the unknown coefficients Φ_j, i.e.,

$$\frac{\delta F}{\delta \Phi_i} = 0 \qquad i = 1, 2, 3, \ldots, N \tag{2.24}$$

a system of linear equations, similar to that of Equation (2.19), is obtained. The variational method requires that the property (2.14) is valid. As a consequence the matrix vector [SS] is symmetrical and the system is identical to that obtained by the residual Galerkin's method.

2.5.1 A Field Problem Solution by Means of the Variational Method

Let us consider a field problem described by the differential equation (2.1), with homogeneous boundary conditions, i.e., of the kind given in Equation (2.4), Equation (2.6), or Equation (2.7). Then, the operation of the inner product is defined as in Equation (2.10). The differential operator is assumed to satisfy both the properties [Equation (2.13) and Equation (2.14)]. The aim is to demonstrate that the solution of Equation (2.1) corresponds to the minimum of the functional defined in Equation (2.23), i.e., to $\delta F = 0$ with $\delta(\delta F) > 0$.

It is

$$\begin{aligned} F + \delta F &= F(\phi + \delta\phi) \\ &= \frac{1}{2}\left\langle L(\phi + \delta\phi), \phi + \delta\phi\right\rangle - \frac{1}{2}\left\langle \phi + \delta\phi, \mathrm{f}\right\rangle - \frac{1}{2}\left\langle f, \phi + \delta\phi\right\rangle \end{aligned} \tag{2.25}$$

Thanks to the additive property, neglecting the infinitesimal of higher order, the following is obtained:

$$\delta F = \frac{1}{2}\left\langle L\phi, \delta\phi\right\rangle + \frac{1}{2}\left\langle L\delta\phi, \phi\right\rangle - \frac{1}{2}\left\langle \delta\phi, \mathrm{f}\right\rangle - \frac{1}{2}\left\langle f, \delta\phi\right\rangle \tag{2.26}$$

Since it has been assumed that L satisfies Equation (2.14), it is

$$\delta F = \frac{1}{2}\left\langle \delta\phi, L\phi - \mathrm{f}\right\rangle + \frac{1}{2}\left\langle L\phi - \mathrm{f}, \delta\phi\right\rangle \tag{2.27}$$

Thus, from the definition of the inner product (2.10), the following results:

$$\delta F = \mathrm{Re}\left(\left\langle \delta\phi, L\phi - \mathrm{f}\right\rangle\right) \tag{2.28}$$

where Re means the real part. By imposing the stationary condition $\delta F = 0$, it is

$$\mathrm{Re}\left(\left\langle \delta\phi, L\phi - \mathrm{f}\right\rangle\right) = 0 \tag{2.29}$$

Since $\delta\phi$ is an arbitrary variation of the function ϕ, one can immediately conclude that ϕ has to satisfy the condition (2.1), in order that Equation (2.29) is satisfied for any $\delta\phi$. Then, to verify that it is a point of minimum, it is found that

$$\delta(\delta F) = \delta F(\phi + \delta\phi) - \delta F(\phi) = \mathrm{Re}\left(\left\langle \delta\phi, L\delta\phi\right\rangle\right) > 0 \tag{2.30}$$

since it has been assumed that L is positive defined, as in Equation (2.13).

It is worth observing that the property (2.14) is essential in order that the stationary point of the functional corresponds to the solution of the field problem. Conversely, to satisfy the condition (2.13) means that this stationary point corresponds to a minimum. Such a condition can be removed, since the aim is to find a solution for Equation (2.1), no matter whether this solution corresponds to a minimum, to a maximum, or to a flex point of the functional.

2.5.2 Definition of the Modified Variationals

When the field problem is described by Equation (2.1) and by nonhomogeneous boundary conditions, i.e., of the kind (2.5) and (2.8), it is essential to modify the definition of the functional F given in Equation (2.23) in order to apply again the method described above. In fact, in this case, the differential operator does not match the property (2.14).

In order to remedy this drawback, a new unknown function $\phi' = \phi - \psi$ is introduced instead of ϕ, where ψ is any function that satisfies the prefixed nonhomogeneous conditions. Therefore, the function ϕ' satisfies the homogeneous conditions, and then the differential operator L satisfies the property (2.14). Substituting ϕ' instead of ϕ and $f' = f - L\psi$ instead of f in the definition of the functional (2.23), it results in

$$F(\phi') = \frac{1}{2}\langle L\phi', \phi'\rangle - \frac{1}{2}\langle \phi', f'\rangle - \frac{1}{2}\langle f', \phi'\rangle \tag{2.31}$$

From Equation (2.31) the modified functional is defined, taking into account the nonhomogeneous boundary conditions, as

$$F(\phi) = \frac{1}{2}\langle L\phi, \phi\rangle - \frac{1}{2}\langle L\phi, \psi\rangle + \frac{1}{2}\langle \phi, L\psi\rangle - \frac{1}{2}\langle \phi, f\rangle - \frac{1}{2}\langle f, \phi\rangle \tag{2.32}$$

In this equation, the second and the third term in the second member, containing the function ψ, can be changed in integrals along the domain boundary, by means of the Stokes theorem and the divergence theorem. Thus, the function ψ disappears when the fixed boundary conditions are applied.

Further modifications can be applied to both the definition of the functional and the definition of the inner product, which allow the method to be generalized. In particular, it is possible to adapt the method to complex differential operators (used in computation of media with losses) or to operators that do not satisfy the property (2.13) and (2.14). The reader who wants to know more about these techniques may refer to the tests in the References.

2.5.3 Natural Boundary Conditions

In the variational method, the boundary conditions may be distinguished between natural and essential conditions. The natural boundary conditions are those that are automatically satisfied in the stationary point of the functional. The essential boundary conditions have to be added to the definition of the functional in order to obtain the exact solution of the field problem.

With the common definition of the functional, the Neumann conditions, which are applied to the derivative of the function ϕ, are natural conditions, i.e., are directly satisfied by the obtained solution. Conversely, the Dirichlet conditions, which are applied directly to the value of the function ϕ, are essential conditions, i.e., they have to be forced in the field solution.

2.6 The Finite Element Method

The finite element method is essentially based on the subdivision of the whole domain in a fixed number of subdomains. Despite of the classical methods described above, where the interpolating functions v_i are defined on the whole domain D, in the finite element method they are defined only on each subdomain. It follows that, because of the small dimension of these subdomains, the function ϕ is approximated by simple interpolating functions whose coefficients are the unknown quantities. The solution of the field problem is obtained when these unknown coefficients are found.

The finite element analysis is organized in the following steps:

1. *Partition of the domain*: The domain is divided in subdomains; they are characterized by reduced dimensions.
2. *Choice of the interpolating functions*: The functions v_i are chosen. As said earlier, with the small dimension of the subdomains, these functions can be very simple.
3. *Formulation of the system to resolve the field problem*: The system of equations, representing the field solution, is developed indifferently by means of Galerkin's method or the Rayleigh-Ritz method.
4. *Solution of the problem*: The solution is obtained by solving the resulting system of equations.

2.6.1 Partition of the Domain

The first step of the finite element method is to divide the domain. The whole domain D is subdivided in N_m elements D_m ($m = 1, 2, 3, \ldots, N_m$). The way to achieve such a subdivision greatly affects the solution accuracy. Moreover, it influences the memory space required to the computer.

In one-dimensional problems, the domain is a curve and each subdomain is a segment, as shown in Figure 2.1(a). The connection of the different segments forms the original curve. In two-dimensional problems, the domain is a surface and each subdomain is a polygon, usually a triangle or a rectangle, as shown in Figure 2.1(b). In three-dimensional problems, the domain is a volume and each subdomain is a tetrahedron, a triangular prism, or a rectangular solid, as shown in Figure 2.1(c).

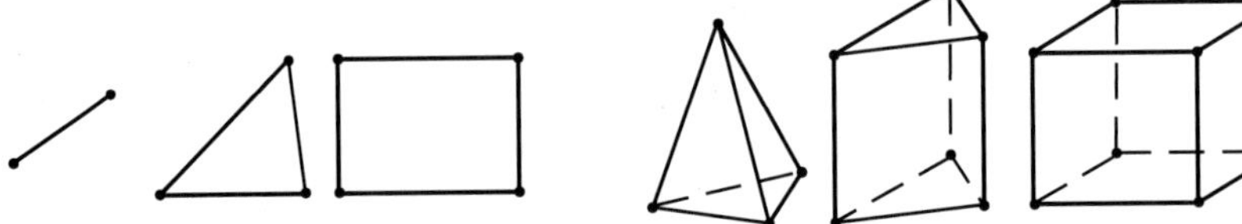

FIGURE 2.1
Elements for the partition of the domain.

2.6.2 Choice of the Interpolating Function

The second step consists of the choice of the interpolating function for approximating the unknown function in each m-th element. If a first-order polynomial is chosen, a linear interpolation is achieved. With a second-order polynomial, a quadratic interpolation is achieved. Also a higher-order polynomial can be chosen; however, although they yield to a higher accuracy in the interpolation, they require a more complex formulation and, thus, are barely adopted. Once the order of the polynomial is chosen, the unknown solution in each m-th element is written as

$$\phi_m{}^*(x,y,z,t) = \sum_{j=1}^{n} \Phi_{mj}\, v_{mj}(x,y,z,t) \tag{2.33}$$

where n is the number of the nodes of the element, Φ_{mj} is the value of ϕ in the j-th node of the m-th element. Finally, v_{mj} is the interpolating function referred to the j-th node of the m-th element. The highest order of the function defines also the order of the element.

2.6.3 Formulation of the System

To solve the field problem, the values of Φ_{mj} have to be computed in the nodes of each element. It is necessary to prepare a system of equations, whose solution corresponds to the values Φ_{mj}. To develop this system, both the variational method and the residual method may be adopted, in a similar way as explained above, but applied to each element separately.

In the case of Galerkin's method, Equation (2.16) is applied to each element. The residual integral is put to zero. In the m-th element, the n integrals given by

$$\begin{aligned} R_{im} &= \int_\tau v_i \left(L\phi_m{}^* - f_m\right) d\tau \\ &= \int_\tau v_i\, L\left(\sum_{j=1}^{n} \Phi_{mj}\, v_{mj}\right) d\tau - \int_\tau v_i\, f_m\, d\tau \quad i = 1, 2, \ldots, n \end{aligned} \tag{2.34}$$

are posed equal to zero. A system of n equations with the n unknown Φ_{mj} is obtained. Applying Equation (2.34) to all the N_m elements that form the domain, and considering the relationships that link the adjacent elements, a system of this kind is obtained:

$$[SS][\phi] - [T] = 0 \tag{2.35}$$

which is formed by N_n equations, with N_n unknown Φ_j.

In the case of the Rayleigh-Ritz method, the functional is given by

$$\begin{aligned} F(\phi^*) &= \sum_{m=1}^{M} F(\phi_m^*) \\ &= \sum_{m=1}^{M} \left[\frac{1}{2} \int_\tau \phi_m^* \, L\phi_m^* \, d\tau - \frac{1}{2} \int_\tau f_m \, \phi_m^* \, d\tau \right] \end{aligned} \tag{2.36}$$

that in matrix form results in

$$F(\phi^*) = \frac{1}{2} [\phi]^t [SS][\phi] - [\phi]^t [T] \tag{2.37}$$

The system is achieved by imposing the stationary condition to the functional, which is to put to zero all the partial derivatives of Equation (2.37) with respect to Φ_j, which is

$$\frac{\partial F}{\partial \Phi_j} = 0 \qquad j = 1, 2, 3, \ldots, N \tag{2.38}$$

A system of N_n equations of the same form of Equation (2.35) is obtained. However, using the variational method, the matrix [SS] is symmetrical, since the condition (2.14) has to be satisfied. This is not verified in general with Galerkin's method.

2.6.4 Solution of the Problem

Once that the system (2.35) is developed, it is possible to compute the values Φ_i in the N_n nodes of the domain. The system (2.35) is a system of equations that may be solved by means of common numerical algorithms. Since the matrix [SS] contains several zeros, suitable algorithms exist to resolve such a system rapidly.

References

1. O.C. Zienkiewicz, *The Finite Element Method in Engineering Science*, McGraw Hill, London, 1971.
2. H.C. Martin and G.F. Carey, *Introduction to Finite Element Analysis: Theory and Application*, McGraw-Hill, New York, 1973.
3. D.H. Norrie and G. de Vries, *The Finite Element Method: Fundamentals and Applications*, Academic Press, New York, 1973.

4. M.V.K. Chari and P.P. Silvester, *Finite Elements in Electrical and Magnetic Field Problem*, New York, John Wiley & Sons, 1980.
5. R.D. Cook, *Concepts and Applications of Finite Element Analysis*, Wiley, New York, 1981.
6. S.S. Rao, *The Finite Element Method in Engineering*, Pergamon Press, Oxford, 1982.
7. R.K. Livesley, *Finite Element: an Introduction for Engineers*, Cambridge University Press, Cambridge, 1983.
8. P.P. Silvester and R.L. Ferrari, *Finite Element Analysis and Design of Electromagnetic Devices*, Cambridge University Press, Cambridge, England, 1983.
9. J.N. Reddy, *An Introduction to Finite Element Method*, McGraw Hill, New York, 1984.
10. R. Wait and A.R. Mitchell, *Finite Element Analysis and Applications*, Wiley, New York, 1985.
11. D.A. Lowther and P.P. Silvester, *Computer Aided Design in Magnetics*, Springer Verlag, New York, 1986.
12. H. Grandin, *Fundamentals of the Finite Element Method*, Macmillan, New York, 1986.
13. D.S. Burnett, *Finite Element Analysis: from Concepts to Applications*, Addison-Wesley Publishing Company, Reading, MA, 1987.
14. S.R.H. Hoole, *Computer-Aided Analysis and Design of Electromagnetic Device*, Elsevier, New York, Amsterdam, London, 1989.
15. W.B. Bickford, *A First Course in the Finite Element Method*, Richard D. Irwin, Homewood, 1990.
16. J.M. Jin, *The Finite Element Method in Electromagnetics*, John Wiley & Sons, New York, 1992.
17. N. Ida and J.P.A. Bastos, *Electromagnetics and Calculation of Fields*, Springer-Verlag, New York, 1992.
18. S.J. Salon, *Finite Element Analysis of Electrical Machine*, Kluwer Academic Publishers, Boston, MA, 1995.
19. A. Reece and T. Preston, *Finite Element Method in Electric Power Engineering*, Oxford University Press, UK, 2000.

3

Applications of the Finite Element Method to Two-Dimensional Fields

Although the three-dimensional codes are increasingly used, the majority of the field problems concerning the analysis of electrical machines can be carried out by a two-dimensional (2D) analysis. This provides several advantages and results in an appreciable reduction of the computation time. The phenomena omitted in the 2D analysis that are not negligible have to be taken into account by means of suitable corrections to the obtained solution.

This chapter deals with the application of the finite element method to 2D field problems. The latter will be described using Cartesian coordinates.

3.1 Introduction

3.1.1 Statement of the Two-Dimensional Field Problem

In 2D field problems, the considered domain is a surface *S*, and its boundary is a curve. Let ϕ be the unknown function that is to be determined. It is a scalar function of the space coordinates x and y, i.e., $\phi = \phi(x,y)$. The time dependence is omitted. Let f be the forcing function, which is also function of x and y, and independent of the time. The 2D field problem is defined by the differential equation of second order

$$-\text{div}\left(\alpha \cdot \text{grad}\phi\right) + \beta\phi = -\frac{\partial}{\partial x}\left(\alpha_x \frac{\partial \phi}{\partial x}\right) - \frac{\partial}{\partial y}\left(\alpha_y \frac{\partial \phi}{\partial y}\right) + \beta\phi = f \qquad (3.1)$$

together with the boundary conditions that are imposed on the boundary Γ of the domain. They are Dirichlet's boundary conditions on the portion Γ_1 of the boundary:

$$\phi = \phi_f \qquad \text{on } \Gamma_1 \qquad (3.2)$$

together with Neumann's boundary conditions on the remaining portion Γ_2 of the boundary:

$$\left(\alpha_x \frac{\partial\phi}{\partial x}\mathbf{u}_x + \alpha_y \frac{\partial\phi}{\partial y}\mathbf{u}_y\right)\cdot\mathbf{n} + k\phi = \phi_g \qquad \text{on } \Gamma_2 \tag{3.3}$$

In Equation (3.1) and Equation (3.3), α_x, α_y, and β are known parameters that are related to the physical property of the materials in the domain. In Equation (3.2) and Equation (3.3), k, ϕ_f, and ϕ_g are known parameters that are related to the physical property of the boundary. In particular, ϕ_f and ϕ_g are functions describing the sources along the boundary curves.

The differential notation of Equation (3.1) is of a general nature, whose Laplace's, Poisson's, and Helmholtz's equations are the particular forms. It corresponds to that given in Equation (2.1), where the differential operator corresponds to

$$L = -\frac{\partial}{\partial x}\left(\alpha_x \frac{\partial}{\partial x}\right) - \frac{\partial}{\partial y}\left(\alpha_y \frac{\partial}{\partial y}\right) + \beta \tag{3.4}$$

3.1.2 Application of the Variational Method

It is hereafter verified that the differential operator L, defined by Equation (3.1) satisfies the property (2.14), if the following conditions are fulfilled:

1. The parameters α_x and α_y in Equation (3.1) are real, whether they are numbers or functions.
2. The functions ϕ and φ satisfy the homogeneous Dirichlet's condition or the homogeneous Neumann's condition.

From the inner product operation, defined in Equation (2.10), in a 2D domain it is possible to write

$$\begin{aligned}\langle L\phi,\varphi\rangle &= \int_S \left[-\mathrm{div}\left(\alpha\cdot\mathrm{grad}\phi\right) + \beta\phi\right]\tilde{\varphi}\ dS \\ &= \int_S -\mathrm{div}\left(\alpha\cdot\mathrm{grad}\phi\right)\tilde{\varphi}\ dS + \beta\int_S \phi\tilde{\varphi}\ dS\end{aligned} \tag{3.5}$$

Applying the second Green's theorem (1.23) results in

$$\langle L\phi,\varphi\rangle = \int_S \phi\left[-\mathrm{div}\left(\alpha\cdot\mathrm{grad}\tilde{\varphi}\right)\right]dS - \int_\Gamma \alpha\cdot\left(\tilde{\varphi}\frac{\partial\phi}{\partial n} - \phi\frac{\partial\tilde{\varphi}}{\partial n}\right)dl + \beta\int_S \phi\tilde{\varphi}\ dS \tag{3.6}$$

The second addendum in the second member is null, since it has been assumed that both the functions ϕ and φ satisfy homogeneous conditions on the boundary Γ. Then, it is

$$\begin{aligned}\langle L\phi,\varphi\rangle &= \int_S \phi\left[-\operatorname{div}\left(\alpha\cdot \operatorname{grad}\tilde{\varphi}\right)+\beta\tilde{\varphi}\right]dS \\ &= \langle \phi, L\varphi\rangle\end{aligned} \tag{3.7}$$

With the same assumptions, it is

$$\begin{aligned}\langle L\phi,\phi\rangle &= \int_S\left[-\operatorname{div}\left(\alpha\cdot \operatorname{grad}\phi\right)+\beta\phi\right]\tilde{\phi}\; dS \\ &= \int_S\left[-\operatorname{div}\left(\alpha\cdot \operatorname{grad}\phi\right)\right]\tilde{\phi}\; dS+\beta\int_S \phi\,\tilde{\phi}\; dS\end{aligned} \tag{3.8}$$

Applying the first Green's theorem (1.22) results in

$$\langle L\phi,\phi\rangle = \int_S \alpha\cdot \operatorname{grad}\phi\cdot \operatorname{grad}\tilde{\phi}\; dS-\int_\Gamma \alpha\cdot\tilde{\phi}\;\frac{\partial\phi}{\partial n}\; dl+\beta\int_S \phi\,\tilde{\phi}\; dS \tag{3.9}$$

Substituting the homogeneous boundary conditions (2.4) and (2.7) results in

$$\begin{aligned}\langle L\phi,\phi\rangle &= \int_S \alpha\left|\operatorname{grad}\phi\right|^2 dS+\int_{\Gamma_2} k\left|\phi\right|^2 dl+\beta\int_S \left|\phi\right|^2 dS \\ &= \int_S \alpha\left|\operatorname{grad}\phi\right|^2 +\beta\left|\phi\right|^2 dS+\int_{\Gamma_2} k\left|\phi\right|^2 dl\end{aligned} \tag{3.10}$$

It is possible to see that, with positive α and k, the differential operator is positive defined, i.e., the property (2.13) is satisfied: $\langle L\phi,\phi\rangle > 0$ if $\phi \neq 0$, and $\langle L\phi,\phi\rangle = 0$ if $\phi = 0$.

The functional $F(\phi)$, defined in Equation (2.23), results in

$$\begin{aligned}F(\phi) &= \frac{1}{2}\langle L\phi,\phi\rangle-\frac{1}{2}\langle\phi,f\rangle-\frac{1}{2}\langle f,\phi\rangle \\ &= \frac{1}{2}\int_S \alpha\left|\operatorname{grad}\phi\right|^2 +\beta\left|\phi\right|^2 dS+\frac{k}{2}\int_{\Gamma_2}\left|\phi\right|^2 dl-\frac{1}{2}\int_S\left(\phi\tilde{f}+f\tilde{\phi}\right)dS\end{aligned} \tag{3.11}$$

In case of nonhomogeneous boundary conditions, the technique illustrated in Section 2.5.2 of Chapter 2 is applied. Always in this case the functional becomes

$$F(\phi) = \frac{1}{2}\int_S \alpha|\text{grad}\phi|^2 + \beta|\phi|^2 \, dS + \int_{\Gamma_2} \frac{k}{2}|\phi|^2 - \phi\phi_g \, dl - \frac{1}{2}\int_S \left(\phi\tilde{f} + f\tilde{\phi}\right) dS \quad (3.12)$$

Given that ϕ and f are real functions, it is possible to obtain the following formulation:

$$F(\phi) = \frac{1}{2}\int_S \alpha_x \left(\frac{\partial\phi}{\partial x}\right)^2 + \alpha_y \left(\frac{\partial\phi}{\partial y}\right)^2 + \beta\phi^2 \, dS + \int_{\Gamma_2} \frac{k}{2}\phi^2 - \phi\phi_g \, dl - \int_S f\,\phi \, dS \quad (3.13)$$

It follows that the solution of the field problem, described by Equation (3.1), Equation (3.2), and Equation (3.3), corresponds to that function ϕ that satisfies the boundary conditions (3.2) and corresponds to a stationary point of the functional (3.13).

The variation of $F(\phi)$ with respect to ϕ is

$$\begin{aligned} \delta F(\phi) = \frac{1}{2}\int_S \alpha_x \frac{\partial\phi}{\partial x}\frac{\partial\delta\phi}{\partial x} + \alpha_y \frac{\partial\phi}{\partial y}\frac{\partial\delta\phi}{\partial y} + \beta\phi\delta\phi \, dS + \\ \int_{\Gamma_2} \left(k\phi - \phi_g\right)\delta\phi \, dl - \int_S f\,\delta\phi \, dS \end{aligned} \quad (3.14)$$

The two addenda of the first integral in the second member can be rewritten as

$$\begin{aligned} \alpha_x \frac{\partial\phi}{\partial x}\frac{\partial\delta\phi}{\partial x} &= \frac{\partial}{\partial x}\left(\alpha_x \frac{\partial\phi}{\partial x}\delta\phi\right) - \frac{\partial}{\partial x}\left(\alpha_x \frac{\partial\phi}{\partial x}\right)\delta\phi \\ \alpha_y \frac{\partial\phi}{\partial y}\frac{\partial\delta\phi}{\partial y} &= \frac{\partial}{\partial y}\left(\alpha_y \frac{\partial\phi}{\partial y}\delta\phi\right) - \frac{\partial}{\partial y}\left(\alpha_y \frac{\partial\phi}{\partial y}\right)\delta\phi \end{aligned} \quad (3.15)$$

Let α_x and α_y be continuous in the whole domain. Then, by means the divergence theorem (1.20) and the relationships (3.15), Equation (3.14) can be rewritten as

$$\begin{aligned} \delta F(\phi) = \int_S \left[-\frac{\partial}{\partial x}\left(\alpha_x \frac{\partial\phi}{\partial x}\right) - \frac{\partial}{\partial y}\left(\alpha_y \frac{\partial\phi}{\partial y}\right) + \beta\phi - f\right]\delta\phi \, dS + \\ \oint_\Gamma \left[\left(\alpha_x \frac{\partial\phi}{\partial x}\mathbf{u}_x + \alpha_y \frac{\partial\phi}{\partial y}\mathbf{u}_y\right)\cdot\mathbf{n}\right]\delta\phi \, dl + \int_{\Gamma_2}\left[k\phi - \phi_g\right]\delta\phi \, dl \end{aligned} \quad (3.16)$$

Since ϕ presents a fixed value on Γ_1, equal to ϕ_f, as expressed in Equation (3.2), then $\delta\phi$ becomes zero on the line Γ_1. Consequently, the corresponding integral on Γ_1 becomes zero as well. The variation of the functional becomes

$$\delta F(\phi) = \int_S \left[-\frac{\partial}{\partial x}\left(\alpha_x \frac{\partial \phi}{\partial x}\right) - \frac{\partial}{\partial y}\left(\alpha_y \frac{\partial \phi}{\partial y}\right) + \beta\phi - f \right] \delta\phi \, dS + \int_{\Gamma_2} \left[\left(\alpha_x \frac{\partial \phi}{\partial x}\mathbf{u}_x + \alpha_y \frac{\partial \phi}{\partial y}\mathbf{u}_y\right) \cdot \mathbf{n} + k\phi - \phi_g \right] \delta\phi \, dl \tag{3.17}$$

The condition to have a stationary point of the functional corresponds to force the variation (3.17) to be zero. This must be verified for each variation of $\delta\phi$, so that the surface integral and the line integral must be zero independently. Therefore the results are

$$\begin{gathered} -\frac{\partial}{\partial x}\left(\alpha_x \frac{\partial \phi}{\partial x}\right) - \frac{\partial}{\partial y}\left(\alpha_y \frac{\partial \phi}{\partial y}\right) + \beta\phi - f = 0 \\ \left(\alpha_x \frac{\partial \phi}{\partial x}\mathbf{u}_x + \alpha_y \frac{\partial \phi}{\partial y}\mathbf{u}_y\right) \cdot \mathbf{n} + k\phi - \phi_g = 0 \end{gathered} \tag{3.18}$$

It is recognized that both the differential equation (3.1), which describes the field problem, and the boundary condition (3.3), which forces the derivative of the function ϕ on the boundary, are satisfied. As stated in Chapter 2, this condition is a natural condition, i.e., it is automatically satisfied by the solution. Conversely the boundary condition (3.2) is an essential condition, which must be forced explicitly in solving the problem.

3.2 Linear Interpolation of the Function ϕ

The 2D domain is subdivided in a finite and sufficiently high number of elements. In the simplest case, they are elements of the triangular form, not necessary equal, but not intersecting each other. Each vertex is called a node, and all of them set up the mesh. Let us assume that the structure has been divided in N_m finite elements, and the total number of node is N_n. Each of them assumes the value Φ_i of the potential function ϕ.

Thanks to the small dimensions of the elements, the interpolating functions $v_i(x,y)$ may be simple. In the following, a linear interpolation of the function ϕ is assumed for each m-th triangular element, given by

$$\phi_m(x,y) = a + bx + cy \tag{3.19}$$

In particular, in the three nodes of the triangle, the three i-th values are given by

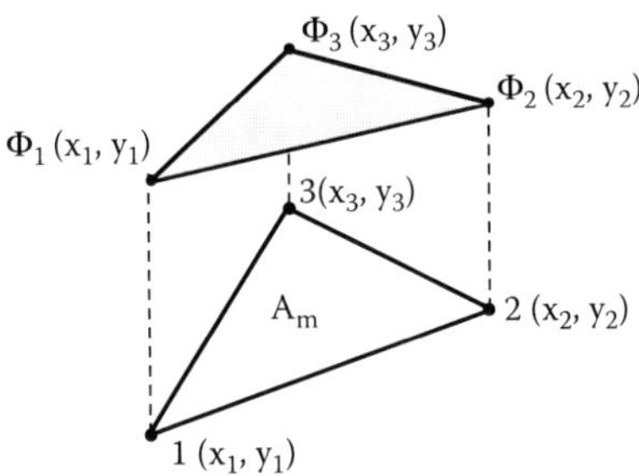

FIGURE 3.1
Linear interpolation of the potential function ϕ_m in the m-th triangular element.

$$\begin{cases} \Phi_1 = a + bx_1 + cy_1 \\ \Phi_2 = a + bx_2 + cy_2 \\ \Phi_3 = a + bx_3 + cy_3 \end{cases} \tag{3.20}$$

From the knowledge of the value of the function in the nodes of each finite element, i.e., Φ_1, Φ_2, and Φ_3, by means of Equation (3.19), it is possible to compute the potential function in any other point of the element. This is represented in Figure 3.1.

If the three values of the potential are given in the three nodes of the element, it is possible to solve the system (3.20) in the three unknowns a, b, and c. At first, it is posed

$$\begin{aligned} A_m &= \frac{1}{2}\begin{vmatrix} 1 & x_1 & y_1 \\ 1 & x_2 & y_2 \\ 1 & x_3 & y_3 \end{vmatrix} \\ &= \frac{1}{2}\left[(x_2y_3 - x_3y_2) + (x_3y_1 - x_1y_3) + (x_1y_2 - x_2y_1)\right] \end{aligned} \tag{3.21}$$

that represents the area of the m-th triangular element, as shown in Figure 3.1. By Cramer's rule, it results that

$$\begin{aligned} a &= \frac{1}{2A_m}\left[\Phi_1(x_2y_3 - x_3y_2) + \Phi_2(x_3y_1 - x_1y_3) + \Phi_3(x_1y_2 - x_2y_1)\right] \\ b &= \frac{1}{2A_m}\left[\Phi_1(y_2 - y_3) + \Phi_2(y_3 - y_1) + \Phi_3(y_1 - y_2)\right] \\ c &= \frac{1}{2A_m}\left[\Phi_1(x_3 - x_2) + \Phi_2(x_1 - x_3) + \Phi_3(x_2 - x_1)\right] \end{aligned} \tag{3.22}$$

It is

$$a = \frac{1}{2A_m}\left[\Phi_1 p_1 + \Phi_2 p_2 + \Phi_3 p_3\right]$$

$$b = \frac{1}{2A_m}\left[\Phi_1 q_1 + \Phi_2 q_2 + \Phi_3 q_3\right] \quad (3.23)$$

$$c = \frac{1}{2A_m}\left[\Phi_1 r_1 + \Phi_2 r_2 + \Phi_3 r_3\right]$$

where

$$\begin{array}{lll} p_1 = x_2 y_3 - x_3 y_2 & q_1 = y_2 - y_3 & r_1 = x_3 - x_2 \\ p_2 = x_3 y_1 - x_1 y_3 & q_2 = y_3 - y_1 & r_2 = x_1 - x_3 \\ p_3 = x_1 y_2 - x_2 y_1 & q_3 = y_1 - y_2 & r_3 = x_2 - x_1 \end{array} \quad (3.24)$$

Equation (3.19) expresses the unknown function $\phi_m(x,y)$, inside the m-th triangular finite element. It can be rewritten as

$$\begin{aligned} \phi_m(x,y) &= \left(\frac{1}{2A_m}\sum_{i=1}^{3} p_i\Phi_i\right) + \left(\frac{1}{2A_m}\sum_{i=1}^{3} q_i\Phi_i\right)\cdot x + \left(\frac{1}{2A_m}\sum_{i=1}^{3} r_i\Phi_i\right)\cdot y \\ &= \sum_{i=1}^{3} \frac{(p_i + q_i x + r_i y)}{2A_m}\cdot\Phi_i \\ &= \sum_{i=1}^{3} v_i(x,y)\cdot\Phi_i \end{aligned} \quad (3.25)$$

It is worth noticing that is possible to define the function $\phi_m(x,y)$ in each point of the triangle as linear combination of the values Φ_1, Φ_2, and Φ_3 in

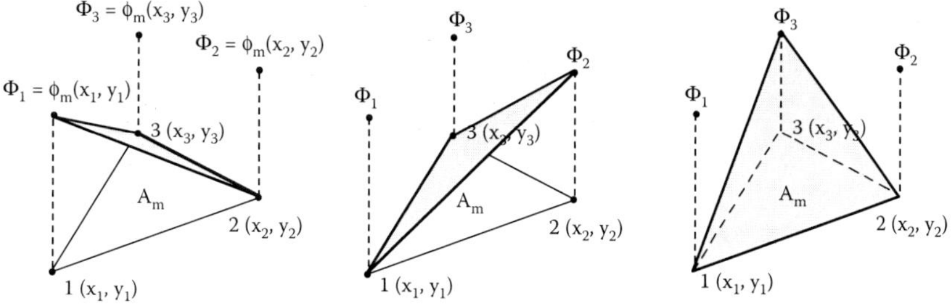

FIGURE 3.2
Representation of the linear interpolating functions of the function ϕ_m in the m-th triangular element drawn in Figure 3.1.

the nodes of the triangle itself. It is then evident that, to get the field, it is sufficient to compute the values of the unknown function $\phi_m(x,y)$ in the nodes of each element that form the whole domain. In Equation (3.25), the interpolating functions $v_i(x,y)$, with i = 1, 2, 3, that interpolate the function $\phi_m(x,y)$ have been highlighted. Each of them can be represented graphically, as illustrated in Figure 3.2.

3.2.1 Computation on the Function $\phi_m(x,y)$

Before going on, let us consider some operations on the function $\phi_m(x,y)$ defined in Equation (3.19), which will be useful in the following.

If we let ϕ_m be a scalar function, its gradient is given by

$$\begin{aligned} \text{grad}\phi_m &= \text{grad}(a + bx + cy) \\ &= \left[\frac{\partial(a+bx+cy)}{\partial x}, \frac{\partial(a+bx+cy)}{\partial y}, 0\right] \\ &= (b,\ c,\ 0) \end{aligned} \tag{3.26}$$

By analogy, letting ϕ_m be a vector function, and with a z-axis component only, as commonly occurs in 2D problems, its curl is given by

$$\begin{aligned} \text{curl}\phi_m &= \text{curl}\left[0,\ 0,\ (a + bx + cy)\right] \\ &= \left[\frac{\partial(a+bx+cy)}{\partial y}, -\frac{\partial(a+bx+cy)}{\partial x}, 0\right] \\ &= (c,\ -b,\ 0) \end{aligned} \tag{3.27}$$

The values $(\text{grad}\phi_m \cdot \text{grad}\phi_m)$ and $(\text{curl}\phi_m \cdot \text{curl}\phi_m)$, inside each triangular element, are constant, since they do not depend on the coordinates x and y. In fact, the results are

$$\text{grad}\phi_m \cdot \text{grad}\phi_m = b^2 + c^2 \tag{3.28}$$

$$\text{curl}\phi_m \cdot \text{curl}\phi_m = b^2 + c^2 \tag{3.29}$$

3.3 Application of the Variational Method

The field problem is solved by means of the integral formulation, computing a functional F that is related to the field problem. It is expressed as a function of the N_n unknown values Φ_i of the potential ϕ. The field problem becomes a N_n-dimensional vector of Φ_i that minimizes the functional and satisfies Dirichlet's boundary conditions (Neumann's boundary conditions, if any, are naturally satisfied, as seen in Section 3.1.2). In other words, it is necessary to put to zero all the derivative of the functional F with respect to Φ_i in the N_n nodes of the domain, that is

$$\frac{\partial F}{\partial \Phi_i} = 0 \quad \text{with} \quad i = 1, 2, 3, \ldots, N_n \tag{3.30}$$

Since the functional F is given by the sum of the functionals F_m associated to all the N_m finite elements that form the domain

$$F = \sum_{m=1}^{N_m} F_m \tag{3.31}$$

the solution of the field problem, expressed by Equation (3.30), can be expressed as

$$\frac{\partial F}{\partial \Phi_i} = \sum_{m=1}^{N_m} \frac{\partial F_m}{\partial \Phi_i} = 0 \quad \text{with} \quad i = 1, 2, 3, \ldots, N_n \tag{3.32}$$

3.3.1 Functional Referred to the m-th Element

The functional associated to the m-th finite element, say F_m, depends only on the values Φ_i in the three nodes of the element. In fact, F_m depends on the function ϕ_m in the element and the latter is a linear combination of the potential in the nodes, as defined by Equation (3.25). Thus, all the derivatives of F_m are null with respect to the unknown values Φ_i in the nodes that do not belong to the m-th element.

For the sake of simplicity, the most general case of Helmholtz's equation (3.1) is disregarded, and we focus on Poisson's equation (or Laplace's equation in specific cases), that cover almost all of the problems about the electrical machines. As a consequence, $\beta = 0$ in Equation (3.1) and in Equation (3.4). In addition, on the boundary part Γ_2 only the homogeneous Neumann's condition is considered, yielding to $k = 0$ and $\phi_g = 0$ in Equation (3.3). Finally, referring to Equation (3.1), it is assumed that in each finite element the forcing term f is constant and equal to f_m, and the values α_x and α_y are constant as well.

By means of these simplifications, the functional F of Equation (3.13), referred to the m-th element, becomes

$$F_m = \frac{1}{2} \int_{A_m} \alpha_x \left(\frac{\partial \phi_m}{\partial x} \right)^2 + \alpha_y \left(\frac{\partial \phi_m}{\partial y} \right)^2 dS + f_m \int_{A_m} \phi_m \, dS \tag{3.33}$$

With the interpolation of ϕ_m given in Equation (3.25), the partial derivatives of ϕ_m are

$$\frac{\partial\phi_m}{\partial x}=\frac{1}{2A_m}\sum_{i=1}^{3}q_i\Phi_i$$
$$\frac{\partial\phi_m}{\partial y}=\frac{1}{2A_m}\sum_{i=1}^{3}r_i\Phi_i \tag{3.34}$$

so that

$$\left(\frac{\partial\phi_m}{\partial x}\right)^2=\frac{1}{4A_m^{\,2}}\sum_{i=1}^{3}\sum_{j=1}^{3}q_iq_j\Phi_i\Phi_j$$
$$\left(\frac{\partial\phi_m}{\partial y}\right)^2=\frac{1}{4A_m^{\,2}}\sum_{i=1}^{3}\sum_{j=1}^{3}r_ir_j\Phi_i\Phi_j \tag{3.35}$$

Both the terms in Equation (3.35) are constant inside the elements. They can be put outside the integral of Equation (3.33). After some manipulations, the functional F_m of the m-th element results in

$$F_m=\frac{1}{8A_m}\sum_{i=1}^{3}\sum_{j=1}^{3}\left(\alpha_xq_iq_j+\alpha_yr_ir_j\right)\Phi_i\Phi_j+f_m\frac{A_m}{3}\sum_{i=1}^{3}\Phi_i \tag{3.36}$$

Using the vector notation, it is

$$F_m=\frac{1}{2}\begin{bmatrix}\Phi_1 & \Phi_2 & \Phi_3\end{bmatrix}\begin{bmatrix}s_{11} & s_{12} & s_{13}\\ s_{21} & s_{22} & s_{23}\\ s_{31} & s_{32} & s_{33}\end{bmatrix}\begin{bmatrix}\Phi_1\\ \Phi_2\\ \Phi_3\end{bmatrix}+\begin{bmatrix}\Phi_1 & \Phi_2 & \Phi_3\end{bmatrix}\begin{bmatrix}t_1\\ t_2\\ t_3\end{bmatrix} \tag{3.37}$$
$$=\frac{1}{2}\left[\Phi_{123}\right]^t\left[S_m\right]\left[\Phi_{123}\right]+\left[\Phi_{123}\right]^t\left[T_m\right]$$

where the terms of the matrix $[S_m]$ and the vector $[T_m]$ are

$$s_{ij}=\frac{1}{4A_m}\left(\alpha_xq_iq_j+\alpha_yr_ir_j\right)$$
$$t_i=f_m\frac{A_m}{3} \tag{3.38}$$

It is observed that the terms s_{ij} depend only on the geometry and the material characteristic (expressed by the coefficients α_x and α_y). If the material is linear, the terms s_{ij} do not depend on the value of the potentials in the nodes (since α_x and α_y are constant with the field). By analogy, the terms t_i are only functions of the geometry of the element and of the forcing quantity f_m.

Since the research of the stationary point of the functional F_m is carried out by posing to zero the derivative of F_m with respect to the values Φ_1, Φ_2, and Φ_3 in the three nodes of the m-th element. Deriving F_m in Equation (3.37) with respect Φ_1, Φ_2, and Φ_3 and equating the result to zero, the results are

$$\left[\frac{\partial F_m}{\partial \Phi_1}, \frac{\partial F_m}{\partial \Phi_2}, \frac{\partial F_m}{\partial \Phi_3}\right]^t = \left[S_m\right]\left[\Phi_{123}\right] - \left[T_m\right] = \left[0\right] \tag{3.39}$$

Equation (3.39) shows that, for the generic m-th triangular element, a system of three equations is obtained, with the three unknown potential values Φ_1, Φ_2, and Φ_3 in the nodes of the element. When this system of equations is satisfied, the field problem solution is obtained in the m-th element.

3.3.2 Functional Referred to the Whole Domain

The complete solution of the field problem consists of the solution of the system (3.39) for each finite element that belongs to the domain. It is necessary to pass from the local matrix $[S_m]$ that refers to the single m-th element (i.e., to the three values Φ_1, Φ_2, and Φ_3 only) to the global matrix [SS] that refers to the complete domain (i.e., to all the values Φ_i of the N_n nodes). This operation is accomplished by combining all the functionals ϕ_m of the adjacent elements.

As an example, let us consider two triangles that share the nodes 2 and 3, where the function Φ assumes the values Φ_2 and Φ_3, as shown in Figure 3.3.

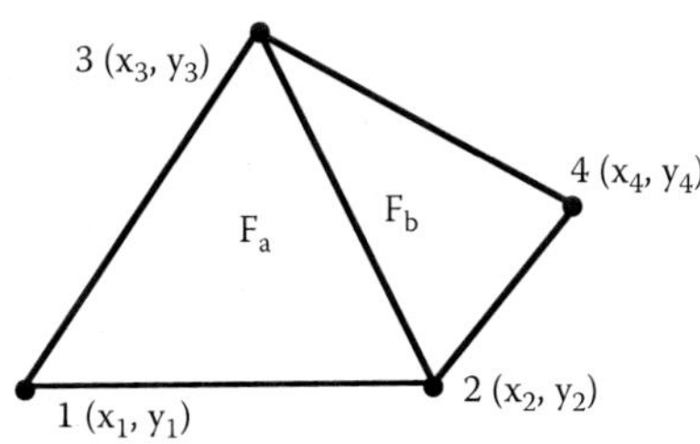

FIGURE 3.3
Adjacent elements with shared nodes 2 and 3.

The two triangles are characterized by the local functionals F_a and F_b, given by

$$F_a = \frac{1}{2}\left[\Phi_{123}\right]^t\left[S_a\right]\left[\Phi_{123}\right] - \left[\Phi_{123}\right]^t\left[T_a\right]$$
$$F_b = \frac{1}{2}\left[\Phi_{234}\right]^t\left[S_b\right]\left[\Phi_{234}\right] - \left[\Phi_{234}\right]^t\left[T_b\right] \tag{3.40}$$

Then, the sum of the two functionals, $F_a + F_b$, is obtained as

$$F_a + F_b = \frac{1}{2}\left[\Phi_{1234}\right]^t\left[S_{ab}\right]\left[\Phi_{1234}\right] - \left[\Phi_{1234}\right]^t\left[T_{ab}\right] = \tag{3.41}$$

$$\frac{1}{2}\begin{bmatrix}\Phi_1\\ \Phi_2\\ \Phi_3\\ \Phi_4\end{bmatrix}^t \begin{bmatrix} s_{a(11)} & s_{a(12)} & s_{a(13)} & 0 \\ s_{a(21)} & s_{a(22)}+s_{b(11)} & s_{a(23)}+s_{b(12)} & s_{b(13)} \\ s_{a(31)} & s_{a(32)}+s_{b(21)} & s_{a(33)}+s_{b(22)} & s_{b(23)} \\ 0 & s_{b(31)} & s_{b(32)} & s_{b(33)} \end{bmatrix} \begin{bmatrix}\Phi_1\\ \Phi_2\\ \Phi_3\\ \Phi_4\end{bmatrix} - \begin{bmatrix}\Phi_1\\ \Phi_2\\ \Phi_3\\ \Phi_4\end{bmatrix}^t \begin{bmatrix} t_{a(1)} \\ t_{a(2)}+t_{b(1)} \\ t_{a(3)}+t_{b(2)} \\ t_{b(3)} \end{bmatrix}$$

Proceeding in the same way for each triangular element with shared nodes, the complete functional of the whole domain is obtained as

$$F = \frac{1}{2}\left[\Phi\right]^t\left[SS\right]\left[\Phi\right] - \left[\Phi\right]^t\left[T\right] \tag{3.42}$$

where $[\Phi]$ is the column vector of all the N_n values of the function ϕ in all the nodes of the domain. The dimension of the matrix [SS] is $N_n \times N_n$. It is called the "stiffness" matrix, due to the analogy with the structural analysis to which the finite element method has been originally applied. Finally, [T] is the column vector of the known terms, and its dimension is N_n.

The stiffness matrix is a sparse matrix, i.e., with several zeros. As observed in Equation (3.38), in linear conditions, this matrix is built and is not modified during the process of search of the solution. The stationary point of the functional F (3.42) corresponds to

$$\left[SS\right]\left[\Phi\right] - \left[T\right] = \left[0\right] \tag{3.43}$$

It is a linear system of N_n equations with N_n unknown Φ_i in the nodes of the triangular finite elements. Thanks to the interpolation (3.25) valid in each element, once the values Φ_i are computed, it is possible to go back to the value of ϕ in each point of the domain.

3.3.3 Assigning the Essential Boundary Conditions

The exact field solution has to satisfy the boundary conditions. Whereas Neumann's boundary conditions may be not imposed, since they are natural conditions, it is necessary to impose Dirichlet's boundary conditions, which are essential conditions.

To impose Dirichlet's boundary conditions, the stiffness matrix [SS] and the column vector [T] of the known terms in Equation (3.43) must be modified. As an example, let us suppose that the potential value Φ_k has to be assigned to the k-th node. In the stiffness matrix [SS], each term of the k-th row is put to zero, except the term on the principal diagonal in which unity is assigned. Then the value Φ_k is assigned to the k-th term of the column vector [T].

3.3.4 The Problem's Solution

Once the system is assembled, including the assignment of the essential boundary conditions, it is still in the form (3.43). The system is solved by one of the common numerical methods, among them:

1. The Gauss-Jordan direct method. This method is applied when the physical properties of the materials are linear, which is when all the parameters of the materials are constant.
2. The Newton-Rapson iterative method. This method is applied when the materials are not linear, which is when the parameters of the materials (α_x and α_y) are not constant, but they are functions of the value of the magnitude of the magnetic fields. The common case is the B-H curve of the ferromagnetic materials.

3.4 Simple Descriptions of Electromagnetic Fields

This section deals with some simple applications of the finite element method to 2D electromagnetic fields, expressed in Cartesian coordinates. For all the field problems, a convenient potential function ϕ is singled out and the corresponding functional F is achieved. In particular, the various expressions of the functional F_m will be reported, referred to as the m-th element.

3.4.1 An Electrostatic Field

Starting from Maxwell's equations, reported in Chapter 1, the electrostatic field problem is described by the relations div$\mathbf{D} = \rho$, curl$\mathbf{E} = \mathbf{0}$, and $\mathbf{D} = \varepsilon\mathbf{E}$.

All the time derivatives are null and the current density **J** is null as well. The unique specific electric force is Coulomb's electric field, $\mathbf{E} = \mathbf{E}_c$. The second equation highlights that the electric field is irrotational in the whole domain, which is conservative. Thus, it is possible to define a scalar electric potential V such as $\mathbf{E}_c = -\text{grad}V$. From these equations, the field problem is described by the following almost harmonic scalar equation:

$$\text{div}\mathbf{D} = \text{div}(\varepsilon\mathbf{E}_c) = \text{div}\varepsilon(-\text{grad}V) = \rho \tag{3.44}$$

In case of a 2D field, using Cartesian coordinates, this equation becomes

$$\frac{\partial}{\partial x}\varepsilon_x\frac{\partial V}{\partial x} + \frac{\partial}{\partial y}\varepsilon_y\frac{\partial V}{\partial y} = -\rho \tag{3.45}$$

If the medium is homogeneous and isotropic, i.e., the electric permittivity ε is constant, Equation (3.45) is reduced to the scalar Poisson's equation

$$\frac{\partial^2 V}{\partial x^2} + \frac{\partial^2 V}{\partial y^2} = -\frac{\rho}{\varepsilon} \tag{3.46}$$

The functional related to the 2D electrostatic field, referred to the surface S of the domain *D*, is given by

$$F = \int_S \left(\frac{1}{2}\varepsilon E_c^{\,2} - \rho V\right) dS \tag{3.47}$$

With reference to the m-th element, substituting the interpolation given in Equation (3.19) in Equation (3.47) and using the relation (3.28), the functional F_m is given by

$$\begin{aligned} F_m &= \frac{1}{2}\varepsilon\int_{A_m}(\text{grad}V\cdot\text{grad}V)dS - \rho\int_{A_m} V\,dS \\ &= \frac{1}{2}\varepsilon\int_{A_m}(b^2+c^2)dS - \rho\int_{A_m}(a+bx+cy)\,dS \\ &= \frac{1}{2}\varepsilon A_m(b^2+c^2) - \rho A_m\left(a + b\frac{x_1+x_2+x_3}{3} + c\frac{y_1+y_2+y_3}{3}\right) \end{aligned} \tag{3.48}$$

The coefficients a, b, and c are functions of the potential V_1, V_2, and V_3 of the nodes of the m-th triangular element, where the functional is computed. Substituting the expressions (3.22), that have been obtained for a, b, and c as a function of the potentials, into Equation (3.48), F_m can be written as

$$F_m = \frac{1}{2}[V_{123}]^t[S_m][V_{123}] - [V_{123}]^t[T_m] \tag{3.49}$$

where $[V_{123}]$ is the column vector of the potentials V_1, V_2, V_3 of the three nodes of the triangular element, and $[S_m]$ is the stiffness matrix, with dimension 3×3. The generic component corresponding to the i-th row and j-th column of $[S_m]$ is given by

$$s_{ij} = \frac{\varepsilon}{4A_m}(q_i q_j + r_i r_j) \tag{3.50}$$

Finally $[T_m]$ is a column vector of known terms, which is $[T_m] = [t_1, t_2, t_3]^t$, where

$$t_i = -\frac{A_m}{3}\rho \tag{3.51}$$

Of course, Equation (3.50) and Equation (3.51) coincide with those given in Equation (3.38), once equating $f_m = -\rho$ and $\alpha_x = \alpha_y = \varepsilon$.

3.4.2 A Stationary Current Field

The equations describing the stationary current field are div**J** = 0, rot**E** = **0**, and **J** = σ**E**. All the time derivatives are null and the unique specific electric force is Coulomb's electric field, $\mathbf{E} = \mathbf{E}_c$. The study may be set out in two ways.

3.4.2.1 *By Means of the Scalar Potential*

Since the field is stationary, the electric field **E** is irrotational in the whole domain; thus it is conservative. It is possible to define a scalar electric potential V such as **E** = −gradV. The field problem is described by the following, almost harmonic scalar equation

$$\text{div}\mathbf{J} = \text{div}(\sigma\mathbf{E}) = \text{div}\sigma(-\text{grad}V) = 0 \tag{3.52}$$

that, in a 2D field using Cartesian coordinates, becomes

$$\frac{\partial}{\partial x}\sigma_x\frac{\partial V}{\partial x} + \frac{\partial}{\partial y}\sigma_y\frac{\partial V}{\partial y} = 0 \tag{3.53}$$

Furthermore, if the electric conductivity σ is constant, Laplace's equation is achieved:

$$\frac{\partial^2 V}{\partial x^2} + \frac{\partial^2 V}{\partial y^2} = 0 \tag{3.54}$$

The functional F_m, referred to the surface A_m of the m-th element, is

$$\begin{aligned} F_m &= \int_{A_m} \left(\frac{1}{2} \sigma E^2 \right) dS \\ &= \frac{1}{2} \sigma \int_{A_m} (\text{grad}V \cdot \text{grad}V) dS \\ &= \frac{1}{2} \sigma A_m (b^2 + c^2) \end{aligned} \tag{3.55}$$

Since the coefficients a, b, and c are functions of the potentials V_1, V_2, and V_3 of the nodes of the triangular element, the functional F_m can be expressed as

$$F_m = \frac{1}{2} [V_{123}]^t [S_m][V_{123}] \tag{3.56}$$

where $[V_{123}] = [V_1, V_2, V_3]^t$, and the generic component of the matrix $[S_m]$ is given by

$$s_{ij} = \frac{\sigma}{4A_m} (q_i q_j + r_i r_j) \tag{3.57}$$

3.4.2.2 By Means of the Vector Potential

Alternatively, in place of the scalar potential V, the stationary current field can be studied by means of a suitable vector potential. Since the current density vector **J** is solenoidal, it is possible to define an electric vector potential **N** such as **J** = curl**N**, always being div(curl**N**) = 0. By handling these equations, the field problem is described by the almost harmonic vector equation

$$\text{curl}\mathbf{E} = \text{curl}\left(\frac{\mathbf{J}}{\sigma} \right) = \text{curl}\frac{1}{\sigma}\left(\text{curl}\mathbf{N} \right) = \mathbf{0} \tag{3.58}$$

The coordinate system is chosen so that the electric vector potential **N** has only a z-axis component, which is **N** = $[0, 0, N_z]$. In this way, there are only x-axis and y-axis components of the current density vector **J**:

$$\mathbf{J} = \text{curl}\left[0,\ 0,\ N_z\right] = \left[\frac{\partial N_z}{\partial y},\ -\frac{\partial N_z}{\partial x},\ 0 \right] \tag{3.59}$$

With a constant conductivity σ Laplace's equation is achieved:

$$\frac{\partial^2 N_z}{\partial x^2} + \frac{\partial^2 N_z}{\partial y^2} = 0 \tag{3.60}$$

The functional of the m-th triangular element is

$$\begin{aligned} F_m &= \int_{A_m} \left(\frac{1}{2\sigma} J^2 \right) dS \\ &= \frac{1}{2\sigma} \int_{A_m} (\text{curl}\mathbf{N} \cdot \text{curl}\mathbf{N}) dS \\ &= \frac{1}{2\sigma} A_m (b^2 + c^2) \end{aligned} \tag{3.61}$$

Since a, b, and c are functions of the potential values N_1, N_2, and N_3 of the three nodes of the triangular element, the functional F_m becomes

$$F_m = \frac{1}{2} [N_{123}]^t [S_m][N_{123}] \tag{3.62}$$

where $[N_{123}] = [N_1, N_2, N_3]^t$, and $[S_m]$ is the stiffness matrix. Its generic component is

$$s_{ij} = \frac{1}{4\sigma A_m} (q_i q_j + r_i r_j) \tag{3.63}$$

3.4.3 A Magnetostatic Field

Starting from Maxwell's equations given in Chapter 1, the magnetostatic field problem is described by the equations div**B** = 0, curl**H** = **J**, and **B** = μ**H**. Since div(curl**B**) = 0, a magnetic vector potential **A** is defined such as **B** = curl**A**. The field problem is described by the almost harmonic vector equation

$$\text{curl}\mathbf{H} = \text{curl}\frac{\mathbf{B}}{\mu} = \text{curl}\frac{1}{\mu}\text{curl}\mathbf{A} = \mathbf{J} \tag{3.64}$$

In the 2D field, using Cartesian coordinates, the current density vector **J** has only a component normal to the plane (x, y), which is only a z-axis component. Consequently, the magnetic vector potential **A** has only a z-axis component, i.e., the vector **A** is parallel to the vector **J**. Then these vectors can be expressed as **J** = $[0, 0, J_z]$ and as **A** = $[0, 0, A_z]$. Equation (3.64) becomes

$$\frac{\partial}{\partial x}\frac{1}{\mu_x}\frac{\partial A_z}{\partial x}+\frac{\partial}{\partial y}\frac{1}{\mu_y}\frac{\partial A_z}{\partial y}=-J_z \tag{3.65}$$

The magnetic flux-density vector **B** has components only on the plane (x, y), because

$$\mathbf{B}=\text{curl}\left[0,0,A_z\right]=\left[\frac{\partial A_z}{\partial y},-\frac{\partial A_z}{\partial x},0\right] \tag{3.66}$$

In homogeneous media, with magnetic permeability $\mu_x = \mu_y = \mu$, Equation (3.66) is reduced to a Poisson's equation:

$$\frac{\partial^2 A_z}{\partial x^2}+\frac{\partial^2 A_z}{\partial y^2}=-\mu J_z \tag{3.67}$$

By means of the linear interpolation of Equation (3.19), the functional referred to the m-th finite element is given by

$$\begin{aligned} F_m &= \int_{A_m}\left(\frac{1}{2}\mathbf{B}\cdot\mathbf{H}-\mathbf{J}\cdot\mathbf{A}\right)dS \\ &= \frac{1}{2\mu}\int_{A_m}(\text{curl}\mathbf{A}\cdot\text{curl}\mathbf{A})dS - J_z\int_{A_m}A_z dS \\ &= \frac{1}{2\mu}A_m(b^2+c^2)-J_zA_m\left(a+b\frac{x_1+x_2+x_3}{3}+c\frac{y_1+y_2+y_3}{3}\right)\end{aligned} \tag{3.68}$$

Since a, b, and c are functions of the potentials A_1, A_2, and A_3 of the three nodes of the triangular element, F_m becomes

$$F_m=\frac{1}{2}\left[A_{123}\right]^t\left[S_m\right]\left[A_{123}\right]-\left[A_{123}\right]^t\left[T_m\right] \tag{3.69}$$

where $[A_{123}] = [A_1, A_2, A_3]^t$, and $[S_m]$ is the stiffness matrix, whose generic component is

$$s_{ij}=\frac{1}{4\mu A_m}(q_iq_j+r_ir_j) \tag{3.70}$$

The column vector $[T_m]$ is a function of the current density, which is imposed as the source. Its generic component is

$$t_i=-\frac{A_m}{3}J_z \tag{3.71}$$

3.4.4 A Magnetostatic Field without Current

Let us consider a magnetostatic field problem characterized by a null current density vector $\mathbf{J} = \mathbf{0}$ in the whole domain. In such a case, the equations describing the field are div$\mathbf{B} = 0$, curl$\mathbf{H} = \mathbf{0}$, and $\mathbf{B} - \mu\mathbf{H}$. Since the magnetic field strength $\mathbf{H}$ is irrotational, a magnetic scalar potential Ψ can be defined such as $\mathbf{H} = -\text{grad}\Psi$, as $\text{curl}(\text{grad}\Psi) = \mathbf{0}$. From these equations, the field problem is described by the almost harmonic scalar equation

$$\text{div}\mathbf{B} = \text{div}(\mu\mathbf{H}) = \text{div}\mu(-\text{grad}\Psi) = 0 \tag{3.72}$$

In the 2D field, described using the Cartesian coordinates, and with a constant magnetic permeability μ, Equation (3.72) becomes Laplace's equation given by

$$\frac{\partial^2\Psi}{\partial x^2} + \frac{\partial^2\Psi}{\partial y^2} = 0 \tag{3.73}$$

The functional F_m of the m-th triangular element is given by

$$\begin{aligned} F_m &= \int_{A_m}\left(\frac{1}{2}\mathbf{B}\cdot\mathbf{H}\right)dS \\ &= \frac{1}{2}\mu\int_{A_m}(\text{grad}\Psi\cdot\text{grad}\Psi)dS \\ &= \frac{1}{2}\mu A_m(b^2 + c^2) \end{aligned} \tag{3.74}$$

Since a, b, and c depend on the potentials Ψ_1, Ψ_2, and Ψ_3 of the three nodes of the triangular element, Equation (3.74) becomes

$$F_m = \frac{1}{2}\left[\Psi_{123}\right]^t\left[S_m\right]\left[\Psi_{123}\right] \tag{3.75}$$

where $[\Psi_{123}]=[\Psi_1, \Psi_2, \Psi_3]^t$, and $[S_m]$ is the stiffness matrix whose generic component corresponds to

$$S_{ij} = \frac{\mu}{4A_m}(q_i q_j + r_i r_j) \tag{3.76}$$

3.4.5 A Magnetic Field with Permanent Magnets

The equations that describe the field are curl$\mathbf{H} = \mathbf{0}$, div$\mathbf{B} = 0$, and $\mathbf{B} = \mathbf{B}_{res} + \mu\mathbf{H}$, where $\mathbf{B}_{res}$ is the residual flux density vector of the permanent magnet.

However, since div(curl**B**) = 0 and div**B** = 0, it is possible to define a magnetic vector potential **A** so that **B** = curl**A**. From these equations, it results that

$$\begin{aligned}\text{curl}\mathbf{H} &= \text{curl}\left(\frac{\mathbf{B}-\mathbf{B}_{\text{res}}}{\mu}\right) = \text{curl}\left(\frac{\mathbf{B}}{\mu}\right) - \text{curl}\left(\frac{\mathbf{B}_{\text{res}}}{\mu}\right) \\ &= \text{curl}\frac{1}{\mu}(\text{curl}\mathbf{A}) - \text{curl}\left(\frac{\mathbf{B}_{\text{res}}}{\mu}\right) = \mathbf{0}\end{aligned} \tag{3.77}$$

The vector **M** = curl($\mathbf{B}_{res}/\mu$) is the magnetization vector. The differential equation that describes the field problem is

$$\text{curl}\frac{1}{\mu}(\text{curl}\mathbf{A}) = \text{curl}\left(\frac{\mathbf{B}_{\text{res}}}{\mu}\right) \tag{3.78}$$

The functional that is associated with the field, referred to as the area A_m of the m-th triangular element, is given by

$$\begin{aligned}F_m &= \int_{A_m} \frac{1}{2\mu}(B - B_{res})^2\, dS \\ &= \frac{1}{2\mu}\int_{A_m} (B^2 - 2BB_{res} + B_{res}{}^2)dS \\ &= \frac{1}{2\mu}\int_{A_m} (\text{curl}\mathbf{A}\cdot\text{curl}\mathbf{A} - 2B_{res}\left|\text{curl}\mathbf{A}\right| + B_{res}{}^2)dS\end{aligned} \tag{3.79}$$

Substituting the linear interpolation given in Equation (3.19) in $\mathbf{A} = A_z\mathbf{u}_z$, the functional F_m (3.79) becomes

$$F_m = \frac{1}{2}\left[A_{123}\right]^t\left[S_m\right]\left[A_{123}\right] - \left[A_{123}\right]^t\left[T_m\right] \tag{3.80}$$

where $[A_{123}] = [A_1, A_2, A_3]^t$, and the generic component of the stiffness matrix $[S_m]$ is

$$s_{ij} = \frac{1}{4\mu A_m}(q_i q_j + r_i r_j) \tag{3.81}$$

The generic component of the column vector $[T_m]$ is a function of the residual flux density, and it is expressed as

$$t_i = \frac{1}{\mu}\left(B_{res,y}q_i - B_{res,x}r_i\right) \tag{3.82}$$

3.5 Appendix: Integration in Triangular Elements

Let us consider a triangular element defined in the plane (x, y) by the three points 1:(x_1,y_1), 2:(x_2,y_2), and 3:(x_3,y_3), so that its center of gravity is in the origin of the coordinate system, i.e.,

$$\frac{x_1 + x_2 + x_3}{3} = \frac{y_1 + y_2 + y_3}{3} = 0$$

Some surface integrals on the triangle area A_m are reported as follows:

$$\int_{A_m} x \, dx \, dy = 0$$

$$\int_{A_m} y \, dx \, dy = 0$$

$$\int_{A_m} dx \, dy = \frac{1}{2} \begin{vmatrix} 1 & x_1 & y_1 \\ 1 & x_2 & y_2 \\ 1 & x_3 & y_3 \end{vmatrix} = A_m$$

$$\int_{A_m} x^2 \, dx \, dy = \frac{A_m}{12} \left(x_1^2 + x_2^2 + x_3^2 \right)$$

$$\int_{A_m} y^2 \, dx \, dy = \frac{A_m}{12} \left(y_1^2 + y_2^2 + y_3^2 \right)$$

$$\int_{A_m} xy \, dx \, dy = \frac{A_m}{12} \left(x_1 y_1 + x_2 y_2 + x_3 y_3 \right)$$

References

See references in Chapter 2.

4

The Analysis Procedure Using the Finite Element Method

4.1 Introduction

This chapter deals with the analysis of the electrical machines by means of the finite element method. Instead of a generic problem, the magnetic field problem is investigated. Such a problem is the most common in the analysis of electrical machines, including transformers, rotating machines, and actuators.

In any case, the study of the magnetic field alone is not sufficient for a complete analysis of the machine. For instance, the evaluation of the electric field between the turns of the coils requires an electrostatic field analysis; the computation of the Joule losses in conductors with restricted sections, especially in the terminals, requires a current field analysis. However, the extension of the finite element analysis to vector fields of a nature different from the magnetic field is left to the reader.

4.2 Reduction of the Field Problem to a Two-Dimensional Problem

The electrical machine to be analyzed is obviously a three-dimensional (3D) structure. However, a 3D analysis is required to subdivide the whole structure by means of 3D finite elements (as seen in Chapter 2, Figure 2.1(c)) and requires heavy processing and long computation time. That is why, if possible, the field problem should be reduced to a 2D problem. This is carried out by individuating any symmetry of the machine. Such a symmetry can be of two kinds:

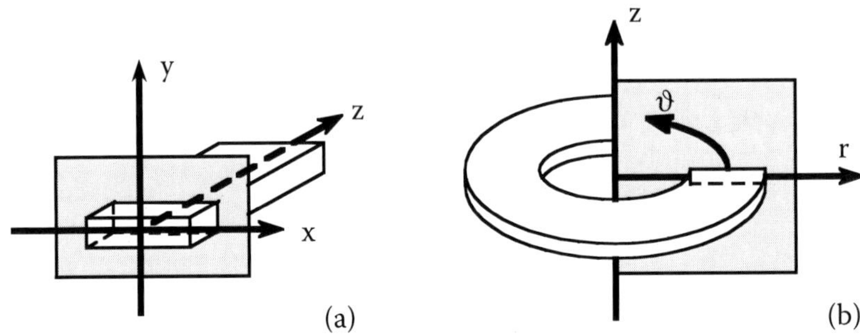

FIGURE 4.1
Planar symmetry (a) and axial symmetry (b).

1. *Planar symmetry (or xy symmetry)*: The magnetic phenomena are assumed to be identical on each plane (x, y) normal to the z-axis, as shown in Figure 4.1(a). The effects of the ending edges of the machines are therefore neglected.
2. *Axial symmetry (or rz symmetry)*: It is assumed that the magnetic phenomena are repeated identical on each semiplane (r, z) obtained as a rotation around the z-axis, which is called the symmetry axis, as shown in Figure 4.1(b).

In this chapter, a field problem with planar symmetry is considered. The study will be carried out in the plane (x, y) perpendicular to the z-axis, as in Figure 4.1(a). In the next chapter, the finite element analysis of an axial-symmetric structure will be presented and discussed.

As described in Chapter 3, the study of a 2D magnetostatic field problem with planar symmetry is simplified as follows:

1. The current density vector **J** has a z-axis component only, which is $\mathbf{J} = [0, 0, J_z]$.
2. The magnetic vector potential **A** is parallel to the vector **J**, thus it has a z-axis component only, i.e., $\mathbf{A} = [0, 0, A_z]$. As regards the divergence of the magnetic vector potential, Coulomb's position (1.68) is generally adopted, which is $\text{div}\mathbf{A} = 0$.
3. The flux density vector **B** has components only on the plane (x, y), as stated by Equation (3.66).
4. With a constant magnetic permeability μ, the field problem is described by Poisson's equation (3.67), which is reported again:

$$\frac{\partial^2 A_z}{\partial x^2} + \frac{\partial^2 A_z}{\partial y^2} = -\mu J_z \tag{4.1}$$

4.3 Boundary Conditions

The magnetic vector potential A_z can be determined from Poisson's equation (4.1), in each point of the domain *D*, once the current density J_z is fixed and the value of A_z is known on the boundary Γ of the domain itself. Going into details, in a part of the boundary, say Γ_1, the value of A_z is assigned (Dirichlet's condition), and in the remanent part of the boundary, say Γ_2, the value of the derivative of A_z normal to the boundary line is assigned (Neumann's condition). See Chapter 2, Section 2.2.

Therefore, the assignment of the boundary conditions is a thorough operation, which assumes a fundamental importance in the solution of the field problem. The choice of the boundary conditions not only influences the final solution, but also can further reduce the domain under study.

The boundary conditions are gathered in a group hereafter.

4.3.1 Dirichlet's Condition

This condition corresponds to assign the value of the magnetic vector potential A_z on a given part of the boundary. Generally, the value that is assigned is constant, so that the boundary line assumes the same value of magnetic vector potential A_z. It follows that the flux lines are tangential to the boundary itself, and no flux line crosses that boundary. It is common to assign the homogeneous Dirichlet's condition (2.4), fixing the magnetic vector potential $A_z = 0$ along all or part of the boundary.

Such a condition is equivalent to considering an external material with null magnetic permeability, which is a magnetic insulating material just outside the domain.

As an example, let us refer to a synchronous generator with salient poles. By neglecting the edge effects (essentially due to the end windings), a planar symmetry is recognized, so that only the section of the machine represented in Figure 4.2 can be analyzed. Since the flux lines are confined within the

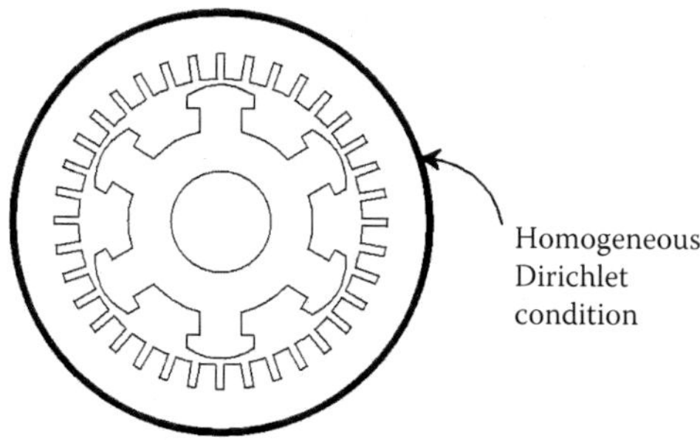

FIGURE 4.2
Homogeneous Dirichlet's condition along the external circumference of a synchronous generator.

stator back iron, the condition $A_z = 0$ is assigned along all the external circumference of the synchronous machine. Figure 4.2. highlights this homogeneous Dirichlet's condition, by means of a bold line.

4.3.2 Neumann's Condition

This condition corresponds to imposing a given value to the derivative of A_z normal to the boundary, so that the flux lines have a given incidence angle with the boundary. In the case of homogeneous Neumann's condition (4.6), the flux lines are forced to be perpendicular to the boundary line. The condition (4.6) applied to a magnetic field constrains the flux density vector **B** to have components only normal to the boundary line.

This condition is equivalent to having an external material with infinite magnetic permeability just outside the domain.

As an example, let us consider a single-phase reactance, with legs of rectangular form. Neglecting the edge effects, a planar symmetry is considered, so that a section of the structure can be simply analyzed, as shown in Figure 4.3(a). It is easy to notice that the structure of the reactance appears to be symmetric with respect to the axis AA′, both geometrically and magnetically. Since the lower part of the machine is exactly the mirrored image of the upper part, with respect to the axis AA′, and the flux density components cannot be discontinuous, then the flux lines must be normal to the axis AA′. The structure can be simplified as shown in Figure 4.3(b), imposing the homogeneous Neumann's condition along the boundary AA′. Thus the analysis is then carried out on such a portion of the structure only.

4.3.3 Periodic Condition

This condition corresponds to assigning a correspondence between the values of the magnetic vector potential along two (or more) boundary lines of the structure. In general terms, two or more boundary lines are chosen. Among them a principal line is selected, and the potential of the other (secondary) lines is expressed as a function of the potential of the principal line.

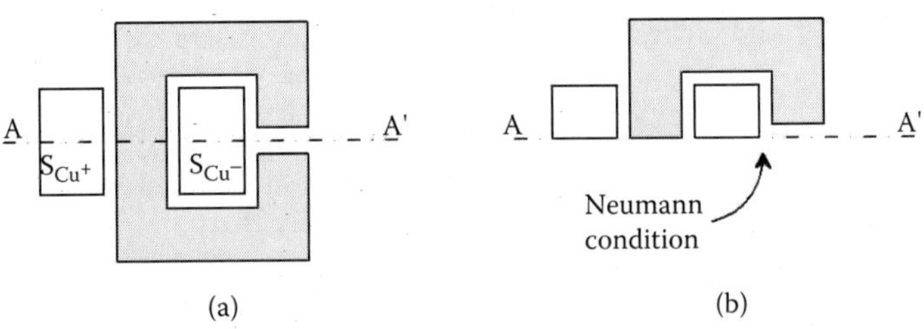

FIGURE 4.3
Reduction of the analysis domain by means of Neumann's conditions.

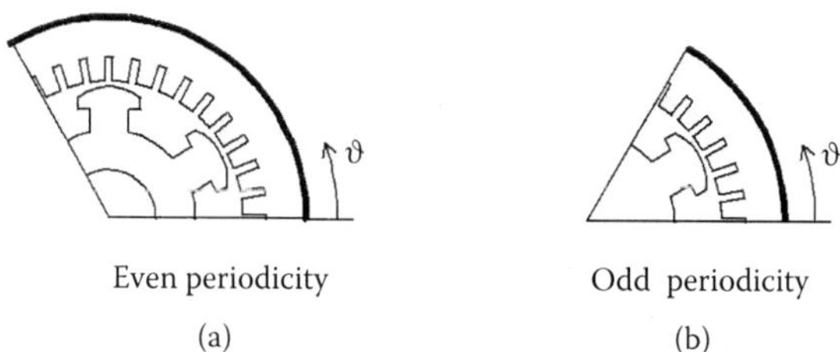

FIGURE 4.4
Reduction of the analysis domain by means of the periodic condition.

These boundary conditions are useful in structures that exhibit a repetition of the electromagnetic fields, but neither Dirichlet's nor Neumann's boundary conditions are appropriate. Assigning the periodic conditions along at least two lines of the boundary, the structure can be reduced again and the analysis carried out only on a part of it, the disregarded parts becoming mirrored images.

As an example, let us refer to the section of the synchronous generator of Figure 4.2. The machine exhibits a number of pole pairs equal to 3, i.e., a recurrence with respect to the azimuthal coordinate ϑ equal to 3. It is possible to study only a third of the machine, as represented in Figure 4.4(a), by imposing on the boundary that

$$A_z(r,\vartheta) = +A_z\left(r,\vartheta + (2k)\frac{\pi}{p}\right) \qquad k = 1, 2, 3, \ldots \tag{4.2}$$

which is named the *even* periodic condition.

In addition, it is possible to study only a sixth of the machine, as shown in Figure 4.4(b), by imposing on the boundary that

$$A_z(r,\vartheta) = -A_z\left(r,\vartheta + (2k-1)\frac{\pi}{p}\right) \qquad k = 1, 2, 3, \ldots \tag{4.3}$$

which is named the *odd* periodic condition.

Finally, along the remaining part of the external circumference, the value of magnetic vector potential $A_z = 0$ is assigned as in Figure 4.2.

As it is evident in the previous example, the process of locating of symmetry lines and periodic lines, together with the corresponding boundary conditions, may reduce the study of a complex structure to only a part of it. This conveys to a reduction of the field problem domain, with the twofold advantages of shorter computing time and of greater accuracy in the analysis of the remaining part.

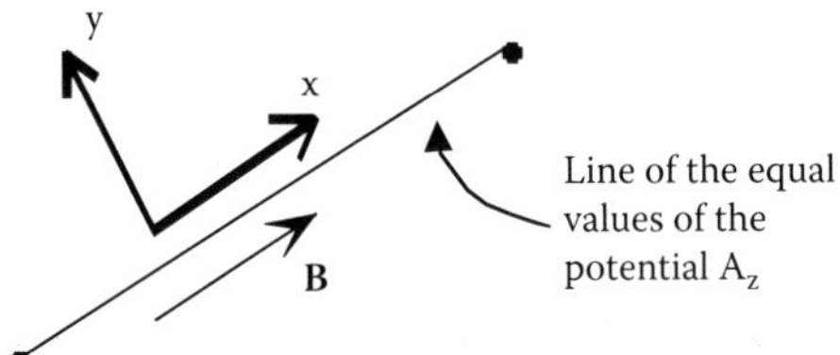

FIGURE 4.5
Flux lines, i.e., lines of the flux density vector **B**.

4.4 Computation of the Solved Structure

Once the structure is drawn and the boundary conditions are assigned, the electric and magnetic properties of the materials are assigned together with the field sources. The latter are current density or permanent magnet magnetization in the magnetostatic field problem. The solution of the field problem (4.1) consists of the knowledge of the magnetic vector potential A_z in each point of the domain. The flux density and magnetic strength vectors are derived from A_z. Some considerations are reported in the following sections.

4.4.1 Drawing the Flux Lines

The flux lines are the lines to which the flux density vector **B** is parallel. They correspond to the equipotential lines of the magnetic vector potential A_z.

In Figure 4.5, the potential A_z is assumed to be constant along a line. Let us refer to a coordinate system Oxy such as the x-axis coincides with the equipotential line. By applying the curl to the magnetic vector potential, the components of the flux density vector are obtained as

$$B_x = \frac{\partial A_z}{\partial y}$$
$$B_y = -\frac{\partial A_z}{\partial x} = 0 \tag{4.4}$$

Equation (4.4) shows that the flux density vector has a component only in the direction of the equipotential line of A_z.

4.4.2 Magnetic Flux and Flux Linkage

The computation of the magnetic flux (1.8), through a surface S oriented by the normal unity vector **n**, is given by

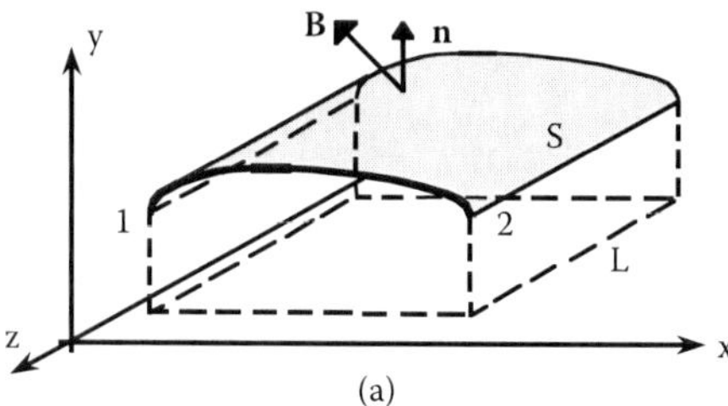

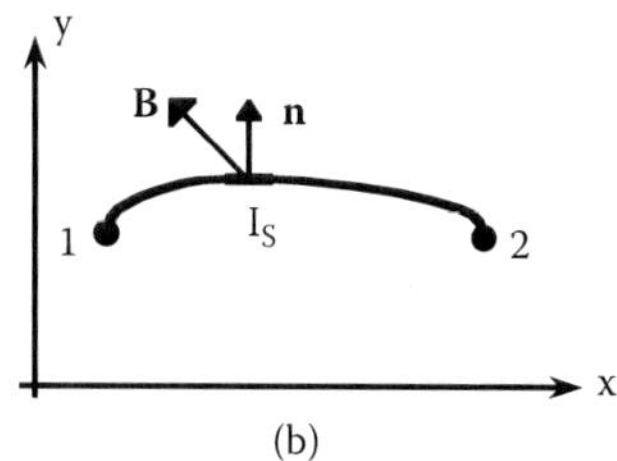

FIGURE 4.6
Surface integral in the 2D problem.

$$\Phi = \int_S \mathbf{B} \cdot \mathbf{n} \, dS = L \int_{l_S} \mathbf{B} \cdot \mathbf{n} \, dl \tag{4.5}$$

In the 2D study, the surface integral is reduced to the computation of a line integral: the magnetic flux is computed by integrating, along the given oriented line l_S, the flux density normal to the line itself; and the result is multiplied by the length L. This is shown in Figure 4.6.

Otherwise, applying the Stokes theorem (1.21), the magnetic flux can be obtained as the loop integral of the magnetic vector potential **A** along the closed line l, bordering the surface S, which is

$$\Phi = \int_S \mathbf{B} \cdot \mathbf{n} \, dS = \int_S (\mathrm{curl}\mathbf{A}) \cdot \mathbf{n} \, dS = \oint_l \mathbf{A} \cdot \mathbf{t} \, dl \tag{4.6}$$

In a 2D problem, such a computation results very simply. Referring to Figure 4.7, it is observed that the line integral is null along the segments 1′2′ and 21, since the magnetic vector potential has the z-axis component only, which is perpendicular to the segments. Along the other two lines, which have length equal to L and are parallel to **A**, the value of A_z is constant. Then the magnetic flux is equal to the difference of A_z in the two points 1 and 2, times the length L, which is

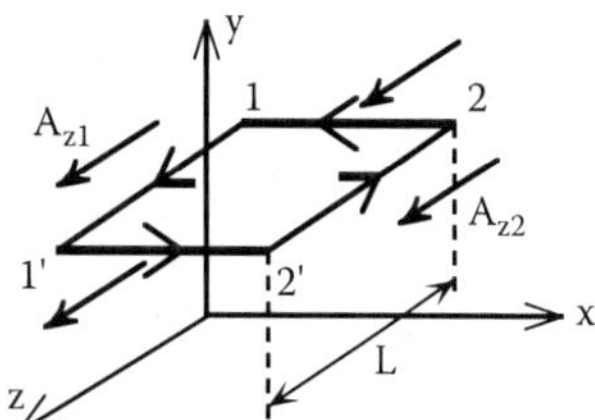

FIGURE 4.7
Computation of the magnetic flux by means of the magnetic vector potential.

$$\Phi = (A_{z1} - A_{z2})L \tag{4.7}$$

In the case of massive conductors with cross-section S_{Cu}, it is convenient to consider the average value of the magnetic vector potential A_z on the conductor surface, which is

$$\Phi = \frac{L}{S_{Cu}} \left(\int_{S_{Cu+}} A_z \, dS - \int_{S_{Cu-}} A_z \, dS \right) \tag{4.8}$$

where S_{Cu+} and S_{Cu-} are the conductor surfaces, positively oriented and negatively oriented, respectively.

If the current is supposed to be uniformly distributed on the conductor surface, the flux linkage Λ is obtained by multiplying the magnetic flux Φ by the number of turns N_t that link the flux.

4.4.3 Joule Power Losses

If the current density is known in each point of the structure, at the time instant t, the instantaneous Joule power losses are given by

$$p_J(t) = \int_\tau \rho J^2 \, d\tau \tag{4.9}$$

Of course, with planar symmetry in which $\mathbf{J} = J_z \mathbf{u}_z$, the volume integral is reduced to

$$p_J(t) = L \int_S \rho J_z^2 \, dS \tag{4.10}$$

where L is the axial length of the structure.

In the case of stationary fields, the current density is not time dependent, so that the power is also constant. In case of fields that are not constant, but are sinusoidal in time, the symbolic representation with complex phasors (1.54) is used. Again the current density has a z-axis component only, but is now represented in complex form as

$$\dot{\mathbf{J}} = \dot{J}_z \mathbf{u}_z = |J_z| e^{j\varphi} \mathbf{u}_z \tag{4.11}$$

The average power that is lost in a period due to Joule effect is given by

$$p_J = \int_\tau \rho \frac{|J_z|^2}{2} \, d\tau = \frac{1}{2} \rho \int_\tau \dot{J} \cdot \tilde{J} \, d\tau \tag{4.12}$$

where the factor 1/2 appears, because $|J_z|$ denotes the maximum of the sinusoidal variation.

4.4.4 Magnetic Energy

The magnetic energy stored in the structure can be computed as

$$W_m = \int_\tau \int_0^B \mathbf{H} \cdot d\mathbf{B} \, d\tau = \int_\tau \int_0^A \mathbf{J} \cdot d\mathbf{A} \, d\tau \tag{4.13}$$

that with linear media becomes

$$\begin{aligned} W_m &= \frac{1}{2}\int_\tau \mathbf{H}\cdot\mathbf{B}\,d\tau = \frac{1}{2}\int_\tau \mu H^2\,d\tau \\ &= \frac{1}{2}\int_\tau \mathbf{J}\cdot\mathbf{A}\,d\tau = \frac{1}{2}\int_\tau J_z A_z\,d\tau \end{aligned} \tag{4.14}$$

In the case of stationary fields, the energy is constant, but in the case of variable fields, this energy has to be considered as instantaneous, at the time t. In particular, if the variable fields have a sinusoidal variation, and they are described by using the symbolic notation, it is possible to compute the average energy stored in a period as

$$W_m = \frac{1}{2}\int_\tau \frac{|B|\cdot|H|}{2}\,d\tau = \frac{1}{2}\int_\tau \frac{|J_z|\cdot|A_z|}{2}\,d\tau \tag{4.15}$$

where 1/2 appears because $|B|$ and $|H|$ represent the maximum value of the flux density and the magnetic field strength, respectively, and also $|J_z|$ and $|A_z|$ represent the maximum value of the current density and magnetic vector potential, respectively.

4.4.5 Magnetic Coenergy

The magnetic coenergy can be computed as

$$\begin{aligned} W'_m &= \int_\tau \mathbf{H}\cdot\mathbf{B}\,d\tau - W_m \\ &= \int_\tau \mathbf{J}\cdot\mathbf{A}\,d\tau - W_m \end{aligned} \tag{4.16}$$

4.4.6 Magnetic Forces

To compute the magnetic forces, which act on an object within the domain, two methods are possible.

4.4.6.1 *Tensor of Maxwell's Strength*

As discussed in Chapter 1, the method of Maxwell's strength tensor allows the magnetic forces acting on an object within the magnetic field to be computed. At first a suitable surface that contains the object is selected. In the 2D problem this surface is reduced to a line, named *l*. The tangential and normal components of the force are computed by means of the integral along the line *l*, as

$$F_t = L\mu_o \oint_l H_t H_n \, dl$$
$$F_n = L\frac{\mu_o}{2} \oint_l \left(H_n^2 - H_t^2\right) dl \tag{4.17}$$

The magnitude of the force and the direction of the force (which is expressed with respect the direction normal to the surface) are given by Equation (1.100) and Equation (1.101), respectively.

As far as the computation time is concerned, it is worth noticing that Maxwell's strength tensor requires only one field solution.

As far as the computation accuracy is concerned, it is observed that with a correct magnetic field distribution, the method of Maxwell's strength tensor provides an exact value of the force, disregarding the choice of the surface (i.e., the line) of integration. However, due to the intrinsic approximation of the adopted numerical method, the continuity of the field components among the adjacent elements is not guaranteed. As a result, the computed forces depend on both the path of integration and the adopted mesh of the domain.

Some preliminary tests are essential, in order to verify the accuracy obtainable with the adopted mesh of the structure and the selected path for the integration. For instance, a possible check may be the evaluation of the force in a circumstance where this force should be null. An example to check the accuracy of the results by using the method of Maxwell's strength tensor is illustrated in Chapter 12.

4.4.6.2 The Virtual Works Method

As an alternative to Maxwell's strength tensor, or just to verify its accuracy in some conditions that are particularly critical, the virtual works method can be adopted. Unfortunately, since this technique is based on the computation of the variation of the magnetic energy between two different positions, it requires the solution of at least two field problems, with an increase of the computation time and with the not negligible problem of the choice of the width of the position variation.

This method is based on the comparison of the energy balance between two different positions, corresponding to a virtual change of position of the object in the direction where the force is computed. As will be discussed in the following chapter, the force is expressed as the derivative of the magnetic coenergy with respect to the virtual movement, with constant electrical sources. This implies the solution of the field problem at least with the structure in two different configurations, with an increase of computation and the problem of the width of the movement.

4.4.7 Determination of the Electrical Parameters

From the field solution, it could be interesting to calculate the parameters of a zero-dimensional electrical network related to the electrical machine that is analyzed. In particular, the electrical resistances, inductances, and capacitors are normally computed.

The resistance is computed from the Joule power losses, as

$$R = \frac{p_J}{I^2} \tag{4.18}$$

In case of linearity, the inductance is obtained from the stored magnetic energy and from the flux linkage, as

$$L = \frac{2W_m}{I^2} = \frac{\Lambda}{I} \tag{4.19}$$

Further observations are reported in the chapters that follow.

References

See references in Chapter 2.

5

Cylindrical Magnetic Devices

In this chapter, a cylindrical magnetic device is considered. A magnetostatic axial-symmetric analysis is carried out. At first, an analytical study is presented. Then the basic concepts about the energy conversion are summarized. Finally, the finite element method is applied for the analysis of the device.

5.1 Introduction

The cylindrical magnetic device is used in applications requiring high thrust but small stroke, e.g., to move power switches, to drive valves, and so on. A section of the cylindrical magnetic device is shown in Figure 5.1. It consists of a fixed magnetic part, i.e., the core, and of a cylindrical magnetic plunger moving axially within the core, guided by a nonmagnetic guide. A coil composed by N_t turns is wound inside the core.

When no currents feed the coil, the plunger is in its rest position, at the lower position, at which the air-gap thickness is maximum, i.e., $g = g_{max}$. When the coil is fed by a current with a sufficiently high amplitude, the plunger pops up, reaching the upper position, at which the air-gap is minimum, i.e., $g = g_{min}$.

The radial gap between the core and the plunger, corresponding to the driver thickness, is constant and equal to t.

5.2 Analytical Study of the Magnetic Device

Before applying the finite element method, an analytical analysis of the magnetic device is carried out. In spite of some convenient assumptions, this study allows a first evaluation of the main electrical and magnetic quantities, in such a way as to achieve their expected value. In the following finite element analysis, it will be easy and rapid to detect any possible errors, if any discrepancy will be found. Furthermore, the analytical and the numerical

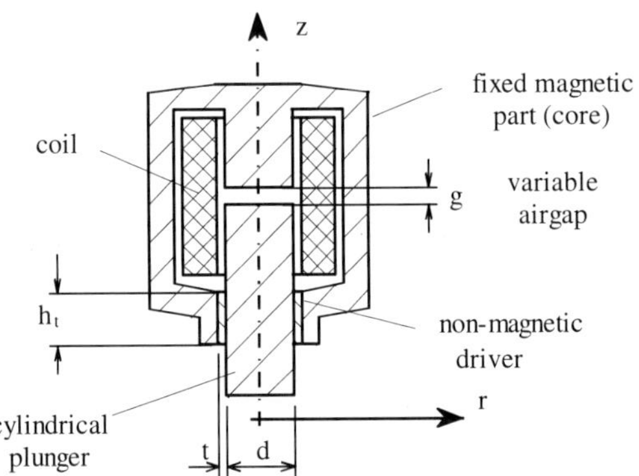

FIGURE 5.1
Sketch of the cylindrical magnetic device.

results can be compared so as to evaluate the effects of the simplifications of the analytical study.

In the analytical analysis, the flux leakages and the air-gap edge effects are neglected. Thus, in correspondence of the air-gap the flux lines are assumed to be normal to the two surfaces.

5.2.1 Computation of the Magnetic Quantities

Assuming that the coil carries a current i is feeding the coil, the flux density in the two air-gaps is computed as follows. Referring to Figure 5.2, using Ampere's law and neglecting the magnetic voltage drop along the iron paths yields

$$N_t i = H_g g + H_t t \tag{5.1}$$

where H_g and H_t are the magnetic field strength in the air-gap g and t, respectively. Because of the small air-gap thickness t, a constant value of H_t is considered.

Gauss's law yields

$$B_g \frac{\pi d^2}{4} = B_t h_t \pi(d+t) \tag{5.2}$$

where B_g and B_t are the flux density in the air-gap g and t, respectively. In Equation (5.2), $\pi(d + t)$ is the average circumference in the middle of the air-gap t. From the two equations (5.1) and (5.2), the flux density B_g is obtained as

$$B_g = \mu_o \frac{N_t i}{g + \dfrac{d^2 t}{4 h_t \pi(d+t)}} = \mu_o \frac{N_t i}{g + t'} \tag{5.3}$$

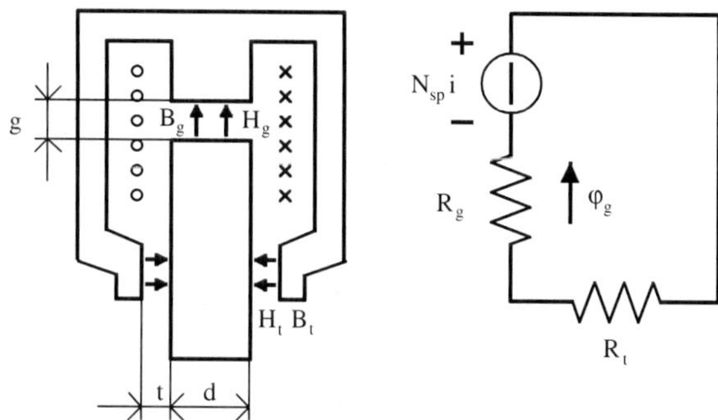

FIGURE 5.2
Definition of the magnetic strength and the flux density in the two air-gaps of the device (a) and equivalent magnetic circuit (b).

with obvious meaning of the term t′.

Alternatively, the value of B_g can be obtained from the equivalent magnetic circuit, shown in Figure 5.2(b). The magnetic reluctance corresponding to the air-gap g is

$$R_g = \frac{g}{\mu_o \frac{\pi d^2}{4}} \tag{5.4}$$

and the magnetic reluctance corresponding to the air-gap t is

$$R_t = \frac{\ln(1+2t/d)}{\mu_o 2\pi h_t} \tag{5.5}$$

With a small ratio t/d, so that ln(1 + 2t/d) is approximated by t/(d + t), Equation (5.5) can be expressed as

$$R_t = \frac{t}{\mu_o \pi h_t (d+t)} \tag{5.6}$$

The magnetic flux φ_g through the base surface of the plunger is computed as

$$\varphi_g = \frac{N_t i}{R_g + R_t} \tag{5.7}$$

The flux density B_g is achieved as the ratio between φ_g and the air-gap surface $\pi d^2/4$, as

$$B_g = \frac{\varphi_g}{\left(\frac{\pi d^2}{4}\right)} \tag{5.8}$$

and therefore it results as in Equation (5.3).

The flux linkage with the coil results as follows:

$$\lambda = N_t \varphi_g = \mu_o \pi d^2 \frac{N_t^2 i}{4(g + t')} \tag{5.9}$$

It is assumed that the magnetic device is without hysteretic phenomena and without losses. If needed, the latter may be considered as external losses. In other words, the system is split in a part without losses (considered in the following study) and a part with losses (not considered here). The magnetic device is then considered to be a conservative system, which can store magnetic energy in a reversible way.

With these assumptions, the current i, the flux linkage λ, and the air-gap thickness g are considered to be the state variables, which define completely the state of the magnetic system. However, Equation (5.9) shows that there is a link among these three quantities (each value of λ corresponds to precise values of i and g). Since hysteresis is neglected, such a correspondence is univocal. Thus, only two state variables are enough to describe completely the state of the system, and they will be chosen on the basis of best convenience.

5.2.2 Magnetic Energy and Coenergy

The magnetic energy can be determined by the system geometry, by the magnetic characteristic of the materials, and by the coil MMF. For a given configuration, with a fixed position of the plunger, i.e., an air-gap $g = g_o$, the energy stored in the magnetic field is computed as the energy furnished by the electric source to bring the current from zero to the value $i = i_o$, to which corresponds the flux linkage $\lambda = \lambda_o = \lambda(i_o, g_o)$. It results that

$$W_m = \int_0^{\lambda_o} i(\lambda) d\lambda \tag{5.10}$$

Such an energy corresponds to the surface that is horizontally hatched in Figure 5.3. With a reversible process, this magnetic energy is completely returned to the electrical circuit, when the current is brought back to zero.

The surface that is vertically hatched, below the curve of Figure 5.3, corresponds to the magnetic coenergy and can be expressed as

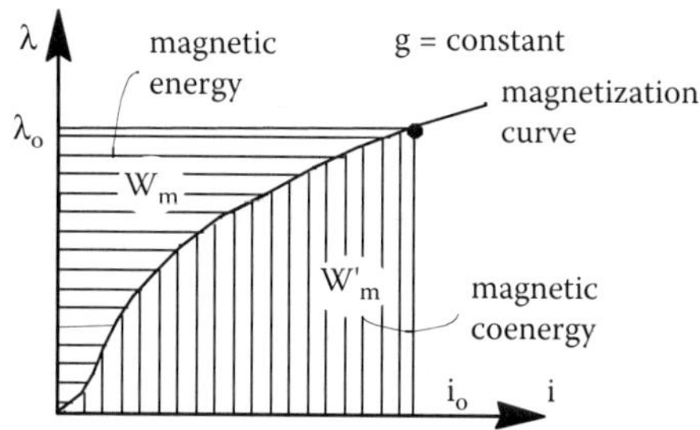

FIGURE 5.3
Magnetization curve, magnetic energy, and coenergy.

$$\begin{aligned} W'_m &= \int_0^{\lambda_o} \lambda(i)di \\ &= \lambda_o i_o - W_m \end{aligned} \tag{5.11}$$

This quantity, which is without an evident physical meaning, proves to be very useful in the computation of the electromechanical quantities. Since the system is conservative, both magnetic energy and coenergy may be considered as state functions. Thanks to the univocal correspondence among g, i, and λ, the magnetic energy and coenergy are usually expressed as functions of the current and the flux linkage, that is $W_m(\lambda, i)$ e $W'_m(\lambda, i)$.

If the system is linear, i.e., the magnetic materials do not reach the saturation, the magnetic energy and coenergy coincide and are computed as

$$W_m = W'_m = \frac{1}{2}\lambda_o i_o \tag{5.12}$$

In the equations above, the magnetic energy is expressed by means of integral quantities that can be measured at the terminals of the device. The magnetic energy can be also expressed as a function of the specific property (or point property) of the magnetic field. Since the flux density has been supposed to be constant in the two air-gaps, the edge effects to be negligible, and the magnetic field strength to be null in the iron paths, the magnetic energy results in

$$W_m = \left(\frac{1}{2\mu_o} B_g{}^2\right)\frac{\pi d^2}{4} g + \left(\frac{1}{2\mu_o} B_t{}^2\right) h_t \pi (d+t) t \tag{5.13}$$

5.2.3 Apparent Inductance and Differential Inductance

The apparent inductance and the differential inductance are functions of the air-gap g. Due to the nonlinear characteristic of the ferromagnetic material

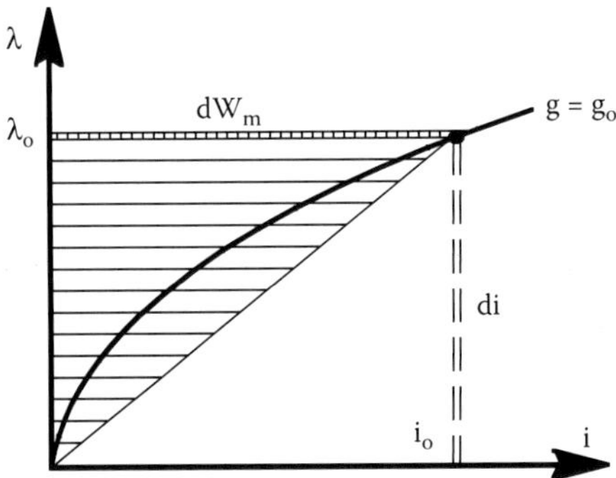

FIGURE 5.4
Current-flux linkage characteristic and surface corresponding to the apparent inductance computation.

(the core and the plunger), they are also functions of the current i flowing through the coil.

Let us consider the operating point (i_o, λ_o), defined by the air-gap g_o, the current i_o, and the flux linkage $\lambda_o = \lambda(i_o, g_o)$. The apparent inductance of the coil is given by the ratio between the flux linkage λ_o and the current i_o, which is

$$L_{app}\left(i_o, g_o\right) = \frac{\lambda_o}{i_o} \tag{5.14}$$

The ratio between an infinitesimal variation of flux linkage, $d\lambda$, and the corresponding variation of current, di, around the operating point (i_o ,λ_o), defines the differential inductance of the coil, which is

$$L_{dif}\left(i_o, g_o\right) = \left.\frac{d\lambda}{di}\right|_{(i_o,\lambda_o)} \tag{5.15}$$

In linear conditions, since the permeability of the materials is constant, the two values of L_{app} and L_{dif} are constant and coinciding: $L_{app} = L_{dif} = L$.

Alternatively, the inductances can be computed from the energy quantity. Referring to Figure 5.4, the apparent inductance can be computed as

$$L_{app}\left(i_o, g_o\right) = \frac{W_m + W'_m}{i_o^{\,2}} = \frac{\lambda_o i_o}{i_o^{\,2}} \tag{5.16}$$

The quantity ($\frac{1}{2}\lambda_o i_o$) has the same dimension of an energy, but it does not have a physical meaning. It corresponds to the surface shown in Figure 5.4. The same figure highlights the magnetic energy variation dW_m corresponding

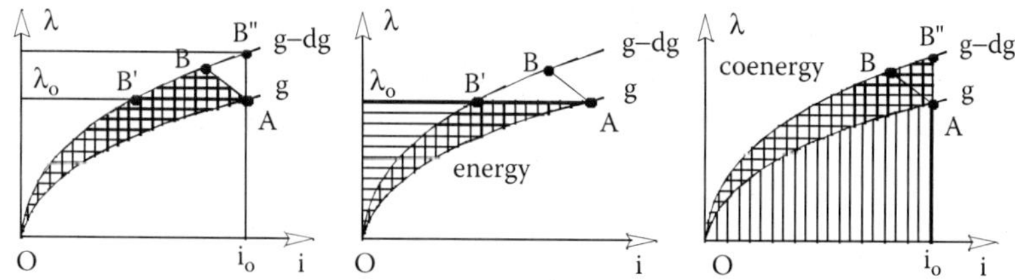

FIGURE 5.5
Energy variations corresponding to the air-gap variation –dg.

to the current variation di, around the working point. From this variation, since $dW_m \approx d\lambda i_o$, the differential inductance can be roughly estimated as

$$L_{dif}\left(i_o, g_o\right) \approx \frac{1}{i_o} \frac{dW_m}{di}\bigg|_{(\lambda_o, i_o)} \tag{5.17}$$

In linear conditions, the two inductances given in Equation (5.16) and Equation (5.17) are simplified as

$$L_{app} = L_{dif} = \frac{2W_m}{i_o^{\,2}} \tag{5.18}$$

5.2.4 Mechanical Forces

Since the magnetic device is a conservative system, the mechanical forces that take effect can be determined by means of the principle of the energy conservation. One has to estimate the variation of the energy corresponding to a infinitesimal movement of the element in the direction along which the force component has to be computed.

The force F_z on the plunger of the magnetic device of Figure 5.1 along the direction of the z-axis is associated to a movement dz = –dg of the plunger itself. According to the movement –dg, there is a variation of both the coil current and the flux linkage. Figure 5.5(a) shows the starting working point A, characterized by an air-gap length g, and the subsequent working point B, characterized by an air-gap length g–dg. Because the movement –dg is infinitesimal, the difference between the two curves is infinitesimal as well.

The force F_z realizes an infinitesimal mechanical work $dW_{mech} = F_z dz = -F_z dg$. It corresponds to the area OAB of Figure 5.5(a), as can be obtained from the following energy balance:

$$dW_{mech} + dW_m = dW_{el} \tag{5.19}$$

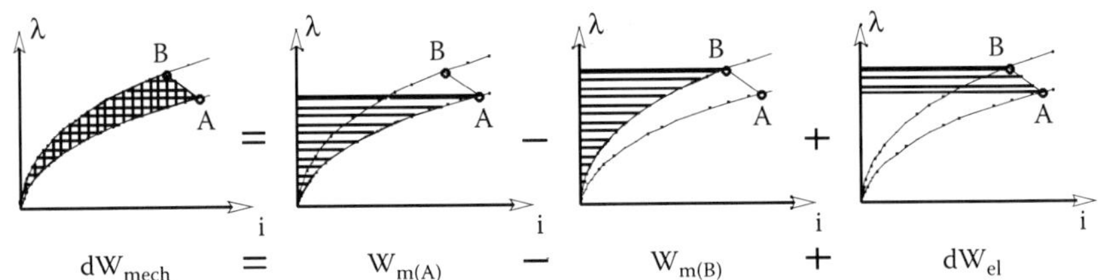

FIGURE 5.6
Energy balance during the air-gap variation -dg.

where dW_m is the variation of the magnetic energy stored in the magnetic field, during the movement of the plunger from A to B, which is

$$\begin{aligned} dW_m &= W_m(B) - W_m(A) \\ &= \int_0^{\lambda_o + d\lambda} i(\lambda, g - dg)d\lambda - \int_0^{\lambda_o} i(\lambda, g)d\lambda \end{aligned} \tag{5.20}$$

and dW_{el} is the electrical energy that is furnished by the electrical source via the coil terminals, which is

$$dW_{el} = \int_{\lambda_o}^{\lambda_o + d\lambda} i(\lambda)d\lambda \tag{5.21}$$

These quantities can be drawn in the (i, λ) plane, as shown in Figure 5.6.

Neglecting the infinitesimal term of second order, the area OAB can be approximated by the area OAB', shown in Figure 5.5(b), which corresponds to the difference between the stored magnetic energy before and after the movement –dg, with a constant flux linkage $\lambda = \lambda_o$. Then the force component is computed as

$$F_z(\lambda_o, g) = -\left.\frac{dW_m}{dz}\right|_{\lambda=\lambda_o} = -\left.\frac{dW_m(\lambda, g)}{-dg}\right|_{\lambda=\lambda_o} \tag{5.22}$$

Equation (5.22) highlights that the energy has to be expressed as a function of λ. Since the flux linkage is constant, the electrical energy furnished to the system is null. The mechanical work $F_z dz = -F_z dg$ is accomplished at the expense of the energy stored in the magnetic field.

In a similar way, the area OAB can be approximated by the area OAB″, shown in Figure 5.5(c). It corresponds to the difference between the coenergy of the system before and after the movement –dg, with a constant current $i = i_o$. In this case, the force is given by

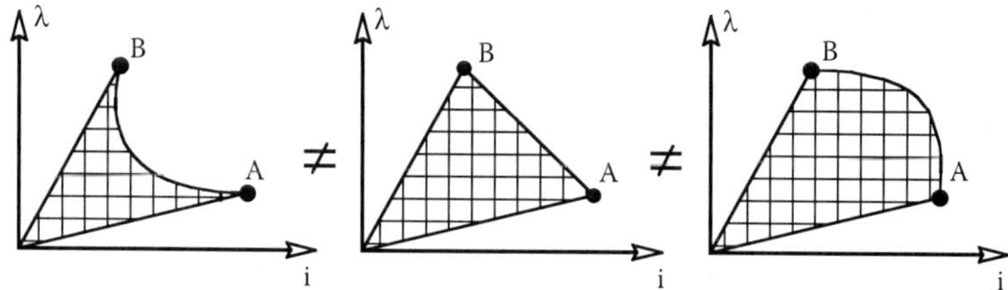

FIGURE 5.7
Mechanical work with different finite movements: although the starting and final points are the same, the dependence on the way of variation of λ and i is evident.

$$F_z(i_o, g) = +\frac{dW'_m}{dz}\bigg|_{i=i_o} = +\frac{dW'_m(i, g)}{-dg}\bigg|_{i=i_o} \tag{5.23}$$

Equation (5.23) highlights that the energy has to be expressed as a function of i. Since the current is constant, i.e., $i = i_o$, the electrical energy furnished to the system is equal to $i_o d\lambda$. The mechanical work $F_z dz = -F_z dg$ is computed as the difference between the electrical energy, furnished by an external source, and the variation of the magnetic energy stored in the magnetic field, which is $½ i_o d\lambda$. This computation corresponds to that adopted using the finite element method, in which the current is set as the source of the field.

Equation (5.22) and Equation (5.23) describe the fundamental relationships of the electromechanical energy conversion. They are used to evaluate the forces of magnetic nature that act on the mobile element. The force results independent of how the current and flux linkage vary during the infinitesimal movement –dg, since these variations are infinitesimal of the second order. In other words, the force depends on the state variables (g, i, and λ) but is independent of the way they change (as physically expected). However, this occurs only if the movement is infinitesimal: with a finite movement, the mechanical work is dependent on the way of variation, as shown in Figure 5.7.

In the particular case of the cylindrical magnetic device of Figure 5.1, supposing linear operations, the magnetic coenergy is computed as $W'_m = ½\lambda i$, where λ is expressed by Equation (5.9), so that

$$W'_m = \frac{1}{2}\mu_o \frac{(N_t i)^2}{g + t'} g \frac{\pi d^2}{4} \tag{5.24}$$

Using Equation (5.23), the force in direction of the infinitesimal movement $dz = -dg$ results in

$$F_z = \frac{1}{2\mu_o}(N_t i)^2 \frac{\pi d^2}{4} \frac{1}{(g + t')^2} \tag{5.25}$$

A positive value of F_z means that the plunger is attracted upwards.

In the equations above, the magnetic force is expressed by means of integral quantities, as energy and coenergy. The same forces can be computed starting from the specific properties of the magnetic field, as an example by means of Maxwell's stress tensor, applied to the magnetic field components. In the particular case of the magnetic device, with the hypothesis that only the z-axis component of the field strength exists in the air-gap g (see the first row of Table 1.2, in Chapter 1), the magnetic pressure normal to the plunger surface is given by

$$p_n = \frac{\mu_o}{2}\left(H_n^{\,2} - H_t^{\,2}\right) = \frac{\mu_o}{2} H_n^{\,2} = \frac{B_g^{\,2}}{2\mu_o} \tag{5.26}$$

where H_n and H_t are the normal and tangential components of the field strength at the surface of the plunger in front of the air-gap. Then the total force on the plunger is

$$F_z = \frac{1}{2\mu_o} B_g^{\,2} \frac{\pi d^2}{4} \tag{5.27}$$

By introducing the expression of B_g given in Equation (5.3) in Equation (5.27), the same expression of the force given in Equation (5.25) is again obtained.

5.3 Finite Element Analysis

5.3.1 Formulation of the Problem for the Finite Element Analysis

The device presents an axial symmetry (see Figure 4.1), thus only a section of the magnetic device may be considered. The analysis is carried out in the (r, z) plane as shown in Figure 5.8.

In fact, some considerations about the symmetry show that only the azimuthal component of the current density vector may exist, i.e., $\mathbf{J} = (0, J_\vartheta, 0)$. As a consequence, the magnetic vector potential is characterized by only the azimuthal component, i.e., $\mathbf{A} = (0, A_\vartheta, 0)$. Then the components of the magnetic flux density $\mathbf{B} = (B_r, 0, B_z)$ are

$$\begin{aligned} B_r &= -\frac{\partial A_\vartheta}{\partial z} \\ B_z &= \frac{1}{r}\frac{\partial (rA_\vartheta)}{\partial r} = \frac{A_\vartheta}{r} + \frac{\partial A_\vartheta}{\partial r} \end{aligned} \tag{5.28}$$

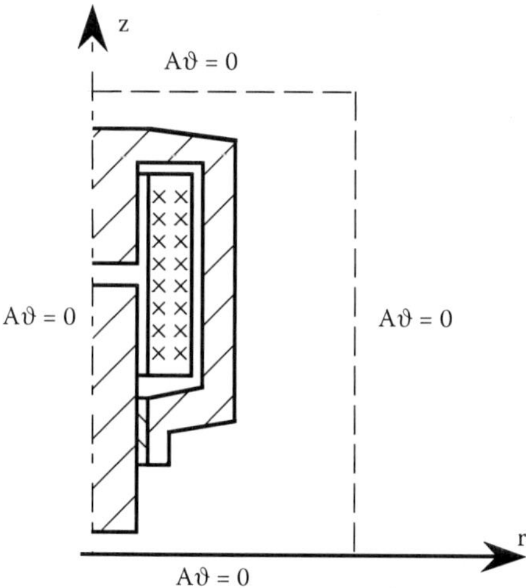

FIGURE 5.8
Sketch of the section of the cylindrical magnetic device, which is used in the finite element analysis and boundary conditions.

In addition, all the field quantities depend on the variables r and z only, i.e., $J_\vartheta = J_\vartheta(r,z)$, $A_\vartheta = A_\vartheta(r,z)$, $B_r = B_r(r,z)$, $B_z = B_z(r,z)$. The components of the magnetic field strength vector $\mathbf{H} = (H_r, 0, H_z)$ are connected to those of the flux density vector by the constitutive law.

The field problem is magnetostatic and is expressed by the Poisson differential equation

$$\nabla^2 A_\vartheta = \frac{1}{r}\frac{\partial}{\partial r}\left(r\frac{\partial A_\vartheta}{\partial r}\right) + \frac{\partial^2 A_\vartheta}{\partial z^2} = -\mu J_\vartheta \tag{5.29}$$

Dirichlet's boundary condition is assigned on the z-axis, with a reference potential $A_\vartheta = 0$. The same value is assigned along a line at a suitable distance from the device. The boundary conditions are shown in Figure 5.8.

A unity relative permeability is considered for nonmagnetic materials, while mild steel B-H curve is used for the magnetic materials. If the flux density is low, a linear behavior may be supposed for the iron paths, with constant magnetic permeability. The typical value of the relative magnetic permeability of nonsaturated iron materials is in the range from $\mu_r = 1000$ to $\mu_r = 10{,}000$.

The coil is formed by N_t turns, but it is modeled as a single conducting bar, as shown in Figure 5.8. Thus, when the coil is fed by a current $i = I$, the overall current N_tI has to be assigned to the equivalent bar.

5.3.2 Computation of the Solved Structure

Once magnetic material characteristics, boundary conditions, and winding current source are assigned, the field problem can be solved. As highlighted by Equation (5.29), the solution consists of computing the azimuthal component of the magnetic vector potential $A_\vartheta(r, z)$ in each point of the domain. All the other magnetic quantities of interest are derived from $A_\vartheta(r, z)$, as described later.

5.3.2.1 Magnetic Flux Density

The two components of the flux density vector are obtained from Equation (5.28). From them, the flux lines are carried out. The magnitude of the flux density results in

$$B = \sqrt{B_r^{\,2} + B_z^{\,2}} \tag{5.30}$$

5.3.2.2 Flux Linkage

At first, a single dimensionless turn is considered. Its radius is r_o and its center is on the z-axis, as illustrated in Figure 5.9(a). The turn flux linkage is computed as the integral of the flux density across the circular surface limited by the turn itself, which is

$$\lambda_{ro} = \int_0^{r_o} \int_0^{2\pi} B_z(r,z)\, r\, d\vartheta\, dz \tag{5.31}$$

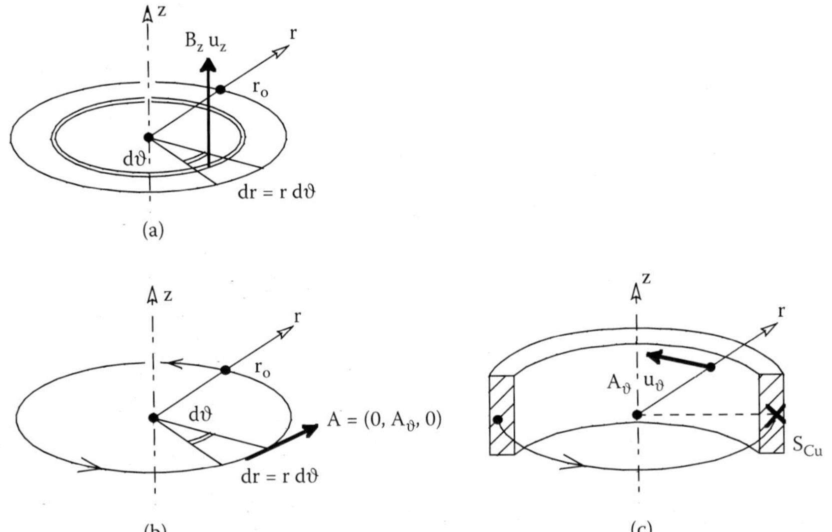

FIGURE 5.9
Integral lines for the computation of the flux linkage.

By means of the Stokes theorem, the flux linkage is conveniently expressed as

$$\begin{aligned}\lambda_{ro} &= \oint_{r_o} A_\vartheta(r,z)\,dr \\ &= \int_0^{2\pi} r_o A_\vartheta(r,z)\,d\vartheta \\ &= 2\pi r_o A_\vartheta(r_o,z)\end{aligned} \tag{5.32}$$

as shown in Figure 5.9(b). If the turn is not dimensionless but has a section S_{Cu}, as shown in Figure 5.9(c), the average flux linkage becomes

$$\lambda = \frac{1}{S_{Cu}} \int_{S_{Cu}} 2\pi r A_\vartheta dS \tag{5.33}$$

At last, if N_t turns are considered that cover the whole surface S_{Cu}, the flux linkage is

$$\lambda = \frac{N_t}{S_{Cu}} \int_{S_{Cu}} 2\pi r A_\vartheta dS \tag{5.34}$$

5.3.2.3 Magnetic Energy and Coenergy

For computing the magnetic energy and coenergy, the equations reported in Section 5.2.2 may be adopted. However, it is convenient to compute the energy from the specific magnetic field quantities: the magnetic field strength **H**, the flux density **B**, the current density **J**, and the magnetic vector potential **A**.

The magnetic energy W_m is computed as

$$W_m = \int_\tau \int_0^B HdB\,d\tau = 2\pi \int_S r \int_0^B HdB\,dS \tag{5.35}$$

where r is the dummy radius. Referring to the current density and the magnetic vector potential, W_m is given by

$$W_m = \int_\tau \int_0^{A_\vartheta} J_\vartheta dA_\vartheta\,d\tau = 2\pi \int_S r \int_0^{A_\vartheta} J_\vartheta dA_\vartheta\,dS \tag{5.36}$$

The magnetic coenergy W'_m, computed from the field quantities, is

$$W'_m = \int_\tau BH\,d\tau - W_m \tag{5.37}$$

In linear conditions, Equation (5.35) becomes

$$W_m = 2\pi \int_S r \frac{1}{2} BH \, dS \tag{5.38}$$

and Equation (5.36) becomes

$$W_m = 2\pi \int_S r \frac{1}{2} J_\vartheta A_\vartheta \, dS \tag{5.39}$$

The core and the plunger of the magnetic device show a nonlinear magnetic characteristic. However, when the magnitude of the flux density is low, they work in the linear part of the B-H curve. Hence, a constant permeability can be assigned even in these objects.

5.3.2.4 Inductances

The apparent inductance is obtained by dividing the flux linkage (5.34) by the current:

$$L_{app} = \frac{\dfrac{N_t}{S_{Cu}} \displaystyle\int_{S_{Cu}} 2\pi r A_\vartheta dS}{i} \tag{5.40}$$

If the magnetic circuit is linear, the inductance L_{app} can be computed from the energy quantities as

$$L_{app} = \frac{2\left(\dfrac{1}{2}\displaystyle\int_\tau BH \, d\tau\right)}{i^2} = \frac{2\pi \displaystyle\int_S rBH \, dS}{i^2} \tag{5.41}$$

The differential inductance has to be computed from two field solutions, corresponding to different coils currents. For instance, let us suppose that the flux linkages $\lambda(i)$ and $\lambda(i + \Delta i)$ correspond to the currents i and $i + \Delta i$, computed as given in Equation (5.34). Then the differential inductance is given by

$$L_{dif} = \frac{\lambda(i+\Delta i) - \lambda(i)}{\Delta i} \tag{5.42}$$

The result depends on the value of the current variation Δi. A more accurate result is obtained if the field problem is solved for two current values around i, which are $i + \Delta i$ and $i - \Delta i$. Adopting the Taylor series expansion of the flux linkage around the current value i, and neglecting the terms of order higher than the second order, it is

$$\begin{aligned} \lambda(i+\Delta i) &= \lambda(i) + \frac{d\lambda(i)}{di}\Delta i + \frac{1}{2}\frac{d^2\lambda(i)}{di^2}\Delta i^2 + \ldots \\ \lambda(i-\Delta i) &= \lambda(i) - \frac{d\lambda(i)}{di}\Delta i + \frac{1}{2}\frac{d^2\lambda(i)}{di^2}\Delta i^2 - \ldots \end{aligned} \tag{5.43}$$

By subtracting the second equation from the first one, the second-order terms disappear and the differential inductance results in

$$L_{dif} = \frac{d\lambda(i)}{di} = \frac{\lambda(i+\Delta i) - \lambda(i-\Delta i)}{2\Delta i} \tag{5.44}$$

5.3.2.5 Electromagnetic Forces

The force acting on the plunger can be evaluated using Maxwell's stress tensor. Let us refer to a surface containing the plunger, which is a line on the (r,z) plane bounding the plunger. The total force F_z on the plunger in the z-axis direction is given as the sum of the force components that are computed on the different parts of the surface. The parts of the whole surface that are normal to the unity vectors $\mathbf{u}_r$ and $\mathbf{u}_z$ are denoted S_r and S_z, respectively. On the surface S_z, whose normal unity vector has a z-axis direction, the force contribution to F_z corresponds to the force component normal to S_z, which is

$$F_n = \int_{S_z} \frac{\mu_o}{2}\left(H_n^{\,2} - H_t^{\,2}\right) dS \tag{5.45}$$

On the surface S_r, whose normal unity vector has an r-axis direction, the force contribution to F_z corresponds to the component of the force tangential to S_r, which is

$$F_t = \int_{S_r} \mu_o H_n H_t dS \tag{5.46}$$

Alternatively, the force in the z-axis direction can be computed from the magnetic coenergy variation, see Equation (5.23), as

$$\begin{aligned} F_z &= -\left.\frac{dW'_m(i,g)}{dg}\right|_{i=const} \\ &= -\left.\frac{W'_m(i,g+\Delta g) - W'_m(i,g-\Delta g)}{2\Delta g}\right|_{i=const} \end{aligned} \tag{5.47}$$

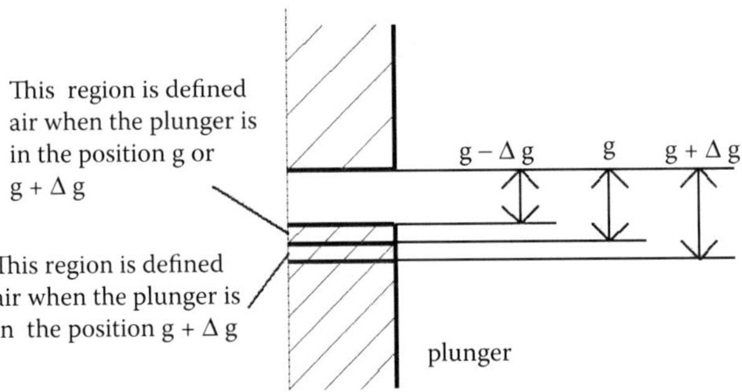

FIGURE 5.10
Definition of the regions that are interesting to the movement of the plunger; they are adopted to avoid errors due to variation of the mesh.

where $W'_m(i, g + \Delta g)$ and $W'_m(i, g - \Delta g)$ are the magnetic coenergy values that are computed from the field solutions, and correspond to the air-gap length $(g + \Delta g)$ and $(g - \Delta g)$, respectively, and to the given current i.

The result of Equation (5.47) depends on the variation of the air-gap length Δg. This has to be neither too small, so that the difference in the numerical solution is not appreciable, nor too large, so that an excessive variation of magnetic coenergy occurs.

In addition, the result may be influenced by the variation of the mesh of the domain, since the geometry changes when the plunger is in the two different positions. In order to avoid this numerical error, it is useful to define the regions involved in the plunger movement, as shown in Figure 5.10. When the plunger occupies these regions, they are considered as iron, assigning the corresponding magnetic property. Conversely, when the plunger does not occupy these regions, they are considered as air.

5.4 Example

This section illustrates the analysis of the cylindrical magnetic device shown in Figure 5.11, together with its main dimensions. The coil is formed by N_t = 1000 turns, whose diameter is d_c = 0.9 mm. The coil is represented by the rectangle with sides 90 mm and 17.5 mm. The nonmagnetic driver, which determines the lateral gap between the core and the plunger, is t = 0.5 mm. When the plunger is in its lower position, the air-gap length is g_{max} = 15 mm, while when the plunger is in its higher position, the air-gap length is g_{min} = 1 mm.

Since the problem is axial-symmetric, the section of the magnetic device of Figure 5.8 is analyzed. The current density J_ϑ is fixed as source, and the magnetic vector potential A_ϑ represents the field solution. The boundary condition $A_\vartheta = 0$ is fixed on the boundary of the domain, as illustrated in Figure 5.8. The mild steel B-H curve is used for the core and the plunger.

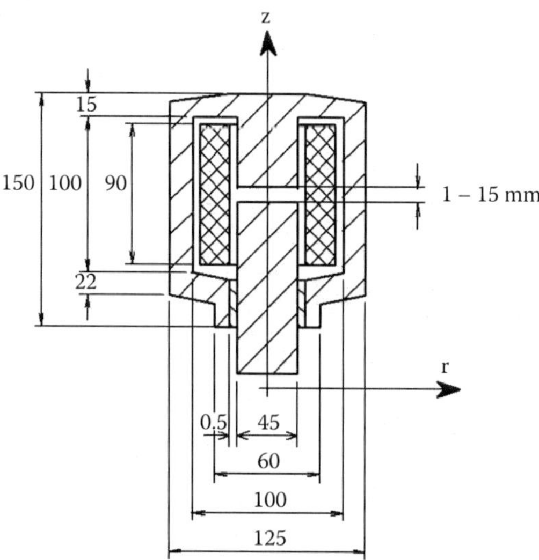

FIGURE 5.11
Main dimension of the cylindrical magnetic device.

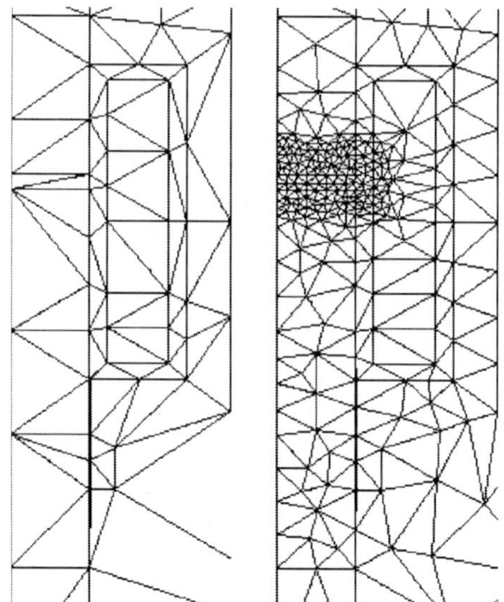

FIGURE 5.12
Mesh of the domain: initial mesh (a) and refined mesh (b).

Figure 5.12 shows the mesh of the device: Figure 5.12(a) shows the initial mesh, while Figure 5.12(b) shows the refined mesh. The mesh was mainly refined in the air-gap region, where the highest field gradients take place.

The field solution consists of the knowledge of the magnetic vector potential A_ϑ. The equipotential lines are shown in Figure 5.13(a). The flux density

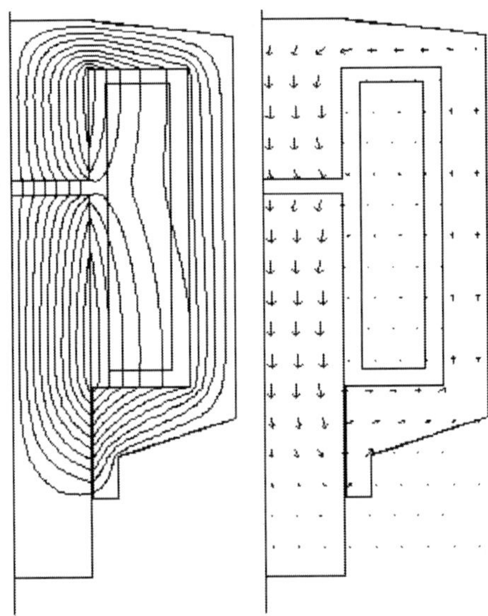

FIGURE 5.13
Equipotential lines of A_ϑ (a) and flux density vectors (b).

vector components are computed from the magnetic vector potential A_ϑ, as reported by Equation (5.28). The vector field **B** is shown in Figure 5.13(b).

As an example, three configurations of the magnetic device are analyzed, corresponding to three air-gap lengths: g = 1 mm, g = 5 mm, and g = 15 mm. The behavior of the z-axis component of the flux density in the middle of the air-gap is shown in Figure 5.14, with a source current equal to i = 5 A. The dependence of the flux density value on the air-gap length is evident. When g = 1 mm, the effect of the plunger and core edges is also manifest.

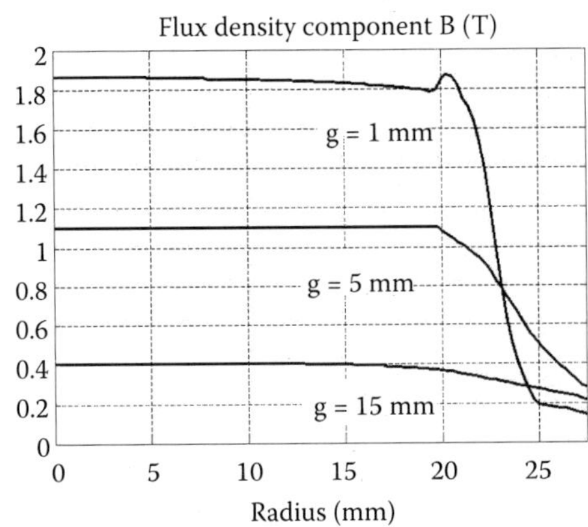

FIGURE 5.14
Behavior of the flux density component B_z in the middle of the air-gap g (with i = 5 A).

TABLE 5.1

Computation on the Solved Structure (with a current i = 5 A)

	g = 15 mm	g = 5 mm	g = 1 mm
$\int_{\tau_{Cu}} A_\vartheta d\tau$	$2.023 \cdot 10^{-6}$	$4.123 \cdot 10^{-6}$	$5.204 \cdot 10^{-6}$
$\int_\tau J_\vartheta A_\vartheta d\tau$	6.422	13.09	16.52
$\int_\tau BH d\tau$	6.415	13.16	16.79
$\int_\tau \int_0^B HdB\, d\tau$	3.217	6.119	3.978
λ (Vs)	1.2843	2.6180	3.3042
W_m (J)	3.217	6.119	3.978
W'_m (J)	3.198	7.041	12.812
L_{app} (H)	0.257	0.524	0.661
L_{dif} (H)	0.257	0.306	0.117

Some magnetic quantities are computed, always considering the three configurations, and reported in Table 5.1. It is worth noting that the magnetic energy obtained by handling the quantities B and H, or the quantities J_ϑ and A_ϑ, is substantially the same. Moreover, the magnetic energy and coenergy are essentially coincident when g = 15 mm, i.e., when the flux density values are low, as reported in Figure 5.14; therefore there is no saturation in any part of the device. Conversely, they are very different when g = 1 mm, i.e., when the flux density values are high, and a high saturation occurs in the iron paths.

In Table 5.2, the analytical and the finite element (FE) results are compared. Concerning the flux density, the results coincide when the air-gap length is high, while they are very different when the air-gap length is low. This is because the analytical model does not consider the iron saturation, though this hypothesis is not acceptable in case of magnetic fields so high. The

TABLE 5.2

Comparison between Analytical and Finite Element (FE) Results

	g = 15 mm		g = 5 mm		g = 1 mm	
	Analytical	FE	Analytical	FE	Analytical	FE
B_g (T)	0.415	0.407	1.219	1.097	5.435 (!)	1.87
λ (Vs)	0.659	1.284	1.939	2.618	—	3.304
W_m (J)	1.648	3.217	4.847	6.119	—	3.978
F_z (N)	217.5	169.4	1881	881.5	—	2086

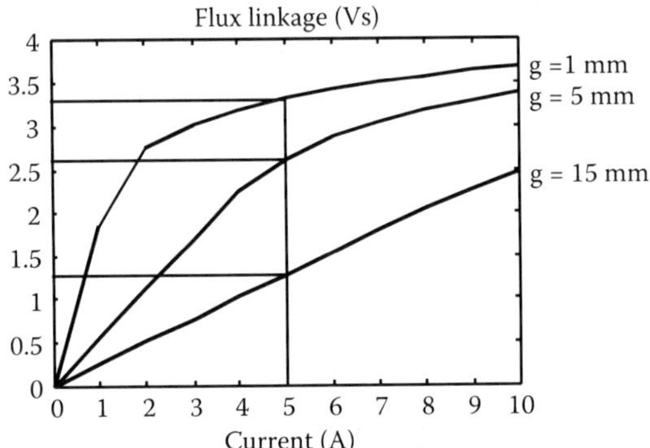

FIGURE 5.15
Flux linkage as a function of the current, with three air-gap lengths.

analytical model should be improved, taking the magnetic voltage drop in the iron paths into account; however, this lies beyond the scope of this example and we will not discuss it in any depth.

Table 5.2 also shows that the analytical values of flux linkage and magnetic energy and those computed by the finite element method are quite different. This is because the analytical model considers only the flux lines crossing the air-gap. This assumption is valid only when the air-gap length is low. Conversely, when the air-gap length is high, the flux lines go through the lateral surface of the plunger, without crossing the air-gap, as highlighted in Figure 5.13. The apparent inductance is computed by using Equation (5.40).

Figure 5.15 shows the flux linkage versus the coil current for the three air-gap lengths. The different saturation level is evident in the three curves. Keeping in mind the expressions for the magnetic energy (5.10) and magnetic coenergy (5.11), and the graphical representation of Figure 5.3, the surfaces representing W_m of Figure 5.15 are compared at the current i = 5 A. The results are reported in Table 5.2.

The knowledge of the flux linkage, at different current amplitudes, allows the differential inductance to be computed, by means of Equation (5.44). In particular, around i = 5 A, with the three different air-gap lengths, the computation result is reported in Table 5.1. With g = 15 mm, which is without saturation, the inductances L_{app} and L_{dif} are effectively coincident. With low air-gap lengths, when saturation occurs in the iron parts, the inductances are quite different.

The force on the plunger is computed with air-gap length g = 5 mm and current i = 5 A. At first the force is computed by means of Maxwell's stress tensor. As shown in Table 5.3, three lines around the plunger are considered, and the force density is integrated along these lines. The line A covers only the upper part of the plunger, so that only the force on this surface is considered. The line B is placed in the middle of the air-gap, covering a

TABLE 5.3

Computation of the Force by Means of Maxwell's Stress Tensor

g = 5 mm i = 5 A	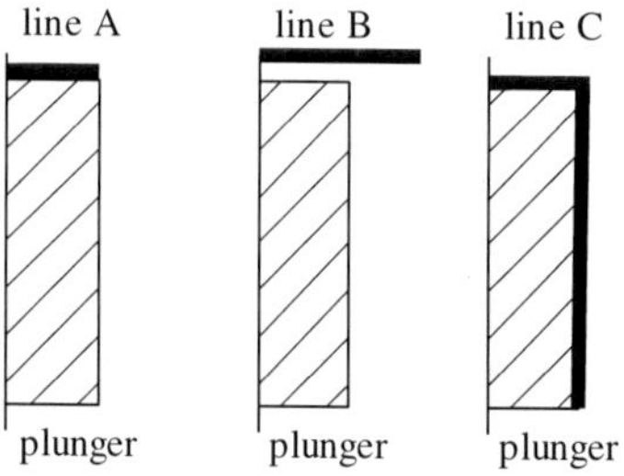 Line	Force (N)		
		Points 100	Points 200	Points 400
line A line B line C	A	782.0	782.1	782.2
	B	829.8	829.8	829.8
plunger plunger plunger	C	856.5	875.1	881.5

line A = on the upper surface of the plunger
line B = in the middle of the air-gap
line C = all around the plunger

surface higher than the plunger surface. Finally, the curve C completely contains the plunger, and the effects of the lateral surface of the plunger are also considered in the force computation. The numerical integration is carried out with a different number of points. It is always a good rule to check the influence of the number of points, to avoid errors due to numerical integration.

By comparing the results of Table 5.3, obtained on the three different lines of integration, the following conclusions arise. The attractive force takes place essentially on the upper surface of the plunger: in fact the force computed on the line A is almost 90% of the total force, considered equal to that computed on the line C. A good approximation of the force, almost 94% of the total force, is obtained computing the force on the line B.

The force acting on the plunger for different air-gap lengths is reported in Table 5.3. These results refer to the integrals along the line C, with 400 points.

The force is also computed by means of the virtual work principle, in the same conditions, i.e., i = 5 A and g = 5 mm. Two further field solutions are carried out, considering a constant current i = 5 A, and two variations of the air-gap length of $\Delta g = 0.5$ mm, the first positive and the second negative. The values obtained from the field solutions are reported in Table 5.4. Using Equation (5.46) yields

$$F_z = -\left.\frac{W'_m(i, g+\Delta g) - W'_m(i, g-\Delta g)}{2\Delta g}\right|_{i=\text{const}} = -\frac{6.579 - 7.477}{2 \cdot 0.5 \cdot 10^{-3}} = 898 \text{ N}$$

which coincides with the force computed by means of Maxwell's stress tensor.

TABLE 5.4

Computation of Some Quantities Around g = 5 mm (current i = 5 A)

	g = 4.5 mm	g = 5 mm	g = 5.5 mm
$\int_{\tau_{Cu}} A_\vartheta d\tau$	$4.278 \cdot 10^{-6}$	$4.123 \cdot 10^{-6}$	$3.961 \cdot 10^{-6}$
$\int_{\tau} J_\vartheta A_\vartheta d\tau$	13.58	13.09	12.58
$\int_{\tau} BH d\tau$	13.63	13.16	12.59
$\int_{\tau}\int_{0}^{B} HdB\, d\tau$	6.153	6.119	6.011
λ (Vs)	2.716	2.618	2.515
W_m (J)	6.153	6.119	6.011
W'_m (J)	7.477	7.041	6.579

References

1. D.C. White and H.H. Woodson, *Electromechanical Energy Conversion*, John Wiley & Sons, New York, 1959.
2. G. Xiong and S.A. Nasar, Analysis of Field and Forces in a Permanent Magnet Linear Synchronous Machine Based on the Concept of Magnetic Charge, *IEEE Trans. on MAG*, vol.25, no.3, pp. 2713–2719, 1989.
3. R. Akmese and J.F. Eastham, Dynamic performance of a brushless dc tubular drive system, *IEEE Trans, MAG*, vol.25, no.5, pp. 3269–3271, 1989.
4. J. Hur, S.B. Yoon, D.Y. Hwang, and D.S. Hyun, Analysis of PMLSM Using Three Dimensional Equivalent Magnetic Circuit Network Method, *IEEE Trans, MAG*, vol.33, no.5, pp. 4143–4145, 1997.
5. I. Boldea and S.A. Nasar, *Linear Electric Actuators and Generators*, Cambridge University Press, Cambridge, UK, 1997.
6. N. Bianchi, S. Bolognani, F. Tonel, Design Considerations for a Tubular Linear PM Motor for Servo Drives, in Proc. *EPE-PEMC 2000 International Conference*, Kosice, Slovak Republic, 5–7 September 2000.
7. N. Bianchi, Analytical Computation of Magnetic Fields and Forces of a Tubular PM Linear Servo Motor, in Proc. di *IEEE IAS Annual Meeting*, Roma, Italy, 8–12 October 2000.
8. J.F. Gieras and Z.J. Piech, *Linear Synchronous Motors. Transportation and Automation Systems*, CRC Press, London–New York, 2000.

6

The Single-Phase Transformer

The aim of this chapter is to describe the finite element analysis of a low-power, single-phase transformer. We start by defining the equivalent circuit of the transformer; then we go on to describe the procedure to obtain its parameters.

6.1 The Single-Phase Transformer

The iron core of the single-phase transformer may be of core type or shell type. Figure 6.1(a) shows a section of the core-type transformer, whose primary and secondary windings are divided in two coils each and wound around the two legs. Figure 6.1(b) shows the shell-type transformer, whose primary and secondary windings are wound around the central leg.

The core of the low-power, single-phase transformer is obtained by stacking silicon-steel laminations. The laminations are insulated to reduce the eddy current losses. Generally a rectangular form of the leg is adopted.

Neglecting the edge effects, e.g., the end-winding flux leakage, the magnetic field is identical on each section of the transformer, along all the net iron length $L_{Fe} = k_{stk}L$, where k_{stk} is the stacking factor and L is the total length of the leg. Thus, the analysis is reduced to a two-dimensional problem analysis.

To reduce mainly the field problem, the analysis of the transformer can be carried out on a part of the section of Figure 6.1. In fact, both structures feature geometric and magnetic symmetries. In the core-type transformer drawn again in Figure 6.2(a), the flux lines must be symmetrical as regards the axis BB′. Then the analysis can be carried out only on the part of the structure drawn in Figure 6.2(b). Similarly, the segment AA′ is a symmetry axis; thus the analysis is reduced to the section part illustrated in Figure 6.2(c). On the symmetry lines, suitable boundary conditions have to be assigned, as will be described in the following.

Similarly, in a shell-type transformer, the analysis can be simplified considering that the segments AA′ and BB′ are symmetry axes for both the

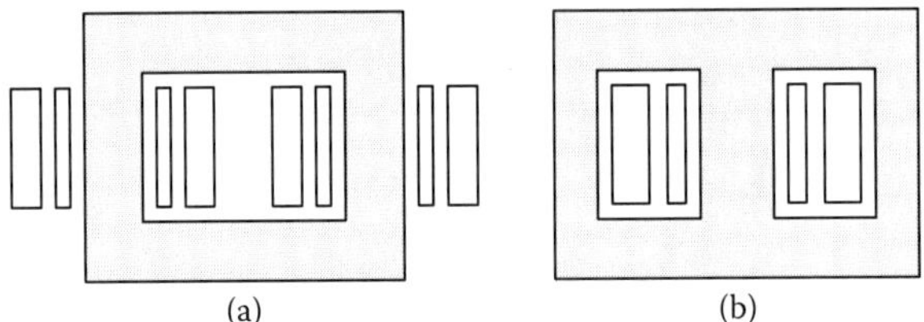

FIGURE 6.1
Single-phase core-type (a) and shell-type (b) transformers.

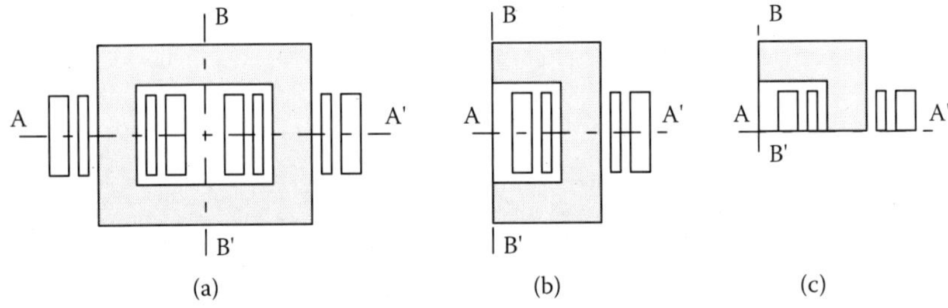

FIGURE 6.2
Core-type transformer: simplification of the structure.

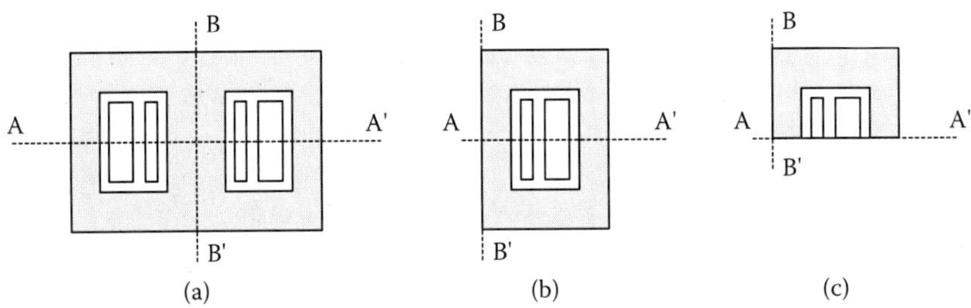

FIGURE 6.3
Shell-type transformer: simplification of the structure.

geometric structure and the flux lines. The section drawn in Figure 6.3(a) can be reduced to the section part drawn in Figure 6.3(b) and then to that drawn in Figure 6.3(c).

6.2 Equivalent Electric Circuit of the Transformer

In the following analysis a lossless transformer will be considered, whereas winding resistances, iron losses, and saturation are not discussed.

The terminal voltage equations of such a transformer are given by

$$\begin{cases} v_1 = \dfrac{d\lambda_1}{dt} = L_1\dfrac{di_1}{dt} + M\dfrac{di_2}{dt} \\ \\ v_2 = \dfrac{d\lambda_2}{dt} = M\dfrac{di_1}{dt} + L_2\dfrac{di_2}{dt} \end{cases} \tag{6.1}$$

where v_1, v_2, i_1, i_2, λ_1, and λ_2 are the voltages, currents, and flux linkages at the two terminal pairs. The sign notation is such as the positive direction of the currents is against the voltage in both terminal pairs. Finally, L_1, L_2, and M are the self- and mutual inductances of the transformer. With the assumption that there is no saturation, these inductances are constant.

The flux linkages can be expressed as

$$\begin{cases} \lambda_1 = L_1 i_1 + M i_2 \\ \\ \lambda_2 = M i_1 + L_2 i_2 \end{cases} \tag{6.2}$$

For the sake of convenience, the self-inductances L_1 and L_2 are split into two addenda, which are

$$\begin{aligned} L_1 &= L_{1\sigma} + L_{1m} \\ L_2 &= L_{2\sigma} + L_{2m} \end{aligned} \tag{6.3}$$

so that the following relationship is satisfied:

$$L_{1m}\, L_{2m} = M^2 \tag{6.4}$$

The transformer formed by L_{1m}, L_{2m}, and M is an ideal transformer characterized by a perfect coupling. The inductances $L_{1\sigma}$ and $L_{2\sigma}$ are the leakage inductances referred to the primary and the secondary winding, respectively.

This split allows the transformer to be represented by means of the equivalent electrical circuit shown in Figure 6.4, where a is the transformation ratio, referred to the perfect link:

$$a = \frac{L_{1m}}{M} = \frac{M}{L_{2m}} \tag{6.5}$$

The voltage v′ and the current i_m are given by

$$\begin{cases} v_1' = a\, v_2'' \\ i_m = i_1 + \dfrac{i_2}{a} \end{cases} \tag{6.6}$$

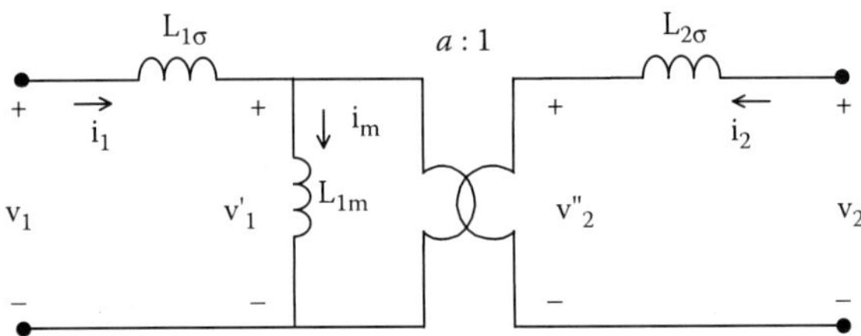

FIGURE 6.4
Equivalent electric circuit of the real transformer, T representation.

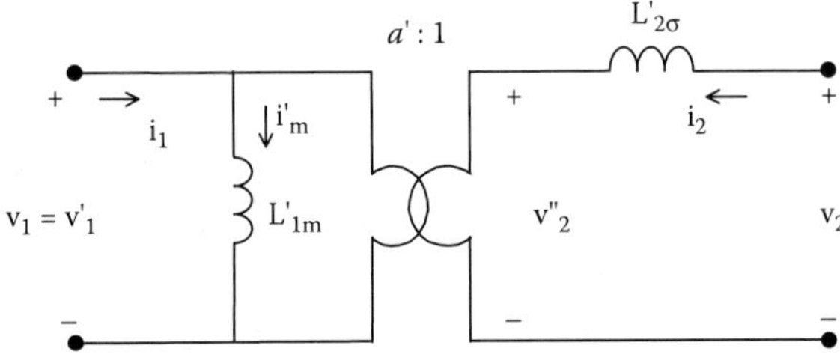

FIGURE 6.5
Equivalent electric circuit, Γ′ representation.

The choice of how the inductances L_1 and L_2 are separated in the two addenda is arbitrary. It is sufficient that Equation (6.2) is satisfied and that the inductances of the model are positive or, at least, null. The latter condition limits the possible values of the transformation ratio:

$$\frac{M}{L_2} \leq a \leq \frac{L_1}{M} \tag{6.7}$$

In the two limit cases, the following results are obtained:

1. With $a' = L_1/M$ the primary leakage inductance is null, so that $v_1 = v_1'$. In this case, the equivalent circuit is shown in Figure 6.5, where

$$\begin{cases} L'_{1m} = L_1 \\ L'_{2\sigma} = \dfrac{L_1 L_2 - M^2}{L_1} \end{cases} \tag{6.8}$$

2. With $a'' = M/L_2$ the secondary leakage inductance is null, so that $v_2 = v_2''$. In this case, the equivalent electric circuit is shown in Figure 6.6, where

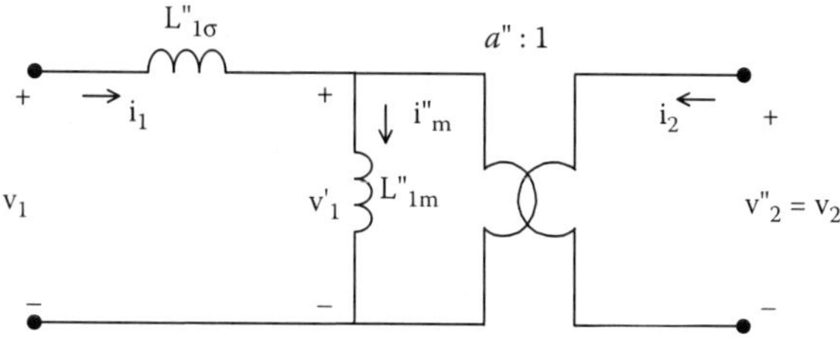

FIGURE 6.6
Equivalent electric circuit, Γ″ representation.

$$\begin{cases} L''_{1m} = \dfrac{M^2}{L_2} \\ L''_{1\sigma} = \dfrac{L_1 L_2 - M^2}{L_2} \end{cases} \tag{6.9}$$

Another common choice is that considering the transformation ratio equal to the ratio between the coils of the transformer, which is

$$a = \frac{N_1}{N_2} \tag{6.10}$$

With this choice, from Equation (6.5) it results that

$$\frac{L_{1m}}{L_{2m}} = \left(\frac{N_1}{N_2}\right)^2 \tag{6.11}$$

Of course the equivalent electric circuit is that in Figure 6.4. As will be hereafter highlighted, the choice of this transformation ratio is useful to impose the current sources in the finite element analysis of the leakage inductances.

When the parameters of the equivalent circuit of Figure 6.4 are known, it is possible to modify the circuit of the transformer using one of the two Γ representations of Figure 6.5 or Figure 6.6, by means of the simple relationships reported in Table 6.1.

6.3 Computation of the No-Load Inductances

In order to compute the magnetizing inductances, a simulation is carried out that corresponds to the no-load test of the transformer. In the test, the

TABLE 6.1

Transformations of the Inductive Parameters of the Transformer

From T to Γ′ Representation	From T to Γ″ Representation
$L'_{1m} = L_{1\sigma} + L_{1m}$	$L''_{1m} = \dfrac{L_{1m}^{2}}{L_{1m} + a^2 L_{2\sigma}}$
$L'_{2\sigma} = \left(L_{2\sigma} + \dfrac{L_{1m}}{a^2}\right)\left(\dfrac{L_{1\sigma} + L_{1m}}{L_{1m}}\right)^2 - \dfrac{1}{a^2}\left(L_{1\sigma} + L_{1m}\right)$	$L''_{1\sigma} = L_{1\sigma} + \dfrac{a^2 L_{2\sigma} L_{1m}}{a^2 L_{2\sigma} + L_{1m}}$

primary winding is fed at the nominal frequency, with voltage equal to the nominal voltage, measuring the primary current and the induced secondary voltage. Similarly, in the simulation, the primary winding current is imposed as the field problem source and the corresponding voltages are computed in the two windings. It is also possible to compute the variation of the induced voltage at no-load, corresponding to different primary current values. Of course, this makes sense when iron saturation occurs.

6.3.1 Statement of the Problem

The magnetic energy is mainly stored in the iron core; thus in the analysis of the no-load operations, it is convenient to increase the mesh in the core of the transformer.

Each part of the coil of the primary winding, concerning the considered section of Figure 6.2(c) or Figure 6.3(c), is composed by a number of conductors:

$N_1/4$, in case of core-type transformer, or

$N_1/2$, in case of shell-type transformer.

A current source I_{1o} is forced in the primary winding. For the sake of simplicity, the latter is simplified as a unique conductive bar, as shown in Figure 6.2(c) and Figure 6.3(c), fed by the total current:

$N_1 I_{1o}/4$, in case of core-type transformer, or

$N_1 I_{1o}/2$, in case of shell-type transformer

We must not forget that the current flows in the conductors and then a free distribution of the current in the equivalent bar is not realistic. It has to be imposed that the current is uniformly distributed in the bar. Of course, if the current assigned in the simulation is constant, a magnetostatic problem arises; then the problem of a nonuniform distribution of the current does not exist. This kind of simulation should be preferred. In this case, the constant current I_{1o} denotes the maximum value of the sinusoidal current waveform.

The secondary winding is open-circuited, and a null conductivity is assigned to its corresponding equivalent bar.

At the beginning, the iron is assumed to be linear, with a constant relative permeability μ_{Fe}. Some considerations of nonlinear characteristics of the iron will be reported in the following.

In order to assign correctly the boundary conditions, it is useful to estimate the flux lines of the transformer. It is possible to believe that the flux density vector **B** has only normal component along the symmetry axis AA′ (see Figure 6.2 and Figure 6.3). Moreover, in case of the core-type transformer, the flux density vector **B** has only a normal component along the symmetry axis BB′ as well. Along these lines, Neumann's boundary conditions are assigned.

In case of the shell-type transformer, the flux density vector **B** has only a tangential component along the symmetry axis BB′. Along this line, Dirichlet's boundary condition is assigned, which is a constant value of the magnetic vector potential, i.e., $A_z = 0$. Along the other lines that delimit externally the analysis domain, it is possible to assign a null magnetic vector potential, i.e., $A_z = 0$.

6.3.2 Computation on the Solved Structure

Once the problem is solved, the following magnetic quantities are computed.

6.3.2.1 *Magnetic Flux Density*

The components of the magnetic flux density in the iron core are obtained by deriving the magnetic vector potential.

6.3.2.2 *Magnetic Energy*

The stored magnetic energy, corresponding to the current feeding the primary winding, can be computed as follows. Assuming linear materials, the magnetic energy density is

$$w_m = \frac{1}{2}\mathbf{B}\cdot\mathbf{H} = \frac{1}{2}\mu H^2 \tag{6.12}$$

Then the magnetic energy stored in the whole transformer is computed by integrating the energy density over all the volume. Since a 2D problem is analyzed, the result is obtained by means of the surface integral, multiplied by the net iron length of the transformer L_{Fe}. Finally, the result has to be multiplied by 4, since only a quarter of the transformer is analyzed. Then the magnetic energy is

$$W_m = 4L_{Fe}\int_S \frac{1}{2}\mu H^2 \, dS \tag{6.13}$$

where S is the total surface of the domain, corresponding to Figure 6.2(c) or Figure 6.3(c).

As an alternative, the magnetic energy can be computed as the energy that is provided to the system by the source, as

$$W_m = 4L_{Fe}\int_S \frac{1}{2}\mathbf{J}\cdot\mathbf{A}\ dS = 4L_{Fe}\int_S \frac{1}{2}J_z\ A_z\ dS \tag{6.14}$$

6.3.2.3 Main Flux

The main flux is obtained by integrating the component of the flux density that is normal to the core section over the same surface. Since a 2D problem is analyzed, the flux density is integrated along a line that crosses the iron core, and then the result is multiplied by the net iron length L_{Fe}. It results that

$$\Phi_o = L_{Fe}\int_l \mathbf{B}\cdot\mathbf{n}\ dl, \quad \text{in case of core-type transformer;} \tag{6.15}$$

$$\Phi_o = 2L_{Fe}\int_l \mathbf{B}\cdot\mathbf{n}\ dl, \quad \text{in case of shell-type transformer.}$$

The ends of the lines are generally chosen within the windings.

6.3.2.4 Flux Linkage

The primary winding flux linkage may be computed by multiplying the main flux by the number of turns of the winding:

$$\Lambda_{1o} = N_1\ \Phi_o \tag{6.16}$$

A more reliable computation of the flux linkage is the integration of the magnetic vector potential A_z over the surfaces of the equivalent bars forming the primary winding. Let S_{Cu1+} be the surface of the bar where the current direction is positive, and S_{Cu1-} be the surface of the bar where the current direction is negative. Of course, the two surfaces are equal, that is $S_{Cu1+} = S_{Cu1-} = S_{Cu1}$. It results that

$$\Lambda_{1o} = N_1 L_{Fe}\frac{1}{S_{Cu1}}\left(\int_{S_{Cu1+}} A_z\ dS - \int_{S_{Cu1-}} A_z\ dS\right) \text{in case of core-type transformer;} \tag{6.17}$$

$$\Lambda_{1o} = 2N_1 L_{Fe}\frac{1}{S_{Cu1}}\left(\int_{S_{Cu1+}} A_z\ dS\right) \quad \text{in case of shell-type transformer.}$$

The same computation is carried out for the secondary winding. The relationships are the same, only subscript 2 will be used in place of 1.

6.3.2.5 Induced FMF

The EMF induced in the two windings are obtained by multiplying the flux linkages by the electric frequency $\omega = 2\pi f$, at which the transformer works. It is obtained that

$$E_{1o} = \omega \Lambda_{1o}$$
$$E_{2o} = \omega \Lambda_{2o} \qquad (6.18)$$

We wish to point out that the simulation is carried out considering the maximum value of the sinusoidal current, say I_{1o}. Consequently, the quantities Φ_o, Λ_{1o}, and E_{1o}, as well as the secondary winding quantities represent the maximum value that is reached. Since the problem has been assumed to be linear, the RMS values of these quantities are obtained by dividing the maximum value by $\sqrt{2}$. This is not possible when the iron saturation occurs.

6.3.2.6 Self- and Mutual Inductances

The primary self-inductance, that is the no-load inductance of the transformer, is obtained from the first part of Equation (6.2), as the ratio between the flux linkage Λ_{1o} and the no-load current I_{1o}. The mutual inductance is obtained from the second part of Equation (6.2), as the ratio between the flux linkage Λ_{2o} and the no-load current I_{1o}. They are

$$L_1 = \frac{\Lambda_{1o}}{I_{1o}}$$
$$M = \frac{\Lambda_{2o}}{I_{1o}} \qquad (6.19)$$

Another way to compute the primary self-inductance is from the magnetic energy, as

$$L_1 = \frac{2W_m}{I_{1o}^2} \qquad (6.20)$$

For obtaining the self-inductance L_2, the secondary winding is fed by the current I_{2o}, and the primary winding is open-circuited (that is, it assigns a null conductivity of the primary winding equivalent bars). Then L_2 is given by the ratio between the secondary flux linkage Λ_{2o} and the current I_{2o}.

6.3.2.7 Joule Power Losses

The Joule power losses of the coils and the corresponding resistances should be computed by the classical analytical relationships, since the coils are modeled as equivalent conducting bars and the simulations are carried out with stationary current.

On the contrary, if the non-uniform current distribution in the winding due to the source frequency has to be estimated, a different procedure must be adopted. The coil is designed in detail, i.e., each turn is drawn in its actual position. The simulation is carried out at the source frequency, always with open-secondary winding (null conductibility). Then the Joule losses are computed in the primary winding.

However, in low-power transformers, which work at the industry frequency, the many turns of small diameter so that the non-uniform current distribution is negligible.

6.3.3 Effect of the Nonlinear B-H Curve

When the flux-density in the iron is quite high, it is no longer possible to consider a constant permeability, but it is necessary to consider the whole B-H curve of the material. Neglecting the magnetic hysteresis, the first magnetization B-H curve is adopted, obtained by the dc test on the material.

A fixed number of simulations is carried out. The values of flux linkage Λ_{1o} are computed at different values of the current I_{1o}. Then the flux linkage Λ is drawn as a function of the current I, as shown in Figure 6.7. This curve is called the *dc magnetizing characteristic.*

From this characteristic the magnetizing characteristic of the transformer can be computed, as described in the following. A voltage is fixed, i.e., a flux

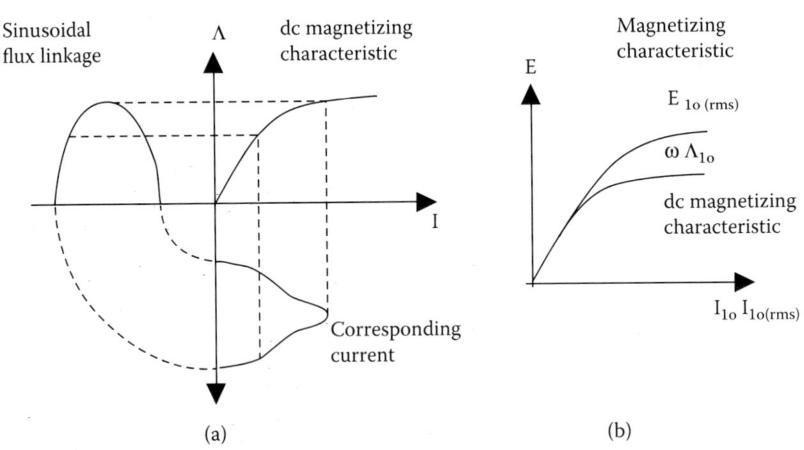

FIGURE 6.7
Drawing of the current waveform (a) and the magnetizing characteristic of the transformer (b).

linkage, with a sinusoidal waveform. The corresponding waveform of the current is achieved, by points, from the dc magnetizing characteristic. This is shown in Figure 6.7(a).

The RMS value of the current waveform is computed, and the relationship between $E_{10(rms)}$ e $I_{10(rms)}$ is achieved. Such a correspondence is less affected by saturation, since the RMS value is computed taking into account the whole current waveform. Such a characteristic is reported in Figure 6.7(b). A simple algorithm for the computation of the characteristic is given at the end of this chapter.

It is worthwhile to observe that the coil resistance is not taken into account. However, the voltage drop on the coil resistance may be high, especially in the low-power transformer. If the primary winding resistance is not negligible, a more accurate computation has to be carried out. The dc magnetizing characteristic of Figure 6.7(a) has to be modified by including the effect of the voltage drop on the resistance.

6.3.4 Estimation of the Iron Losses

Let $p_{s,Fe}$ be the specific losses of the iron core, measured in (W/kg), corresponding to a flux density of sinusoidal waveform, of amplitude $B_M = 1$ T and frequency f = 50 Hz. Then the corresponding losses in an infinitesimal iron volume $d\tau$, where the maximum value of the flux density is given by B_M, are computed by using Steimnetz's relationship:

$$dp_{Fe} = p_{s,Fe}\, B_M^{\alpha} \gamma_{Fe}\, d\tau \qquad (6.21)$$

where γ_{Fe} is the specific weight of the iron(more or less $\gamma_{Fe} = 7700$ kg/m³). The value B_M must be the maximum flux density amplitude, reached in the volume $d\tau$. This is the reason for using the maximum value of the current I_{1o} in the (magnetostatic) simulation. The power α of B_M in Equation (6.21), which ranges from 1.6 to 2.2 in Steimnetz's formula, can be approximated by 2.

Then the iron losses are given by

$$p_{Fe} = \int_{\tau_{Fe}} p_{s,Fe} B_M^2 \gamma_{Fe} d\tau = 4L_{Fe} \int_{S_{Fe}} p_{s,Fe} B_M^2 \gamma_{Fe} dS \qquad (6.22)$$

If the transformer works at a frequency f different from 50 Hz, the iron losses have to be multiplied by $(f/50)^2$.

6.3.5 Example

Let us consider a shell-type transformer, characterized by the following nominal data:

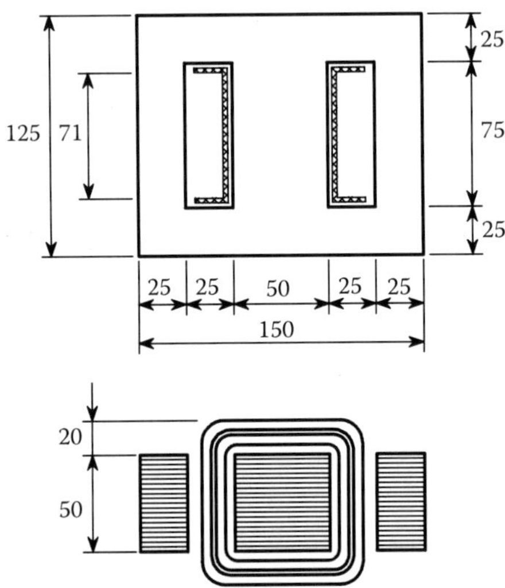

FIGURE 6.8
Main dimensions of the transformer.

$S_n = 200$ VA — rated power, at the secondary in continuous duty operation
$V_{1n} = 220/127$ V — primary voltages
$V_{2n} = 24/18$ V — secondary voltages
f = 50 Hz — frequency
conductor insulation in E class
magnetic material TERNI 1350, with specific iron losses $p_{s,Fe} = 1.3$ W/kg (at $B_M = 1$ T and f = 50 Hz)
rated magnetic flux density in the core $B_M \approx 1.1$ T

The aspect and the dimensions of the iron core are reported in Figure 6.8. The rated currents of the secondary winding are

$$I_{2n} = \frac{S_n}{V_{2n}} = 8.33 \text{ A, at } V_{2n} = 24 \text{ V}$$

$$I_{2n} = \frac{S_n}{V_{2n}} = 11.1 \text{ A, at } V_{2n} = 18 \text{ V}$$

Assuming an efficiency η = 90% and a unity power factor, the rated currents of the primary winding are

$$I_{1n} = \frac{S_n}{\eta V_{1n}} = 1 \text{ A, at } V_{1n} = 220 \text{ V}$$

$$I_{1n} = \frac{S_n}{\eta V_{1n}} = 1.74 \text{ A, at } V_{1n} = 127 \text{ V}$$

Such values have been used for the choice of the diameters of the coils, fixing a current density about 2 A/mm^2. Each coil is characterized by three terminals, as shown in Figure 6.9. The number of conductors have been chosen according to the iron flux density B_M = 1.1 T in the iron core. In addition, in the choice of the number of secondary winding turns, a 4% voltage drop under full load has been assumed. With reference to Figure 6.9, the following number of turns and diameters have been chosen for the secondary winding:

N_2' = 35 turns, with $d_{Cu2'}$ = 2.938 mm
N_2'' = 12 turns, with $d_{Cu2''}$ = 2.488 mm
N_2 = 47 total turns

Assuming a 0.1 mm insulation thickness, the total dimension of the winding results:

2 layers with $d_{Cu2'}$	thickness	6.1 mm
1 layer with $d_{Cu2''}$	thickness	2.6 mm
	thickness	8.7 mm

The following number of turns and diameters have been chosen for the primary winding:

N_1' = 240 turns, with $d_{Cu1'}$ = 1.153 mm
N_1'' = 176 turns, with $d_{Cu1''}$ = 0.885 mm
N_1 = 416 total turns

The dimension of the primary winding results as follows:

5 layers with $d_{Cu2'}$	thickness	6.3 mm
3 layers with $d_{Cu1''}$	thickness	3.0 mm
	thickness	9.3 mm

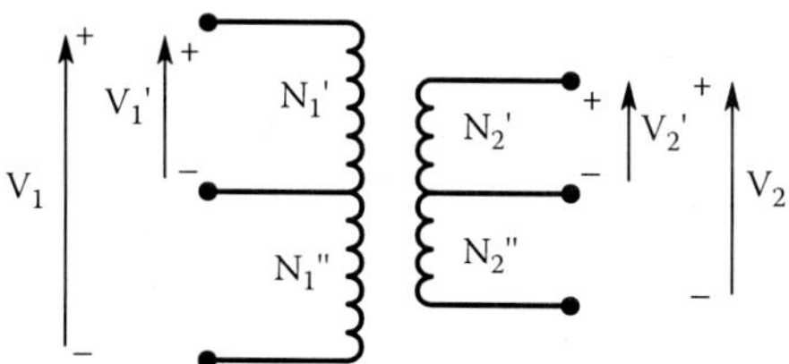

FIGURE 6.9
Sketch of the transformer windings.

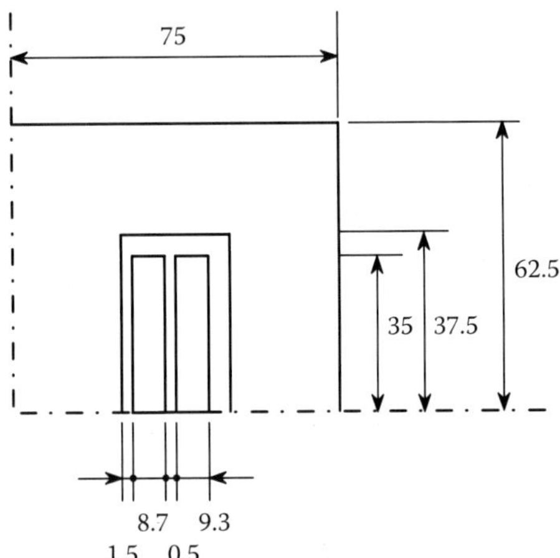

FIGURE 6.10
Section of the transformer that is used in the following simulations.

Finally, an insulation reel with a thickness of 1.5 mm, an insulation between the primary and secondary winding with a thickness of 0.5 mm, and an external insulation with a thickness of 0.5 mm have been adopted. The total thickness of the windings and the corresponding insulation is then equal to 18.7 mm. In Figure 6.10 there are the dimensions of the two coils in a quarter of the section of the transformer. The simulations are carried out referring to this section part.

6.3.5.1 *Computation with Linear Iron*

Let us assume

- a constant current in the primary winding equal to $I_{1o} = 0.1$ A, and null current in the secondary winding. In the equivalent conductive bar results $N_1I_{1o}/2 = 20.8$ A;
- a constant relative permeability of the iron equal to $\mu_{Fe} = 5500$.

Figure 6.11 shows the flux plot, obtained by drawing the equipotential lines of the magnetic vector potential.

Figure 6.12 shows the flux density along the line AA″ (see Figure 6.11) inside the magnetic iron. The average value of the flux density is approximately equal to 1.125 T. The flux density is not exactly constant, due to the different length of the magnetic paths within the iron core. The total magnetic energy stored in the transformer, using Equation (6.13) or Equation (6.14), results in

$$W_m = 58.46 \text{ mJ}$$

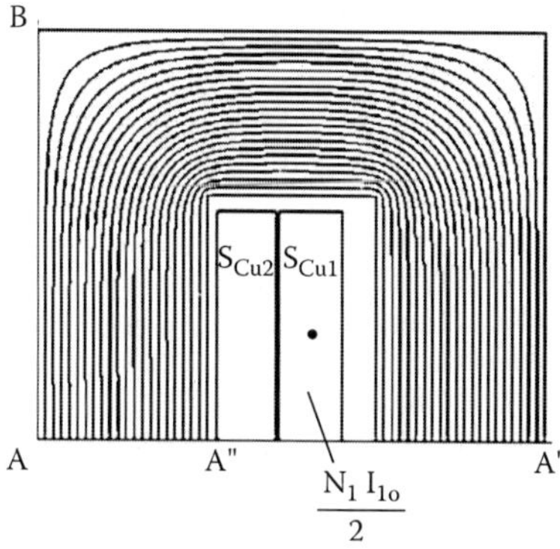

FIGURE 6.11
Flux plot.

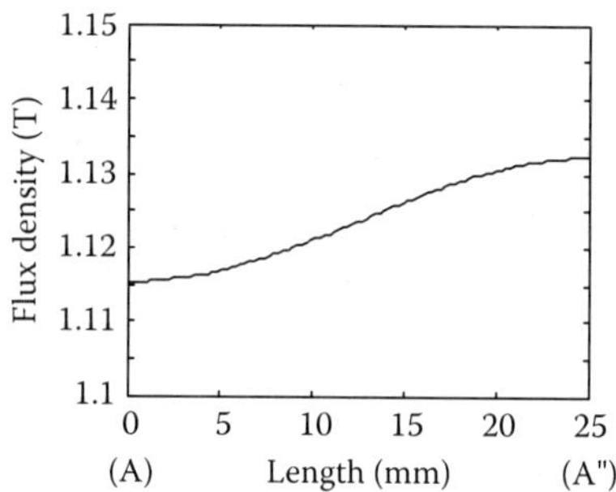

FIGURE 6.12
Flux density within the iron core.

The integral of the magnetic energy density only over the iron core results in

$$W_{m(Fe)} = 58.42 \text{ mJ}$$

which represents almost the totality of the stored magnetic energy. Conversely, by applying Equation (6.14) to the iron core, the integral results null. In fact this equation represents the energy of the transformer as work furnished by the external sources. It is different from zero only in the coils of the primary winding, where the current density is not null.

From the second part of Equation (6.15), the magnetic flux is

$$\Phi_o = 2.7952 \text{ mWb}$$

Then, from Equation (6.16), with $N_1 = 416$ and $N_2 = 47$, the flux linkages are

$$\Lambda_{1o} = 1.1628 \text{ Vs}$$

$$\Lambda_{2o} = 0.1314 \text{ Vs}$$

If Equation (6.17) is directly applied, where $S_{Cu1} = 3.255 \cdot 10^{-4}$ m^2 and $S_{Cu2} = 3.045 \cdot 10^{-4}$ m^2, one obtains

$$\Lambda_{1o} = 1.1696 \text{ Vs}$$

$$\Lambda_{2o} = 0.1321 \text{ Vs}$$

which differ from the previous values just of 0.5%. Finally the voltage induced in the windings are

$$E_{1o} = 367.4 \text{ V}$$

$$E_{2o} = 41.5 \text{ V}$$

Since the system has been supposed to be linear, the RMS values of these voltages result in

$$E_{1o(rms)} = 220 \text{ V}$$

$$E_{2o(rms)} = 24.85 \text{ V}$$

The self-inductance of the primary winding and the mutual inductance, computed using Equation (6.19), are

$$L_1 = 11.69 \text{ H}$$

$$M = 1.321 \text{ H}$$

From Equation (6.20), the same value for L_1 is computed.

6.3.5.2 Computation with Nonlinear Iron

The B-H curve of the TERNI 1350 silicon-steel laminations is now adopted, with a thickness of 0.5 mm and specific iron losses of $p_{s,Fe} = 1.3$ W/kg (at 1 T and 50 Hz), which is not a grain-oriented steel. In the low-power transformer, grain-oriented laminations are not used both to reduce the cost and to simplify the construction (E-shape or U-shape laminations are often used).

Table 6.2 reports the values of the MMF $N_1I_{1o}/2$, the iron flux density B_{Fe}, the primary flux linkage Λ_{1o}, and the induced voltage E_{1o} related to the various primary currents I_{1o}. The magnetizing characteristic is shown in Figure 6.13, highlighting the computed values by dots. In the same figure, the magnetizing characteristic of the transformer is reported, obtained as

TABLE 6.2

Values Corresponding to the Field Analysis

I_{1o} (mA)	$N_1I_{1o}/2$ (A)	B_{Fe} (T)	Λ_{1o} (Vs)	E_{1o} (V)
25.0	5.2	0.36	0.3774	118.5
50.0	10.4	0.73	0.7542	237.0
75.0	15.6	0.94	0.9780	307.2
100.0	20.8	1.05	1.0977	344.9
112.5	23.4	1.09	1.1361	356.9
125.0	26.0	1.12	1.1684	367.1
137.5	28.6	1.15	1.1976	376.2
150.0	31.2	1.17	1.2202	383.3

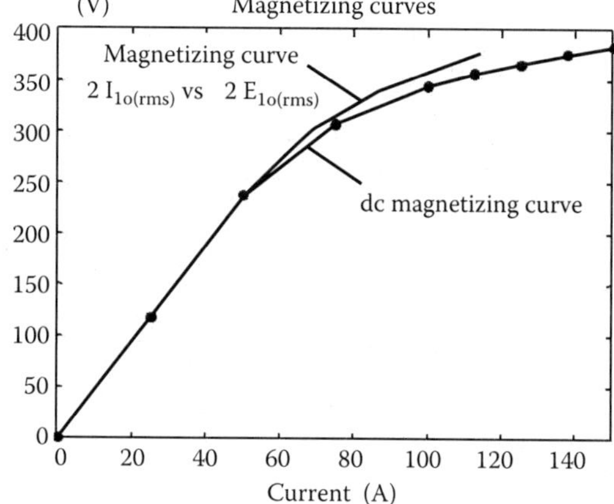

FIGURE 6.13
Current-voltage characteristics.

described in Section 6.3.3. Figure 6.14 shows the sinusoidal voltage waveform, corresponding to the rated value of $E_{1o(rms)} = 220$ V, and the corresponding current waveform. The distortion of the current from the sinusoidal waveform is noticeable.

6.3.5.3 Effect of the Equivalent Air-Gap

To simplify the assembly of the low-power transformers, E-shape laminations are commonly used, as shown in Figure 6.15(a). The dimensions of the lamination are defined by the standardization UNEL 82611, identified from dimension C of the central leg.

In order to consider the small air-gap that occurs between the two parts of the core, some analyses have been carried out with different values of the air-gap. Since the symmetry between the upper and lower parts disappears,

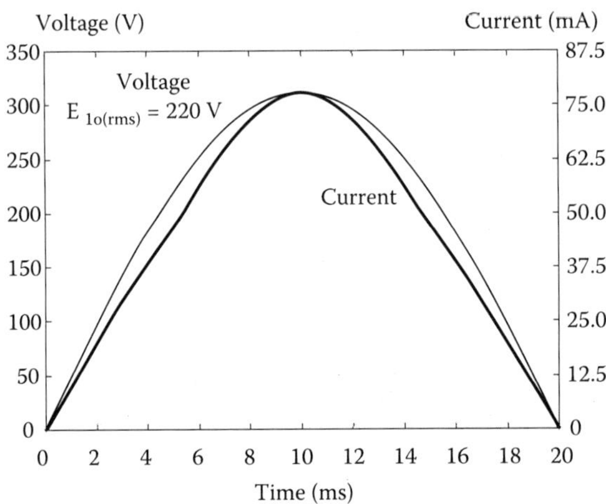

FIGURE 6.14
Voltage and current waveforms corresponding to the rated voltage $E_{1o(rms)} = 220$ V.

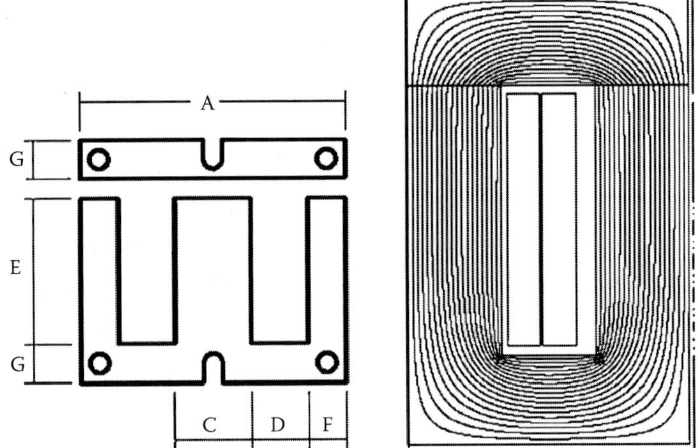

FIGURE 6.15
Standardized lamination for a low-power shell-type transformer (a) and flux plot in half a section of the transformer (b).

half a section of the transformer is considered, as illustrated in Figure 6.3(b). The magnetic flux plot is shown in Figure 6.15(b).

Some magnetic quantities, corresponding to different values of the air-gap thickness, are reported in Table 6.3. All the simulations are carried out with $\mu_{Fe} = 5500$, $N_1I_1 = 208$ A, and $I_2 = 0$. From Table 6.3 the effect of the air-gap thickness is evident, even though it assumes low values.

TABLE 6.3

Effect of the Air-Gap on the Magnetic Quantities

g (mm)	B_{Fe} (T)	$\int_{S_{Cu1+}} A_z dS$ (Vs/m)	$\int_{S_{Cu2+}} A_z dS$ (Vs/m)	$\int_S J_z A_z dS$ (J/m)
0.05	2.13	$17.32 \cdot 10^{-6}$	$16.20 \cdot 10^{-6}$	11.074
0.10	1.19	$9.674 \cdot 10^{-6}$	$9.043 \cdot 10^{-6}$	6.1816
0.15	0.83	$6.745 \cdot 10^{-6}$	$6.0303 \cdot 10^{-6}$	4.3100

6.4 Determination of the Leakage Inductances

The leakage inductances $L_{1\sigma}$ and $L_{2\sigma}$ are determined from the short-circuit test of the transformer, feeding the primary winding and connecting the secondary winding in short circuit. The test is carried out at reduced voltage and with currents almost equal to their rated value. It follows that the iron core works at low magnetic flux density. It is commonly assumed that the iron works in the linear part of its B-H curve.

As regards the equivalent electric circuit of the transformer, Figure 6.4 is considered, with transformation ratio equal to the turn ratio, i.e., $a = N_1/N_2$ as in Equation (6.10). The simulation is carried out by forcing both winding currents, that is, the secondary current is chosen according to the primary current so that the two MMFs result in equal and opposite, i.e., $N_1 i_1 = -N_2 i_2$. Then, from the second part of Equation (6.6), the magnetizing current i_m is null, because of

$$i_m = i_1 + \frac{i_2}{a} = i_1 + \frac{-(N_1/N_2)i_1}{N_1/N_2} = 0 \tag{6.23}$$

Thus the two flux linkages (6.2) correspond to the leakage flux linkages

$$\begin{cases} \lambda_{1\sigma} = L_{1\sigma} i_1 \\ \lambda_{2\sigma} = L_{2\sigma} i_2 \end{cases} \tag{6.24}$$

The current induced in the secondary winding is such as to have a null main flux, so that only leakage magnetic fluxes appear in the transformer.

In practice, the assumption of the transformation ratio $a = N_1/N_2$ is unnecessarily restrictive in order to have a null magnetizing current. In fact, to satisfy this condition, it is sufficient that $i_2 = -i_1/a$, for each value of a. The choice of the transformation ratio equal to the turn ratio is particularly convenient, since this allows the two coil currents to be easily set in the simulation, and a physical meaning to be given to the flux linkages obtained in the solved structure.

Once the leakage inductances $L_{1\sigma}$ and $L_{2\sigma}$ are obtained, the total leakage inductance is computed, referring to the equivalent electric circuit in the Γ' representation (Figure 6.5) or in the Γ'' representation (Figure 6.6).

6.4.1 Statement of the Problem

As far as the geometric symmetry of the structure is concerned, the consideration given above can be repeated: the simulations can be carried out on a quarter of the transformer section.

As far as the boundary conditions are concerned, they depend on the transformer type. In the shell-type transformer (see Figure 6.3), the flux density vectors have a normal component only along the symmetry line AA′, and a tangential component only along the symmetry line BB′. Then the boundary conditions remain the same as those assigned in the no-load simulation (Neumann's condition in AA′ and Dirichlet's condition in BB′).

In the core-type transformer, the flux density vectors have a normal component only along the symmetry line AA′, and a tangential component only along the symmetry line BB′, exactly as for the shell-type transformer. Then the boundary conditions must be modified with respect to those assigned at the no-load.

In this analysis, since only leakage fluxes occur, the flux lines are mainly localized in the air and in the coils. It is appropriate to increase the mesh of the domain in the coils and in the surrounding space.

The iron is assumed to be linear with a constant magnetic permeability μ_{Fe}.

A magnetostatic simulation is conveniently carried out. Constant currents are forced in the two windings: I_1 in the primary coil and $I_2 = -I_1(N_1/N_2)$ in the secondary coil. Such currents are fixed to their rated value, even though, since the problem is linear, the current value does not matter.

As in the no-load simulation, conductive bars are used, equivalent to the whole coil, as indicated in Figure 6.2(c) and Figure 6.3(c), that are fed by the total current. Since the field quantities are not time-dependent, the current is uniformly distributed in the surface of the equivalent bar. Some considerations about simulations with eddy currents will be described in Section 4.3.

The current in the two parts of the primary and secondary coils are imposed as follows. With a core-type transformer, they are

$N_1 I_{1sc}/4$, in the primary coil, and

$N_2 I_{2sc}/4 = -N_1 I_{1sc}/4$, in the secondary coil

while, with a shell-type transformer, they are

$N_1I_{1sc}/2$, in the primary coil, and
$N_2I_{2sc}/2 = -N_1I_{1sc}/2$, in the secondary coil.

6.4.2 Computation on the Solved Structure

Once the field problem is solved, the magnetic quantities are computed as follows.

6.4.2.1 *Magnetic Flux Density*

The components of the magnetic flux density are evaluated by deriving the magnetic vector potential along the coordinate axes, as in Equation (3.66).

6.4.2.2 *Magnetic Energy*

The magnetic fields, corresponding to the assigned currents, are mainly localized in the insulating space between the two windings and the two windings themselves. Remembering that the material has been assumed to be linear, the stored magnetic energy is computed by integrating the magnetic energy density (6.12) over all the volume of the transformer. Since the problem is 2D and only a quarter of the section has been analyzed, the magnetic energy results in

$$W_m = 4 l_{average} \int_S \frac{1}{2} \mu H^2 \, dS \tag{6.25}$$

where S is the total surface of the domain and $l_{average}$ indicates the average length of a conductor (half a length of one turn). Let us highlight that the effect of the end windings has been considered the same of the part of the winding in the section under study. Such an approximation is reasonable, since the fields are mainly localized between the two coils.

Alternatively, the magnetic energy is computed as the energy that is furnished from the source to the system, as

$$W_m = 4 \, l_{average} \int_S \frac{1}{2} J_{1z} A_z + \frac{1}{2} J_{2z} A_z \, dS \tag{6.26}$$

6.4.2.3 *Flux Linkage*

The flux linked with the primary winding is computed by integrating the magnetic vector potential A_z over the surface of the equivalent bar representing the primary coil. It results that

$$\Lambda_{1sc} = N_1 l_{average} \frac{1}{S_{Cu1}} \left(\int_{S_{Cu1+}} A_z \, dS - \int_{S_{Cu1-}} A_z \, dS \right), \text{ with core-type transformer}$$

$$\Lambda_{1sc} = 2N_1 l_{average} \frac{1}{S_{Cu1}} \left(\int_{S_{Cu1+}} A_z \, dS \right), \text{ with shell-type transformer} \tag{6.27}$$

Analogously for the secondary winding, substitute the subscript 2 for subscript 1 in Equation (6.27).

6.4.2.4 *Induced EMF*

The EMFs induced in the two windings are obtained by multiplying the flux linkages by the electric frequency ω. The EMFs are given by

$$\begin{aligned} E_{1sc} &= \omega \, \Lambda_{1sc} \\ E_{2sc} &= \omega \, \Lambda_{2sc} \end{aligned} \tag{6.28}$$

Since the field problem is linear, the quantities I_{1sc}, I_{2sc}, Λ_{1sc}, Λ_{2sc}, and E_{1sc}, E_{2sc} are in proportion. The maximum values are achieved by multiplying the RMS values by the square root of two.

6.4.2.5 *Leakage Inductances*

The short-circuit inductances are obtained by dividing the flux linkages by the respective winding currents. Since $a = N_1/N_2$ and $N_1 I_{1sc} = -N_2 I_{2sc}$, so that $I_m = 0$ from Equation (6.6), then $\Lambda_{1sc} = L_{1\sigma} I_{1sc}$ and $\Lambda_{2sc} = L_{2\sigma} I_{2sc}$. These equations are obtained by examining the electric circuit in Figure 6.4 and substituting Equation (6.3) and Equation (6.23) in Equation (6.2) together with Equation (6.11). In conclusion, it results that

$$\begin{aligned} L_{1\sigma} &= \frac{\Lambda_{1sc}}{I_{1sc}} \\ L_{2\sigma} &= \frac{\Lambda_{2sc}}{I_{2sc}} \end{aligned} \tag{6.29}$$

An alternative way to compute such inductances is to use the energy furnished by the two sources. In the core-type transformer they are

$$\begin{aligned} L_{1\sigma} &= \frac{2}{I_{1sc}^2} \left[4 l_{average} \left(\int_{Scu1+} \frac{1}{2} J_{1z} A_z \, dS - \int_{Scu1-} \frac{1}{2} J_{1z} A_z \, dS \right) \right] \\ L_{2\sigma} &= \frac{2}{I_{2sc}^2} \left[4 l_{average} \left(\int_{Scu2+} \frac{1}{2} J_{2z} A_z \, dS - \int_{Scu2-} \frac{1}{2} J_{2z} A_z \, dS \right) \right] \end{aligned} \tag{6.30}$$

while in the shell-type transformer they are

$$L_{1\sigma} = \frac{2}{I_{1sc}^{\;2}}\left(4l_{average} \int_{Scu1} \frac{1}{2} J_{1z} A_z \; dS \right)$$
$$L_{2\sigma} = \frac{2}{I_{2sc}^{\;2}}\left(4l_{average} \int_{Scu2} \frac{1}{2} J_{2z} A_z \; dS \right) \tag{6.31}$$

that correspond to those given in Equation (6.29), observing that in the core-type transformer

$$J_{1z} = \frac{N_1 I_1}{4S_{Cu1}} \qquad J_{2z} = \frac{N_2 I_2}{4S_{Cu2}}$$

and in the shell-type transformer

$$J_{1z} = \frac{N_1 I_1}{2S_{Cu1}} \qquad J_{2z} = \frac{N_2 I_2}{2S_{Cu2}}$$

The total leakage inductance $L'_{2\sigma}$ (or alternatively $L''_{1\sigma}$) is obtained, applying the equivalences of Table 6.1.

6.4.2.6 *Electric Forces*

The forces on the coils are evaluated by using Maxwell's stress tensor. Because of the 2D problem, the coil is enclosed by a simple line. The tangential and normal forces to each line are computed as

$$F_t = l_{average} \int_l \left(\mu_o H_t H_n \right) dl$$
$$F_n = l_{average} \int_l \frac{\mu_o}{2} \left(H_n^{\;2} - H_t^{\;2} \right) dl \tag{6.32}$$

The x-axis force is the sum of the forces F_t of the vertical lines with the forces F_n of the horizontal lines. Similarly, the y-axis force is the sum of the forces F_t of the horizontal lines with the forces F_n of the vertical lines.

Due to the symmetry with respect to the axis BB′ that has been used in the simulations, the forces refer to half a coil, and then the overall y-axis force component is null while the overall x-axis force component is obtained by doubling the force obtained in half a winding.

Alternatively, the force on the coils can be computed as Lorentz's force. According to Lorentz's specific electric force (1.84) and the definition of the current density vector (1.60), the x-axis and the y-axis that force components in the half coil are

$$F_x = -l_{average} \int_{S_{Cu1}} J_z B_y \, dS$$
$$F_y = l_{average} \int_{S_{Cu1}} J_z B_x \, dS \tag{6.33}$$

6.4.3 Simulation with Induced Currents

Instead of magnetostatic simulations, a variable field problem may be analyzed. The currents in primary and secondary windings are not forced simultaneously: a periodic sinewave current $i_1(t)$ is forced only in the primary winding, while the current $i_2(t)$ results as the solution of the field problem. It is a current induced in the secondary winding. The waveform of the two currents is sinusoidal with a constant frequency, and the complex phasor notation is adopted.

Such an approach is not recommended, however, when it is used it is indispensable to avoid some possible errors. Some remarks are hereafter underlined.

1. If the secondary coil is modeled by the equivalent conductive bar, and eddy currents are computed in this bar, then the distribution of the current in the bar is not uniform. In fact, the current are nonuniformly distributed as the bar size is large and the frequency is high. To have a correct result, it must be imposed that the induced current in the equivalent bar of the secondary winding be uniformly distributed over the whole section. It is not easy to assign such a constraint, if not impossible.

   Of course the same problem occurs for the equivalent bar of the primary winding. However, since the primary current is a current source, the current density can be defined to be uniformly distributed over the bar.

2. In order to obviate the problem described in the point 1, the secondary winding may be subdivided in the effective number of conductors. However, because of the 2D problem, without a specific constraint the induced current in each conductor may be different. It is necessary to impose the constraint of equal current in each conductor that forms the coil (series conductors). Unfortunately, it is not always possible to assign this constraint.
3. With a core-type transformer, and the adopted simplifications, the solution of the variable field problem is carried out as though the secondary conductors are distinguished with a closing path at

the infinite. This is because of the 2D problem. Then, the bar S_{cu2+} links the flux produced by the primary currents, while S_{cu2-} does not link any flux, so that only S_{cu2+} is interested by eddy currents. It is thus compulsory to impose that the sum of the currents induced in S_{cu2+} and S_{cu2-} be null, which is $I_{cu2-} = -I_{cu2+}$. A method to impose this condition is to declare that the bars S_{cu2+} and S_{cu2-} are interested by a total current in any time. This is equivalent to imposing that $I_{cu2+} + I_{cu2-} = 0$.

6.4.4 Example

Let us consider the transformer described in Section 3.6, with the aim of computing the leakage inductances.

The iron relative permeability is fixed to the constant value $\mu_{Fe} = 5500$.

The current $I_{1sc} = 1$ A is forced in the primary winding, and then the total current in the equivalent primary bar of the analyzed section part is

$$\frac{N_1 I_{1sc}}{2} = 208 \text{ A}$$

and the total current in the equivalent secondary bar is

$$\frac{N_2 I_{2sc}}{2} = -\frac{N_1 I_{1sc}}{2} = -208 \text{ A}$$

which is a current $I_{2sc} = -8.851$ A. The flux plot is shown in Figure 6.16.

The flux density within the windings assumes the behavior of Figure 6.17(a) with a maximum value B = 7.44 mT. This is in accord with the theoretical behavior, as reported in Figure 6.17(b). For the sake of comparison, the maximum flux density computed analytically results in

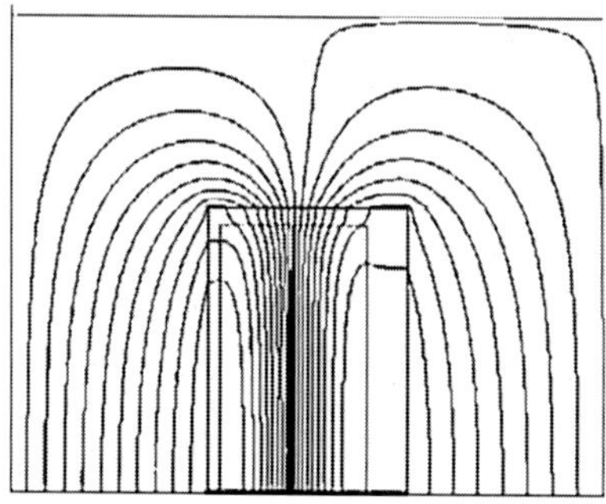

FIGURE 6.16
Flux plot during short-circuit operation.

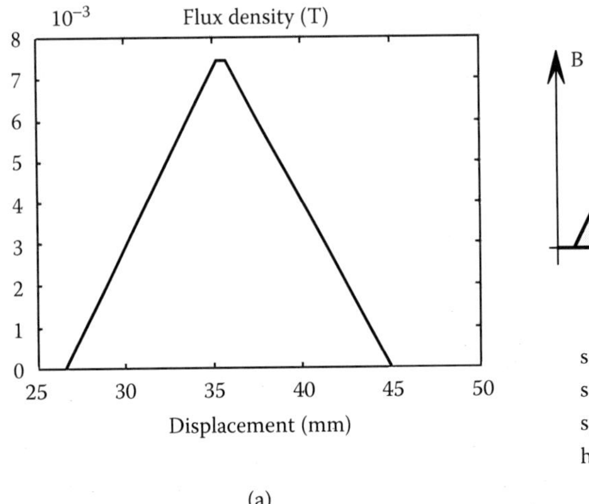

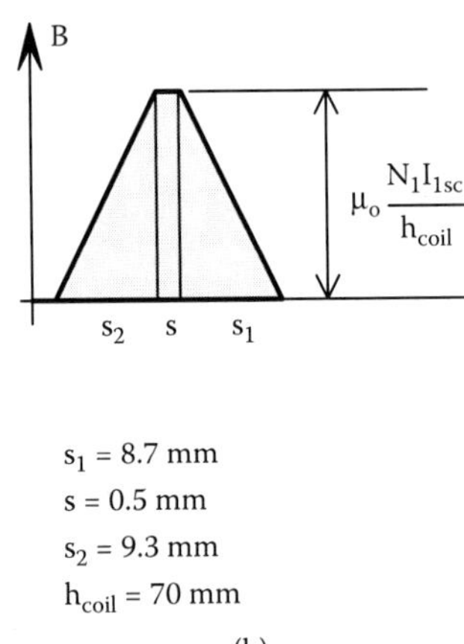

FIGURE 6.17
Flux density behavior within the two windings.

$$B = \mu_o \frac{N_1 I_{1sc}}{h_{coil}} = 7.47 \text{ mT}$$

which corresponds to the value computed from the field solution.

The length $l_{average}$ is computed as the semi-circumference in the middle of the two coils. According to Figure 6.8 and Figure 6.10, $l_{average}$ = 132.8 mm. The stored magnetic energy, computed by Equation (6.25), is

$$W_m = 2.547 \text{ mJ}$$

and results are stored almost completely in the primary winding (45.59%), secondary winding (42.87%), and the surrounding space (11.53%). In the iron, it is lower than 0.01%.

From Equation (6.26) it results always that

$$W_m = 2.547 \text{ mJ}$$

whose 0.9766 mJ in the primary winding and 1.5704 mJ in the secondary winding.

The flux linkages, from Equation (6.27), are

$$\Lambda_{1sc} = 1.9533 \text{ mVs}$$

$$\Lambda_{2sc} = -0.35488 \text{ mVs}$$

The corresponding leakage inductances (6.29) are

$L_{1\sigma} = 1.9533$ mH

$L_{2\sigma} = 0.040076$ mH (that is, 3.1396 mH if referred to the primary winding)

If the total leakage inductance referred to the primary winding is approximated by

$$L_{1\sigma} + \left(\frac{N_1}{N_2}\right)^2 L_{2\sigma} = 5.093 \text{ mH}$$

it is noted that it corresponds to that obtained as

$$L_\sigma = \frac{2W_m}{I_1^2} = 5.094 \text{ mH}$$

The same inductance, computed analytically, is given by

$$L_\sigma = \mu_o \frac{N_1^2}{h_{coil}} 2l_{average}\left(\frac{s_1}{3} + s + \frac{s_2}{3}\right) = 5.365 \text{ mH}$$

The force on the coils is computed by means of Maxwell's stress tensor.

The overall y-axis component of the force (which is in the direction parallel to the axis of the leg) is null, for symmetry reasons. On the contrary, each half a coil is compressed. The value of the compression force on the considered half a coil is $F_y = 2.78$ mN.

The overall x-axis component of the force (which is in the direction normal to the axis of the leg) results in

$$F_x = 0.391 \text{ N}$$

The x-axis force computed analytically is given by

$$F_x = \frac{1}{2}\mu_o \frac{2l_{average}}{h_{coil}}\left(N_1 I_{1sc}\right)^2 = 0.413 \text{ N}$$

which is slightly higher than that obtained from the field solution.

Note: the computation of the x-axis force component on the secondary coil is illustrated as follows. A line surrounding the coil is fixed as shown in Figure 6.18. The x-axis force F_x, is computed in the different paths as

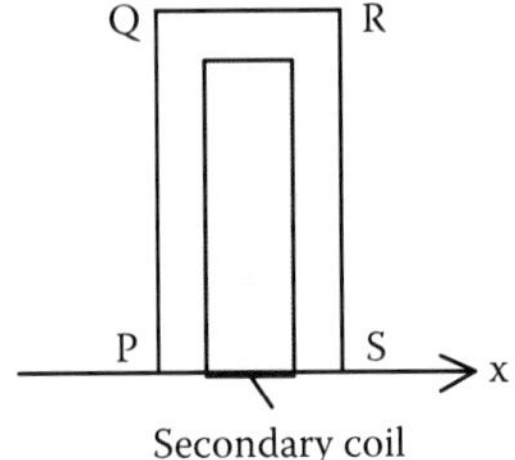

FIGURE 6.18
Line surrounding the secondary coil to compute the x-axis force.

$$F_{xPQ} = 4l_{average} \int_{l_{PQ}} \frac{B_x H_x - B_y H_y}{2} dl = 1.22 \cdot 10^{-3} N$$

$$F_{xQR} = 4l_{average} \int_{l_{QR}} B_x H_y dl = 6.64 \cdot 10^{-3} N$$

$$F_{xRS} = 4l_{average} \int_{l_{RS}} \frac{B_x H_x - B_y H_y}{2} dl = 383.1 \cdot 10^{-3} N$$

The y-axis force component may be computed in a similar way.

6.5 Algorithm for the Construction of the Magnetizing Characteristic of the Transformer

A simple algorithm is described in the following, using the BASIC language, for the construction of the magnetizing characteristic of the transformer. The saturation is considered, using the dc magnetizing characteristic $I_{1o} - \Lambda_{1o}$.

The FUNCTION `current(FlxLnk)` returns the value of the current corresponding to the value of the flux linkage, given as input. It interpolates the value along the dc magnetizing characteristic. For the sake of simplicity, a linear interpolation is used. Better results can be used by means of higher-order interpolations.

```
DECLARE FUNCTION current (FlxLnk)
CLS
CLEAR

CONST pi = 3.14156
a$ = CHR$(34)
DIM f(205), i(205), ang(205), v(205)

omega = 2 * 50 * pi     'electric frequency
Fm = 600                'maximum value of flux linkage
```

```
Vm = Fm * omega        'maximum value of voltage

' computation of the current waveform

 Imax = 0
 FOR k = 1 TO 201
     theta = 2 * pi * (k - 1)/ 200
     ang(k) = theta
     v(k) = Vm * SIN(theta)
     f(k) = Fm * COS(theta)
     i(k) = current(f(k))
     IF i(k) > Imax THEN Imax = i(k)
 NEXT k

' computation of the rms values

 isum = 0
 vsum = 0
 FOR k = 1 TO 200
    isum = isum + i(k) ^ 2
    vsum = vsum + v(k) ^ 2
 NEXT k
 Vrms = SQR(vsum/ 200)
 Irms = SQR(isum/ 200)
```

A possible structure of the FUNCTION `current(FlxLnk)`, supposing Nb=9 points of the dc magnetizing characteristic and a linear interpolation, is as follows. The control of the sign of the input is required if the given characteristic contains only positive values of the current and the flux linkages. Refer to Figure 6.13 and Table 6.2.

```
FUNCTION current(FlxLnk)

'sign evaluation

 sign = 1
 IF FlxLnk < 0 THEN
    sign = -1
    FlxLnk = - FlxLnk
 END IF

'Definition of the dc magnetizing characteristic
'given by a number of Nb points

 Nb = 9
 DIM FluxVect(Nb)
 DIM CurrVect(Nb)

 FluxVect(1) = 0:           CurrVect(1) = 0
 FluxVect(2) = 0.3774:      CurrVect(2) = .025
 FluxVect(3) = 0.7543:      CurrVect(3) = .050
 FluxVect(4) = 0.9780:      CurrVect(4) = .075
 FluxVect(5) = 1.0977:      CurrVect(5) = .100
 FluxVect(6) = 1.1361:      CurrVect(6) = .1125
 FluxVect(7) = 1.1684:      CurrVect(7) = .125
 FluxVect(8) = 1.1976:      CurrVect(8) = .1375
 FluxVect(9) = 1.2202:      CurrVect(9) = .150
```

```
'Interpolation (linear)

 FOR i = 0 TO (Nb - 1)
   IF (FlxLnk >= FluxVect(i) AND FlxLnk < FluxVect(i + 1)) THEN
     Cf =(CurrVect(i+1)-CurrVect(i))/ (FluxVect(i+1)-
      FluxVect(i))
       CurrAux = Cf * (FlxLnk - FluxVect(i)) + CurrVect(i)
   END IF
 NEXT i

   IF (FlxLnk >= FluxVect(Nb)) THEN
      Cf=(CurrVect(Nb)-CurrVect(Nb-1))/(FluxVect(Nb)-
       FluxVect(Nb-1))
      CurrAux = Cf * (FlxLnk - FluxVect(Nb)) +
       CurrVect(Nb)
   END IF

'Sign correction

   current = CurrAux * sign
   FlxLnk = FlxLnk * sign

END FUNCTION
```

References

1. E. Arnold, *Die Wechselstromtechnick. Die Transformatoren,* Vol. 2, Julius Springer, Berlin, 1904.
2. R. Richter, *Elektrische Maschinen. Die Transformatoren,* Vol. 3, Julius Springer, Berlin, 1932.
3. A.S. Longsdorf, *Theory of Alternating Current Machinery,* McGraw-Hill, New York, 1937.
4. T. Bödefeld and H. Segueuz, *Elektrische Maschinen,* Springer-Verlag, Wren, 1942.
5. A. Carrer, *Macchine Elettriche. Parte prima: Trasformatori,* Editrice Universitaria Levrotto & Bella, Torino, 1954.
6. S. Ramo, J.R. Whinnery, and T. Van Duzer, *Fields and Waves in Communication Electronics,* John Wiley & Sons, New York, 1965.
7. G.R. Slemon, *Magnetoelectric Devices: Transducers, Transformers and Machines,* John Wiley & Sons, New York, London, Sydney, 1966.
8. A.E. Fitzgerald, C. Kingsley, Jr., and S.D. Umans, *Electric Machinery,* McGraw-Hill, New York, 1983.
9. I.J. Nagrath and D.P. Kothari, *Electric Machines,* Tata McGraw-Hill Publishing Company Limited, New Delhi, 1985.

7

Single-Phase Variable Reactance

This chapter deals with the finite element analysis of a single-phase variable reactance. In particular the dependence of the reactance on the air-gap length is analyzed. The results obtained by the finite element method are compared with those computed analytically.

7.1 The Single-Phase Variable Reactance

Figure 7.1 shows some structures of single-phase reactance. Figure 7.1(a) shows a shell-type reactance with constant air-gap length; Figure 7.1(b) shows a shell-type reactance with variable air-gap; Figure 7.1(c) shows a core-type reactance with variable air-gap.

Reactances are used in electrical circuits to filter or limit the current. Variable reactances are connected to capacitors to control the reactive power flow and/or to obtain electrical resonance, principally in laboratory tests.

According to the rated power of the reactance, the iron core may be formed by silicon-steel laminations or by grain-oriented laminations. Considering a plane parallel to the plane of the laminations and neglecting the end effects, the study can be reduced to a two-dimensional problem.

In addition, as for the single-phase transformer, the analysis may be reduced to a part of the section of Figure 7.1, after recognizing the symmetry axes of the structure and imposing along them the suitable boundary conditions. For instance, the study of the reactances of Figure 7.1 can be reduced to that of their parts reported in Figure 7.2.

7.2 Computation of the Reactance

The aim of the study is the determination of the value of the self-inductance. If the reactance is characterized by a variable air-gap, it is also important to

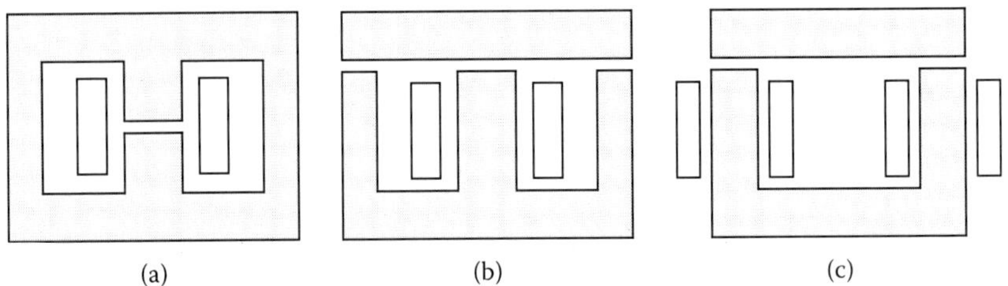

FIGURE 7.1
Magnetic structures of single-phase reactance.

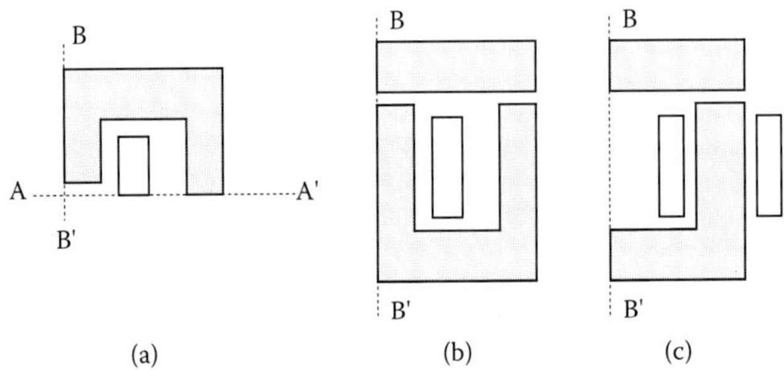

FIGURE 7.2
Reduction of the domain of the structures of Figure 7.1 using their symmetry.

study the dependence of such an inductance on the air-gap length. This dependence is achieved by means of several field analyses with different values of the air-gap. Furthermore, it is possible to assess the attractive forces between the movable part and the fixed part.

7.2.1 Statement of the Field Problem

The magnetic field strength is mainly localized in the air part of the magnetic path. Thus, it is convenient to increase the mesh in the air-gap of the reactance.

A magnetostatic simulation is carried out; a constant current I is forced in the winding as a field source. The flux linkage and the voltage are computed from the field solution. Once again, each section of the primary winding is represented by means of a unique conductive bar, through which the following total current flows:

- $\mathbf{N_tI}$, with shell-type reactance, analyzing the whole structure of Figure 7.1(a) or Figure 7.1(b), or the half of the structure shown in Figure 7.2(b)
- $\mathbf{N_tI/2}$, with shell-type reactance, analyzing a quarter of the structure, as shown in Figure 7.2(a), or with a core-type reactance, as in Figure 7.1(c) or in Figure 7.2(c)

where N_t is the total number of turns and I is the winding current.

Initially, a linear magnetic material is considered, with a constant magnetic relative permeability μ_{Fe}. The effect of the nonlinearity will be investigated later.

In order to assign the boundary conditions correctly, the field lines have to first be estimated along the boundaries. Along the symmetry axis AA′, the magnetic flux density vector **B** presents only a normal component: Neumann's boundary condition is assigned. Along the symmetry axis BB′, with shell-type reactances the magnetic flux density vector **B** presents only a tangential component: Dirichlet's boundary condition is assigned. Conversely, with core-type reactances the magnetic flux density vector **B** presents only a normal component: Neumann's boundary condition is assigned again. Along the other boundaries that delimit the domain (that are suitably distant lines), a null magnetic vector potential is assigned.

7.2.2 Computation on the Analyzed Structure

As the field solution has been obtained, the magnetic quantities of interest can be computed. In the following, the study will be focused on the structure of Figure 7.1(b) only. Thus, in order to extend the analysis to the other structures or parts of structure, the following equations have to be rearranged appropriately.

In addition, a rectangular leg is assumed, whose net iron length is equal to L_{Fe}.

7.2.2.1 Magnetic Flux Density

The flux lines are obtained directly by the magnetic vector potential. The components of the flux density can be assessed by deriving the magnetic vector potential, as described in Equation (3.66).

7.2.2.2 Magnetic Energy

Because of the use of all linear materials, the magnetic energy density is everywhere given by

$$w_m = \frac{1}{2}\mathbf{B}\cdot\mathbf{H} = \frac{1}{2}\mu H^2 \tag{7.1}$$

The magnetic energy stored in the whole reactance is obtained by integrating the magnetic energy density on the whole volume of the reactance. In a 2D problem, this is equivalent to a surface integral multiplied by the length L_{Fe}. If S is the total surface of the domain, it results in

$$W_m = L_{Fe}\int_S \frac{1}{2}\mu H^2 \, dS \tag{7.2}$$

Alternatively, the magnetic energy can be computed as

$$W_m = L_{Fe} \int_S \frac{1}{2} J_z \, A_z \, dS \tag{7.3}$$

which results different from zero only on the surfaces of the coils.

7.2.2.3 *Flux Linkage*

The flux linked with the winding is computed by processing the magnetic vector potential A_z on the surface of the coil. Let S_{Cu+} be the section of the equivalent bar in which the current is positive (with the same direction of the z-axis, i.e., leaving the sheet) and S_{Cu-} be the section of the equivalent bar in which the current is negative (with direction opposite to the z-axis, i.e., going into the sheet). Of course the two values are the same and equal to S_{Cu}. It results that

$$\Lambda = N_t L_{Fe} \frac{1}{S_{Cu}} \left(\int_{S_{Cu+}} A_z \, dS - \int_{S_{Cu-}} A_z \, dS \right) \tag{7.4}$$

7.2.2.4 *Electrical Voltage*

With the assumption of a sinusoidal waveform of the flux linkage, the RMS value of the voltage on the reactance is obtained by multiplying the RMS value of the flux linkage by the electrical frequency ω, which is

$$E_{rms} = \omega \, \Lambda_{rms} \tag{7.5}$$

The voltage at the terminal pairs of the reactance coils is obtained as the geometric sum of the voltage given in Equation (7.5) with the voltage drop on the resistance, computed analytically.

7.2.2.5 *Self-Inductance*

The value of the self-inductance is achieved by dividing the flux linkage by the corresponding current, which is

$$L = \frac{\Lambda}{I} \tag{7.6}$$

Alternatively, always in the linear case, the self-inductance can be achieved from the magnetic energy, resulting in

$$L = \frac{2W_m}{I^2} \tag{7.7}$$

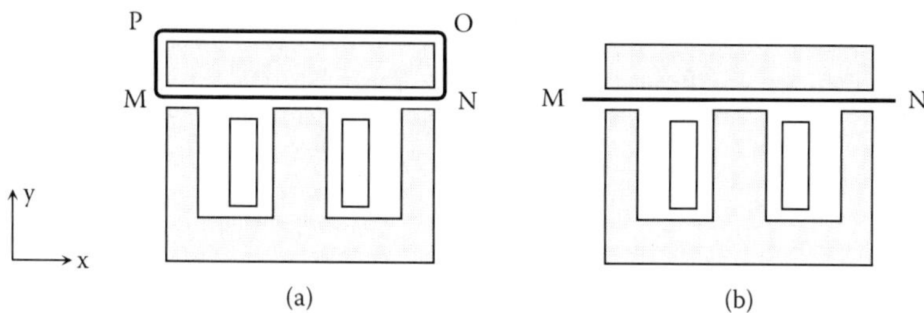

FIGURE 7.3
Lines along which the attractive force is computed.

7.2.2.6 Electric Forces

The force on the movable part of the reactance is assessed by means of Maxwell's stress tensor. A surface is initially chosen that surrounds the movable part on which the force has to be computed. In the 2D problem, the surface is reduced to a line, for instance the line MNOP of Figure 7.3(a). The total force in the y-direction is given by all the force components with direction parallel to the y-axis computed along the four segments of the line, which is

$$F_y = F_{MN} + F_{NO} + F_{yOP} + F_{yPM} \tag{7.8}$$

In Equation (7.8) there are the normal components of the force along the segments MN and OP:

$$\begin{aligned} F_{yMN} &= -L_{fe}\int_{MN}\frac{\mu_o}{2}\left(H_y^{\,2} - H_x^{\,2}\right)dl \\ F_{yOP} &= L_{fe}\int_{OP}\frac{\mu_o}{2}\left(H_y^{\,2} - H_x^{\,2}\right)dl \end{aligned} \tag{7.9}$$

and the tangential components of the force along the segments NO and PM:

$$\begin{aligned} F_{yNO} &= L_{fe}\int_{NO}\left(\mu_o H_y H_x\right)dl \\ F_{yPM} &= -L_{fe}\int_{PM}\left(\mu_o H_y H_x\right)dl \end{aligned} \tag{7.10}$$

Since the magnetic field assumes a negligible value outside the structure, the main addendum is the normal force on the segment MN. The attractive force can be estimated from the integration of the magnetic pressure on the line MN only, as illustrated in Figure 7.3(b).

For computing the force, an alternative of Maxwell's stress tensor is the method of the virtual works. Let $W_m(y)$ be the magnetic energy computed with the movable part in the generic position y. The movable part is shifted a small quantity in the y-axis direction, maintaining constant the source current. Let Δy be the change of position. The magnetic energy $W_m(y+\Delta y)$ is computed with the movable part in the new position $y+\Delta y$. Then, since the system is linear, the force is computed as

$$F_y = \left.\frac{\Delta W_m(y)}{\Delta y}\right|_{i=const} = \left.\frac{W_m(y+\Delta y) - W_m(y)}{\Delta y}\right|_{i=const} \tag{7.11}$$

Of course, this result depends on the chosen value of the movement Δy.

An improvement to the force computation is obtained considering the changes of position $+\Delta y$ and $-\Delta y$. The force yields

$$F_y = \left.\frac{W_m(y+\Delta y) - W_m(y+\Delta y)}{2\Delta y}\right|_{i=const} \tag{7.12}$$

In this way, all the terms of the force due to the second derivative of the energy are eliminated. This is verified by expanding the magnetic energy around the position y via the Taylor series. Neglecting the upper-order addenda, it is

$$W_m(y+\Delta y) = W_m(y) + \frac{dW_m(y)}{dy}\Delta y + \frac{1}{2}\frac{d^2W_m(y)}{dy^2}\Delta y^2 + \ldots \tag{7.13}$$

$$W_m(y-\Delta y) = W_m(y) - \frac{dW_m(y)}{dy}\Delta y + \frac{1}{2}\frac{d^2W_m(y)}{dy^2}\Delta y^2 + \ldots \tag{7.14}$$

By subtracting the second equation from the first one, it results that

$$\frac{dW_m(y)}{dy} = \frac{W_m(y+\Delta y) - W_m(y-\Delta y)}{2\Delta y} \tag{7.15}$$

7.2.3 Considerations with Nonlinear B-H Curve

As for the single-phase transformer, the dc magnetizing characteristic is initially obtained from discrete values of current I and flux linkage Λ. Neglecting the magnetic hysteresis and the resistive voltage drop, assuming a sinusoidal waveform of the flux linkage, the corresponding current waveform is computed. Then, from the current waveform the RMS value of the current is computed. The relationship between E_{rms} and I_{rms} is obtained, as reported in Figure 7.4.

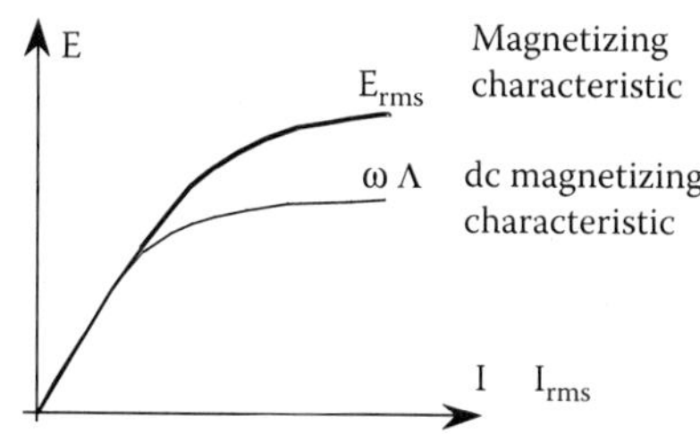

FIGURE 7.4
Magnetizing characteristics.

7.3 Example

A single-phase variable reactance, designed for laboratory tests, is analyzed. It has been designed for giving rise to resonance phenomena, when connected with capacitor loads. The reactance is changed so that a very high voltage is applied to the capacitor load under test, even though the voltage source is lower. The air-gap length is adapted from time to time depending on the capacitor load, in order to obtain the resonance.

According to a capacitor load in the range C = 10 ÷ 200 nF, the variation of the inductance is in the range L = 50 ÷ 1000 H.

The rated data of the single-phase reactance are as follows:

1. f = 50Hz — frequency
2. V_n = 50 kV — RMS voltage in resonance conditions
3. I_n = 3.15 A — maximum RMS current
4. S_n = 160 kVA — apparent power
5. magnetic material — grain-oriented TERNI M5T30, with thickness 0.3 mm and specific iron losses $p_{s,Fe}$ = 0.5 W/kg (at B_M = 1 T, f = 50 Hz)
6. maximum flux density between 1.4 and 1.5 T (to avoid saturation)
7. the reactance is in oil, so that a class A insulation is chosen

The leg is almost circular, with S_{Fe} = 4253 mm^2 and external diameter D = 249.6 mm (the stacking factor is k_{stk} = 0.93, and the utilization factor, due to the steps of the leg, is 0.935). The distance between the iron leg and the winding is 25 mm with two insulating layers of 5 mm thickness each and three oil channels. The distance between the coil and the iron yoke is 100 mm, and between the two external diameters of the coils it is 110 mm. The main dimensions of the reactance are reported in Figure 7.5. The movable part is chosen with a rectangular section with a width equal to the diameter D of the leg and a height equal to 170.4 mm.

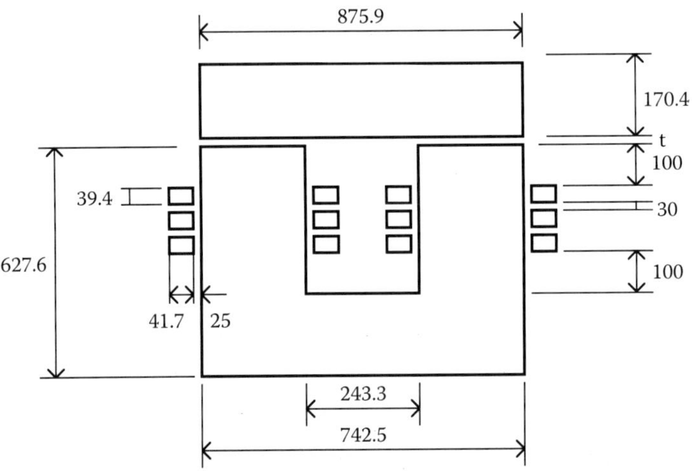

FIGURE 7.5
Main dimensions of the reactance.

7.3.1 Inductance Adjustment

The variation of the inductance is obtained operating on the air-gap length and on the number of turns of the reactance.

Regarding the air-gap variation, the position that has to be analyzed with the utmost care corresponds to the higher length, when the higher leakage flux occurs. The leakage may be so high that may inhibit the regulation of the reactance. The critical value of the air-gap length may be considered more or less equal to 10 mm. The minimum value of the air-gap presents some limitations as well, due to mechanical considerations, mainly caused by nonperfect lamination joints. The minimum air-gap length has been estimated between 1.5 and 2 mm.

As a guideline, an analytical expression of the inductance is displayed. It can be obtained from the study of the magnetic circuit of the reactance. The self-inductance of one coil of the winding, formed by $N_t/2$ turns (the other coil is open-circuited), is given by

$$L = \frac{\left(\frac{N_t}{2}\right)^2}{\frac{2t}{\mu_o S_{Fe}} + \frac{l_{Fe,average}}{\mu_{Fe}\mu_o S_{Fe}}} \approx N_t^2 \frac{\mu_o S_{Fe}}{8t} \tag{7.16}$$

The inductance is essentially inversely proportional to the air-gap length. Thus, the constraints imposed on the variation of the air-gap correspond to constraints on the variation of the inductance itself. With a variation of the air-gap length between 2 mm and 10 mm, it yields a ratio between the maximum and the minimum value equal to 5.

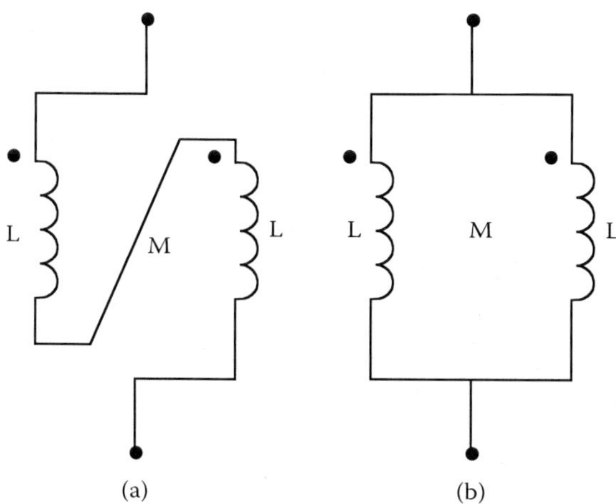

FIGURE 7.6
Connection between the two coils.

Such a variation ratio is increased operating on the number of turns of the reactance, i.e., connecting the two coils of the winding in series or in parallel. Let us assume that each coil is characterized by a self-inductance equal to L, expressed by Equation (7.16), and a mutual inductance $M \approx L$. When the two coils are connected in series, as in Figure 7.6(a), the total inductance is

$$L_s \approx 2(L + M) \approx 4L \approx N_t^2 \frac{\mu_o S_{Fe}}{2t} \tag{7.17}$$

When the two coils are connected in parallel, as in Figure 7.6(b), the total inductance is reduced to a quarter, that is

$$L_p \approx \frac{1}{2}(L + M) \approx L \approx N_t^2 \frac{\mu_o S_{Fe}}{8t} \tag{7.18}$$

The ratio between the maximum inductance, obtained with the minimum air-gap length and with series-connected coils, and the minimum inductance, obtained with the higher air-gap length and with parallel-connected coils, becomes equal to 20. In conclusion, with the parallel connection, the resonance is obtained with capacitor loads in the range C = 40 ÷ 200 nF, while with the series connection, the resonance is obtained with C = 10 ÷ 50 nF.

7.3.2 Definition of the Winding

The maximum flux density is obtained with the parallel connection when the higher voltage V_n is applied to each coil. The voltage applied to the coil is halved with the series connection. The number of turns has been chosen equal to 3780 for each coil, which is a total number of turns $N_t = 7560$.

The maximum value of the current occurs when the reactance is in resonance with the capacitive load C = 200 nF with a current $I = \omega C V_n = 3.15$ A, to which corresponds the apparent power $S_n = 160$ kVA. Such resonance is obtained with the lowest value of inductance, which is with the parallel connection. The current in the conductors is then halved, resulting in 1.575 A.

The diameter of the conductor corresponds to $d_{Cu} = 1.093$ mm with a double insulating thickness (UNEL 01722-3-4).

Each coil is divided in three elementary coils of 1260 turns each. The turns are composed of 35 layers of 36 turns each. Each layer is insulated with a paper sheet of thickness 0.1 mm. The dimensions of each elementary coil result: a 41.7 mm width and a 39.4 mm height, as shown in Figure 7.5.

7.3.3 Analysis

The analysis of the reactance is reduced to a 2D analysis on the plane (x,y). To have a planar symmetry, the circular shape of the leg is modeled as it would be rectangular, but preserving the same cross area. This is shown in Figure 7.7. The circular leg has the diameter D = 249.6 mm, then keeping the width of the rectangular leg equal to D, the z-axis length becomes L_{Fe} = 170.4 mm.

The height of the movable part has to be adapted as well, in order to refer the magnetic quantities at the same machine z-axis length L_{Fe}. Such a height becomes equal to the diameter D, as shown in Figure 7.7. Finally, with the aim of considering the effective iron volume together with the effect of the squared edges of the yoke, the horizontal length of the movable part is increased, as shown by the dotted line in Figure 7.7.

With the required simplifications, the 2D analysis allows a good estimation of the reactance behavior, with the advantage of a faster analysis.

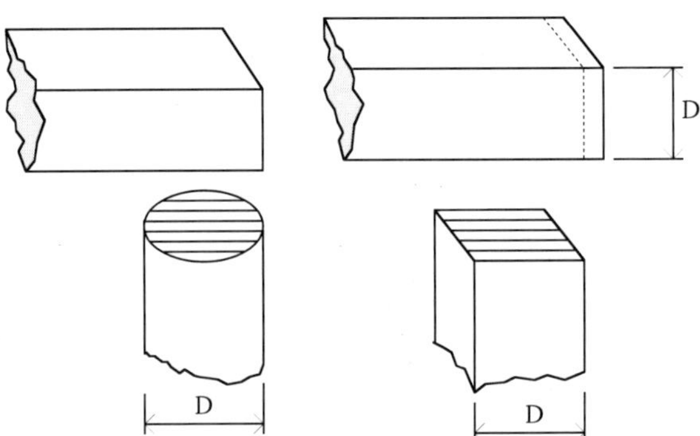

FIGURE 7.7
Equivalent structure of the reactance to apply the planar symmetry.

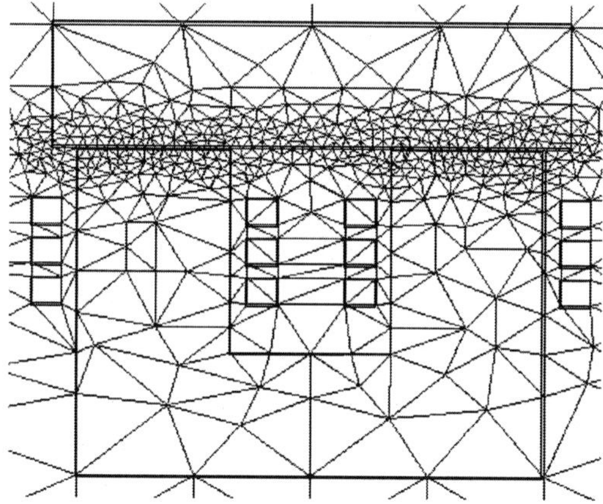

FIGURE 7.8
Mesh of the section of the reactance.

We will carry out the analysis on the whole section of the machine of Figure 7.5. However, the structure presents a symmetry axis, corresponding to the BB′ axis of Figure 7.2(c). Then, only half a structure could be analyzed, assigning Neumann's boundary condition on the BB′ axis.

Since the magnetic field strength is higher in the air-gap, the mesh is increased in this zone, as shown in Figure 7.8. The subdivision of the structure is carried out using some 2000 triangular elements. This choice proved to be appropriate to obtain accurate results: a 20% increase of the number of the finite elements generates a variation of the magnetic energy in the field solution lower than 0.3%.

The magnetic B–H curve Terni M5T30 is used for the iron core.

Various simulations are carried out with different values of air-gap length t and current I. Since the elementary coils are schematized by means of conductive bars, the equivalent current in each coil is obtained as the product of the current I by the number of turns of each elementary coil, which is $N_t/6 = 1260$.

Figure 7.9 shows the flux plot. The detail of Figure 7.9(b) highlights that the flux lines are not confined within the air-gap, but they expand at the extremity of the legs. The existence of such flux lines points out that, at the same current (and then at the same MMF), the flux linkage is higher than that analytically predicted with the assumption of flux lines that are only normal to the air-gap. Consequently, the inductance is higher. It is then necessary to verify that the inductance assumes the required values, especially when the air-gap length is maximum.

Figure 7.10 shows the flux density magnitude in the middle of the air-gap, corresponding to different values of air-gap length and current. The flux density is drawn as a function of the distance from the symmetry axis of the reactance. Figure 7.10(a) corresponds to an air-gap t = 1.5 mm and a current I = 0.154 A. Figure 7.10(b) corresponds to an air-gap t = 9 mm and a current I = 3.19 A.

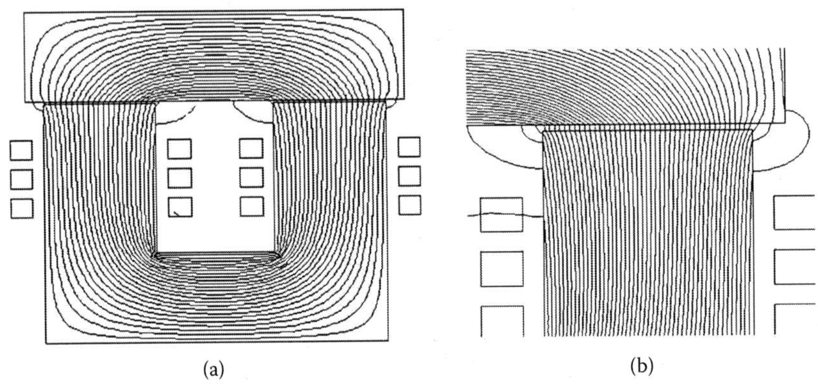

FIGURE 7.9
Flux plots.

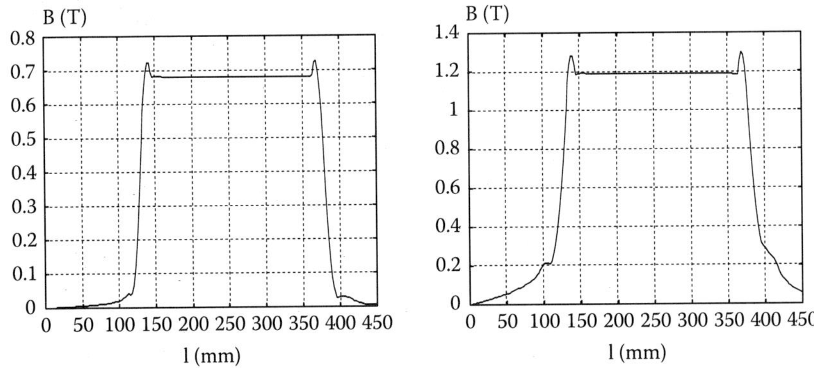

FIGURE 7.10
Flux density in the middle of the air-gap as a function of the distance of the symmetry axis; (a): t = 1.5 mm and I = 0.154 A; (b): t = 9 mm and I = 3.19 A.

In both figures, the edge effect is evident: in the central zone the value of the flux density magnitude is essentially constant; near the edges it increases about 10%.

Table 7.1 contains the values of the inductance obtained varying the air-gap t. Low current is adopted to avoid saturation effect. The inductances are computed by Equation (7.7) using the magnetic energy, given by Equation (7.2) or Equation (7.3), since the problem is essentially linear. The same result is obtained, however, if the inductances are computed by Equation (7.6) using the flux linkages, given by Equation (7.4). The inductances with parallel-connected coil are obtained simply dividing by 4 the inductances with series-connected coils.

For an easy comparison, in the same table the inductances computed analytically are reported as well, as given by Equation (7.17) and Equation

TABLE 7.1

Values of the Self-Inductance, Computed at Low Current

t (mm)	L_s (H)	L_s(analytic) (H)	L_p (H)	L_p(analytic) (H)
1.5	1047.0	1002.3	261.7	250.6
1.6	972.4	940.6	243.4	235.1
1.7	917.6	886.0	229.4	221.5
1.8	870.2	837.4	217.6	209.4
1.9	826.8	793.9	206.7	198.5
2.0	789.8	754.7	197.5	188.7
2.1	752.2	719.2	188.1	179.8
2.2	720.0	686.8	180.0	171.7
2.3	690.0	657.3	172.6	164.3
2.4	662.8	630.1	165.7	157.5
2.5	638.0	605.2	159.5	151.3
3.0	538.4	505.1	134.6	126.3
4.0	412.8	379.6	103.2	94.9
5.0	336.4	304.0	84.1	76.0
6.0	285.8	253.6	71.5	63.4
7.0	249.0	217.5	62.3	54.4
8.0	221.0	190.4	55.3	47.6
9.0	199.5	169.3	49.9	42.3
10.0	182.1	152.4	45.5	38.1

(7.18). As expected, the values obtained from the finite element method are higher than those analytically computed. This is essentially due to the flux lines external to the air-gap, as highlighted in Figure 7.9. The relative difference between the two inductances is as high as the air-gap increases. In fact, with the higher values of t, the flux outer the air-gap increases.

Figure 7.11 illustrates graphically the results reported in Table 7.1.

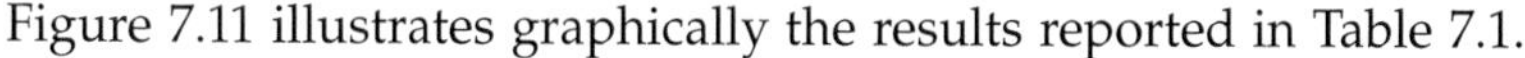

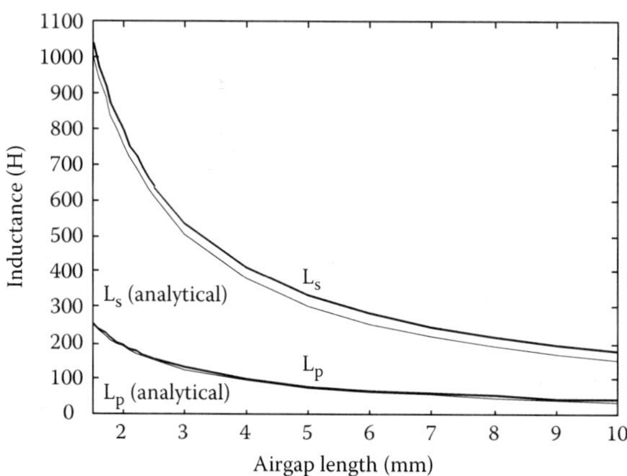

FIGURE 7.11
Behavior of the inductance as a function of the air-gap, with constant current.

TABLE 7.2

Effect of the Saturation (t = 1.5 mm)

I (mA)	W_m (J)	Λ (Vs)	L_s (H)
58	1.76	60.71	1046.6
104	5.66	108.91	1047.2
150	11.78	157.12	1047.5
196	20.12	205.35	1047.7
242	30.68	253.58	1047.9
288	43.40	301.36	1046.4
334	58.19	348.46	1043.3
373	71.91	387.99	1040.2
419	89.41	432.30	1031.7
458	105.30	468.97	1024.0
504	123.10	506.53	1005.0
550	136.00	531.26	965.9
596	144.40	545.89	915.9
642	149.20	553.40	862.0
688	153.60	560.23	814.3
734	158.50	567.33	772.9
780	162.8	573.31	735.0
826	166.50	578.24	700.0
872	169.90	582.62	668.1
918	173.10	586.60	639.0
964	176.10	590.17	612.2

In order to highlight the saturation effect, the air-gap length is fixed to t = 1.5 mm and the current is changed between 0 and 1 A. The analysis is carried out with current higher than the nominal current. In fact, with such an air-gap length, the maximum current is 0.4 A (peak value).

Some results are given in Table 7.2, which reports the magnetic energy, the flux linkage, and the inductance computed by means of Equation (7.6). Since the problem does not remain linear, the inductance assumes the meaning of apparent inductance. Let us remember that the computation by means of the magnetic energy (7.7) is correct only in linear conditions. From Table 7.2 it is possible to verify that, with low currents, Equation (7.6) and Equation (7.7) yield the same result, while they do not with high current.

Figure 7.12 shows the flux linkage Λ versus the current I. Since fixed values have been used, this is the dc magnetizing curve. The same figure also shows the apparent inductance (obtained as the ratio between Λ and I), as a function of the current.

From the figure, it is easy to verify that the reactance works in linearity during the normal operation, since the maximum nominal current is 0.4 A.

A sinusoidal waveform of the flux linkage is assumed, which is a sinusoidal waveform of the voltage with a negligible resistive voltage drop. Then the current waveform is reconstructed from the dc magnetizing characteristic of Figure 7.12.

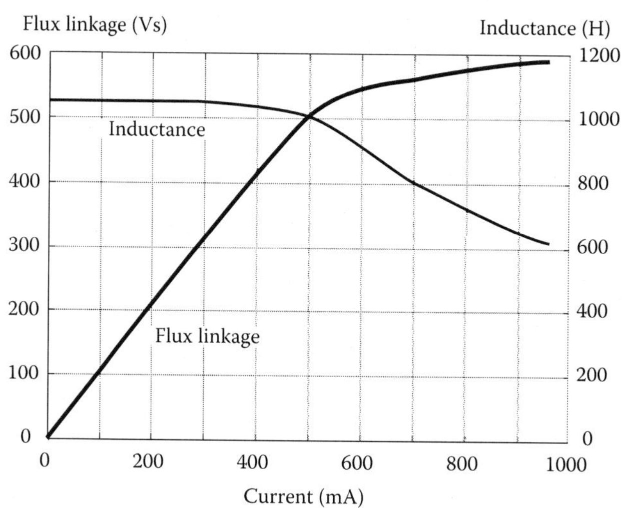

FIGURE 7.12
Flux linkage and inductance as functions of the current (at t = 1.5 mm).

The algorithm for the interpolation is described at the end of Chapter 6. Figure 7.13 shows some waveforms of flux linkage and current, for different peak values of the flux linkage. The saturation effect on the current waveform is evident.

As described in the previous chapter, the magnetizing characteristic of the reactance is obtained, computing the RMS value of the current in correspondence to the flux linkage. As a consequence, a new value for the inductance can be defined, say L*. It is obtained as the ratio between the RMS value of the flux linkage and the RMS value of the current. Referring to Figure 7.13(d), obtained with N_j points, here are the results:

$$\Lambda_{rms} = \frac{\Lambda_M}{\sqrt{2}} = \frac{600}{\sqrt{2}} = 424.3 \text{ Vs}$$

$$I_{rms} = \sqrt{\frac{\omega}{2\pi}\int_0^{2\pi/\omega} i(t)^2 dt} = \sqrt{\frac{1}{N_j}\sum_{j=1}^{N_j} i_j^2} = 0.556 \text{ A}$$

$$L^* = \frac{\Lambda_{rms}}{I_{rms}} = 763.7 \text{ H}$$

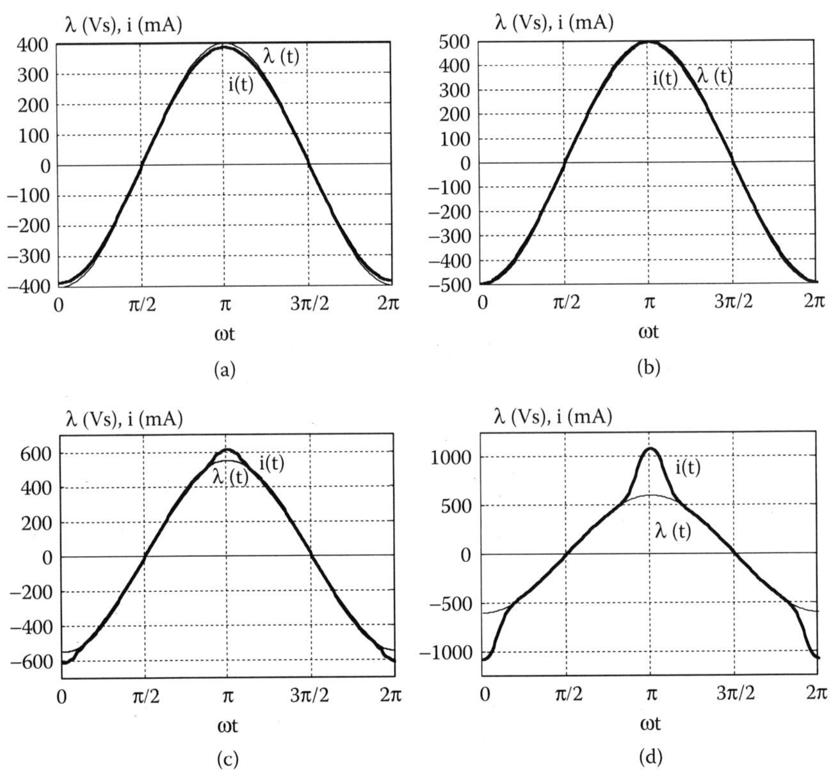

FIGURE 7.13
Current waveforms, corresponding to a sinusoidal flux linkage with peak value (a): 400 Vs; (b): 500 Vs; (c): 550 Vs; (d): 600 Vs.

References

1. G. Kron, "Equivalent Circuits to Represent the Electromagnetic Field Quantities," *The Physical Rev.*, vol. 64, Aug. 1–15, 1943, pp. 126–128.
2. L.F. Blume, A. Boyajian, *Transformer Engineering*, John Wiley, New York, 1951.
3. G. Rago, *Costruzioni Elettromeccaniche e Disegno*, Sansoni, Firenze, 1968.
4. N. Bianchi, F. Dughiero, "Optimal Design Techniques Applied to Transverse-Flux Induction Heating Systems," in *IEEE Trans. on Magnetics*, vol. 31, n. 3, Maggio 1995, pp. 1992–1995.
5. N. Bianchi, F. Dughiero, S. Lupi, "Design of Induction Heating Systems by Optimisation of Field Shape," in *Proceedings of the International Induction Heating Seminar, IHS '98*, Padova, 13–15 Maggio, 1998, pp. 413–423.
6. W.T. McLyman, *Transformer and Inductor Design Handbook*, CRC Press, Boca Raton, FL, 2004.

8

Synchronous Generators

This chapter describes the finite element analysis of a three-phase synchronous generator. The no-load characteristic and the direct and quadrature axis inductances are investigated.

8.1 Introduction

The synchronous generator is composed of a fixed part, the *stator*, and a rotating part, the *rotor*. In both of them, magnetic materials are used to guide the magnetic flux, and windings are employed to carry the current. The structure of a $2p = 2$ pole synchronous generator is represented in Figure 8.1.

The rotor winding, called field winding or magnetizing winding, is fed by a direct current, by means of dragging contacts, i.e., brushes and conductive rings. This is the primary source of the main flux of the machine. The polar axis in the middle of the rotor pole, sketched in Figure 8.1, is called the direct axis, or d-axis. The interpolar axis is 90 electrical degrees in advance with respect to the rotating direction and is called the quadrature axis, or simply q-axis. The rotor rotates at constant mechanical angular speed ω_m (an electrical angular speed $\omega = p\omega_m$), giving rise to a rotating magnetic field.

The stator winding, also called armature winding, links a variable magnetic flux so that a variable EMF is induced. The three phases of the winding are named a, b, and c. Each phase winding is spaced out of 120 electrical degrees. They are sketched in Figure 8.1 by means of a single turn and are identified by the axis normal to the turn itself. Observe that the orientation of the turns agrees with their axis directions, according to the right-handed screw advance (a rotation in direction of the positive current would cause the right-handed screw to advance in the direction of the turn axis). Finally, the rotor position with respect to the stator is pointed out by the electrical angle ϑ, between the polar axis and the a-phase axis. The angle ϑ is $\vartheta = p\vartheta_m$ where ϑ_m is the mechanical angle between rotor and stator.

When the armature windings are open-circuited, no current flows and the induced EMF can be measured at the winding terminals. This is the no-load

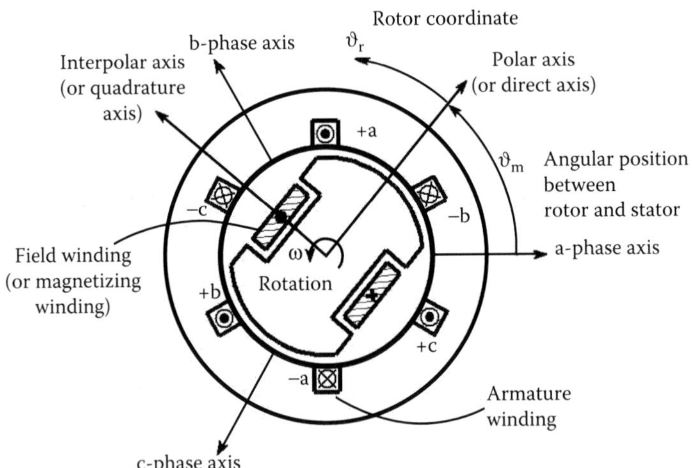

FIGURE 8.1
Structure and references of the three-phase synchronous generator.

voltage of the generator. On the contrary, when the armature windings are connected to the load, variable currents flow in the windings, with the same frequency of the EMF. Such currents cause a magnetic field that rotates at the same speed (called the synchronous speed) of the rotor speed. The electromechanical conversion is obtained by the interaction between the stator and rotor magnetic fields.

Neglecting the edge effects and assuming an identical behavior along the generator, a planar symmetry is supposed: only the plane (x, y) is considered, the magnetic vector potential and the current density have a z-axis component only, which is $\mathbf{A} = (0, 0, A_z)$ and $\mathbf{J} = (0, 0, J_z)$. Then the flux density and magnetic field strength vectors have a component on the plane (x, y) only, which is $\mathbf{B} = (B_x, B_y, 0)$ and $\mathbf{H} = (H_x, H_y, 0)$.

The machine length is chosen equal to the effective length of the iron L_{Fe}, considering the stacking factor k_{stk}, which considers the insulation among the laminations ($k_{stk} = 0.95 - 0.98$) and the possible ventilation channels (e.g., N_{cv} channels, whose wideness is l_{cv}). Then the net length is

$$L_{Fe} = \left(L - N_{cv} l_{cv}\right) k_{stk} \tag{8.1}$$

where L is the total length of the machine.

In addition, a periodicity exists among the pole pairs, so that the study may be carried out only on one pole pair, as in Figure 8.2(a). When a stator symmetry exists among the poles (the number of slots is a multiple of the number of poles), the analysis may be carried out on a single pole centered along the polar axis, as in Figure 8.2(b), or centered along the interpolar axis, as in Figure 8.2(c). Centering the drawing along the d- or the q-axis, as shown in Figure 8.2, is not really necessary, since the periodic conditions may be

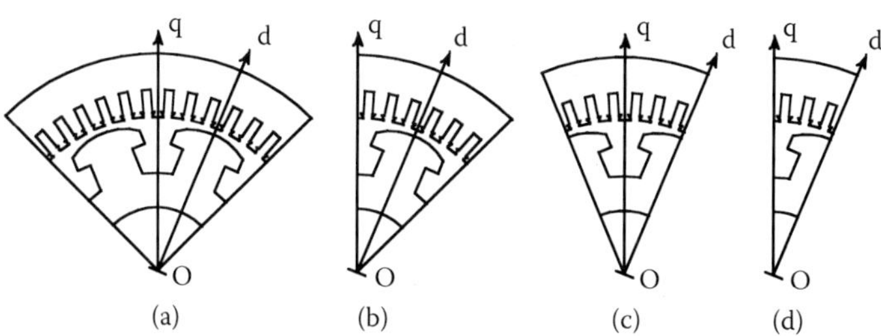

FIGURE 8.2
Sections of the machine and periodicity.

assigned no matter what axis is chosen. However, the choice of the d- or the q-axis is particularly suitable in the assignment of the boundary conditions. In fact, special operating conditions allow Dirichlet's or Neumann's boundary conditions to be assigned, as will be hereafter described.

Finally, some cases exist in which the analysis is reduced to half a pole, as shown in Figure 8.2(d).

8.2 Computation of the No-Load Characteristic

The field winding, formed by N_e turns, is fed by a constant current I_e. The armature windings are open-circuited so that they do not carry any current. They may be considered to have null conductivity. The stator and rotor magnetic material is described by the suitable B-H curve.

The boundary conditions are assigned, considering that the flux lines of half a pole are the mirrored flux lines of the other half of the pole. Then, a value of the magnetic vector potential is fixed along the polar axis (Dirichlet's condition), while a normal flux density vector is assigned along the interpolar axis (Neumann's condition). Since the flux lines do not exit from the yoke, a null magnetic vector potential is assigned, along the external circumference of the stator. Figure 8.3 shows the boundary conditions assigned to the different sections of the generator, corresponding to Figure 8.2(b–d), at the no-load operation.

The field winding, composed by N_e turns and carrying the excitation current I_e, is modeled by an equivalent conductive bar carrying the total current N_eI_e.

The three-phase windings may be displaced into the stator slots, in whatever way with respect to the rotor. However, it is useful to take advantage of the electrical symmetry. Figure 8.4 shows a part of a single-layer, nonchorded winding, referred to one pole of the stator with 6 slots per pole. Figure 8.4(a) and Figure 8.4(b) differ for the winding distribution; in the figures the a-phase axis is highlighted as the stator reference.

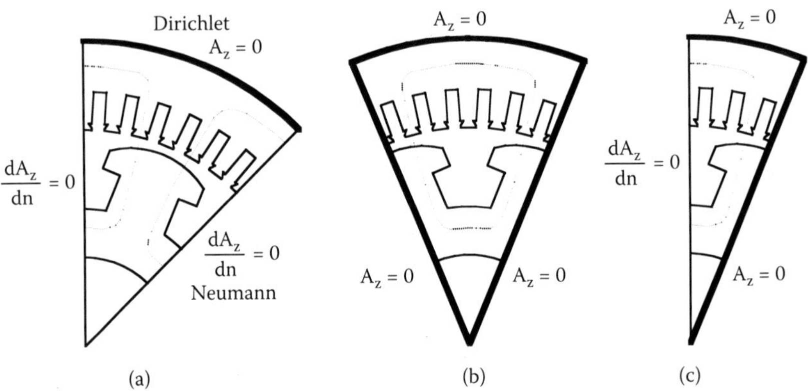

FIGURE 8.3
Boundary conditions in the no-load simulation.

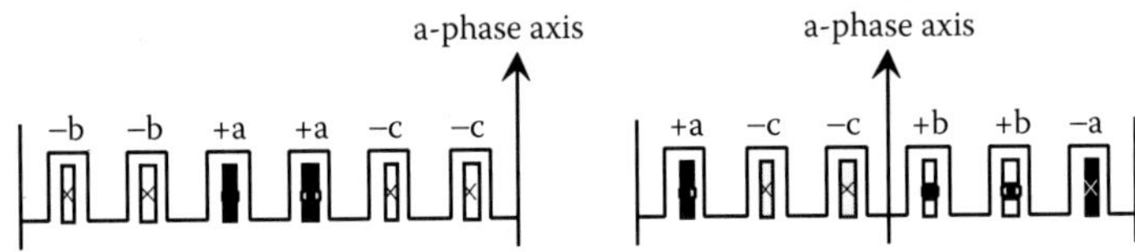

FIGURE 8.4
Distribution in the slots of the three-phase windings: single-layer, nonchorded winding.

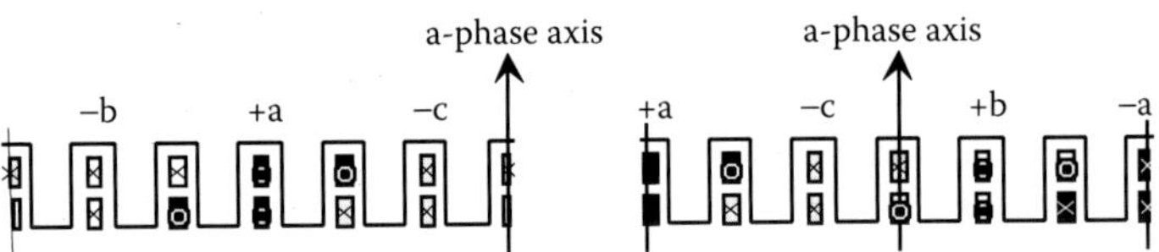

FIGURE 8.5
Distribution in the slots of the three-phase windings: double-layer, one-slot chorded winding.

Figure 8.5 shows a part of a double-layer, one-slot chorded winding, referred to one pole of a stator with 6 slots per pole. In order to have a complete electrical and geometrical symmetry, the stator pole is centered on the middle of the a-phase axis. The stator reference is highlighted by means of the a-phase axis. It is worth noticing that, in this case, the first and the last slots have to be halved.

8.2.1 Computations after the Field Solution

Once the field problem is solved, the z-axis component of the magnetic vector potential $A_z(x, y)$ is known in each point of the domain. The flux linked with the three-phase windings and the induced EMFs are computed. Particular care has to be kept to the turns distribution in the stator slots.

8.2.1.1 *Flux Linkage*

The mean value of the magnetic vector potential in the q-th slot, whose section is S_q, is

$$\frac{1}{S_q}\int_{S_q} A_z dS \tag{8.2}$$

Let n_q be the number of conductors per slot and n_{pp} be the number of parallel paths of the machine. It follows that the total number of conductors of the machine is n_{pp} times the number of series conductors. Considering that the analysis domain is one pole only — see Figure 8.3(a) or Figure 8.3(b) — the j-th phase flux linkage is given by

$$\Lambda_j = 2pL_{Fe}\frac{n_q}{n_{pp}}\sum_{q=1}^{Q/2p} k_{jq}\frac{1}{S_q}\int_{S_q} A_z dS \qquad j = a, b, c \tag{8.3}$$

where 2p is the number of poles and L_{Fe} is given by Equation 8.1; Q is the number of the slots of the generator, so that Q/2p is the number of slots per pole. Finally, k_{jq} is the coefficient taking into account whether the conductors in the q-th slot are of the j-th phase or not, as well as the conductor orientation.

Such coefficient assumes the following values.

Single-Layer Winding (One Coil Side in Each Slot)

$k_{jq} = 0$ — if the coil side in the q-th slot does not belong to the j-th phase;

$k_{jq} = +1$ — if the coil side in the q-th slot belongs to the j-th phase and its orientation is positive with respect to the z-axis direction (leaving the sheet with the references of Figure 8.1);

$k_{jq} = -1$ — if the coil side in the q-th slot belongs to the j-th phase and its orientation is negative with respect to the z-axis direction (going into the sheet with the references of Figure 8.1).

Double-Layer Winding (Two Coil Sides in Each Slot)

$k_{jq} = 0$ — if the coil sides in the q-th slot do not belong to the j-th phase;

$k_{jq} = +0.5$ — if only one coil side in the q-th slot belongs to the j-th phase and its orientation is positive with respect to the z-axis direction;

$k_{jq} = -0.5$ — if only one coil side in the q-th slot belongs to the j-th phase and its orientation is negative with respect to the z-axis direction;

$k_{jq} = +1$ — if both two coil sides in the q-th slot belong to the j-th phase and their orientation is positive with respect to the z-axis direction;

$k_{jq} = -1$ — if both two coil sides in the q-th slot belong to the j-th phase and their orientation is negative with respect to the z-axis direction.

TABLE 8.1

Values of k_{jq} Referring to the Distribution of Figure 8.4(a)

slot (q)	phase (j) a	b	c
	0	−1	0
	0	−1	0
	+1	0	0
	+1	0	0
	0	0	−1
	0	0	−1

TABLE 8.2

Values of k_{jq} Referring to the Distribution of Figure 8.5(a)

slot (q)	phase (j) a	b	c
1	0	−0.25	+0.25
2	0	−1	0
3	+0.5	−0.5	0
4	+1	0	0
5	+0.5	0	−0.5
6	0	0	−1
7	0	+0.25	−0.25

In other words, the magnitude of k_{jq} specifies the relative filling of the q-th slot by the j-th phase conductors. The sign of k_{jq} considers the sign of the scalar product $\mathbf{A}\cdot\mathbf{t}$ of the loop integral: the vector $\mathbf{A}$ has only the component A_z while the tangential unity vector $\mathbf{t}$ (which defines the turn's orientation) coincides with $\mathbf{u}_z$ when a positive current flows in the z-axis direction, or with $-\mathbf{u}_z$ when the same positive current flows opposite to the z-axis direction. Thus, $\mathbf{A}\cdot\mathbf{t} = k_{jq}A_z$ is equal to $+A_z$ or to $-A_z$.

For instance, with the single-layer winding of Figure 8.4(a), the values reported in Table 8.1 are obtained, while with the double-layer winding of Figure 8.5(a), the values reported in Table 8.2 are obtained. In the latter case, the value 0.25 has been used in the slot q = 1 and in the slot q = 7, because only half a slot is considered. Finally, it is worth noticing that the sum of the magnitudes of the coefficients of each row (with whole slot) is always equal to one.

8.2.1.2 Induced EMF

For the sake of convenience, let us consider the case of maximum flux linked with the a-phase winding. This means choosing the a-phase axis parallel to

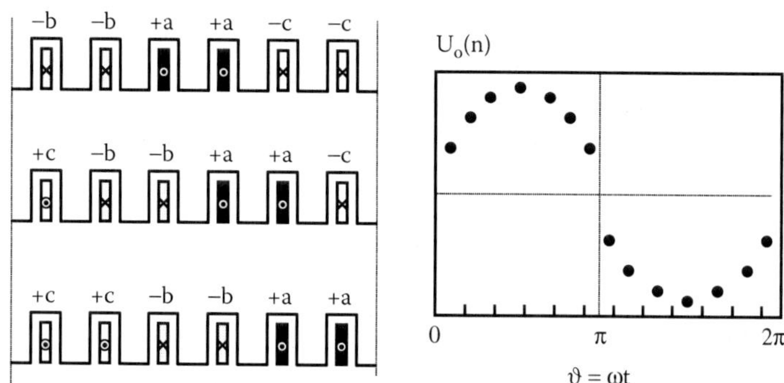

FIGURE 8.6
Construction of the voltage waveform by dots.

the d-axis. Then, indicating Λ_{ma} the maximum flux linkage, with constant rotor speed, the RMS value of the induced EMF is

$$E_a = \frac{1}{\sqrt{2}}\omega\Lambda_{ma} \tag{8.4}$$

which is the no-load line voltage with a wye winding connection. Conversely, with a star winding connection, the no-load line voltage is

$$U_0 = \sqrt{3}E_a \tag{8.5}$$

From the same field solution, supposing a step rotation of the coils of one slot pitch, it is possible to achieve some points of the no-load voltage waveform. Assuming this rotation, different coil sides are associated to each slot, as shown in Figure 8.6, the coefficients k_{jq} are suitably modified. This is possible because only the rotor is fed, which means the flux lines do not depend on the distribution of the conductors into the slots.

The waveform $U_o(n)$ is built by N_c points (e.g., it is common that $N_c = Q/p$). The harmonic content can be approximately estimated. From the discrete Fourier series expansion, the fundamental value is obtained as

$$\begin{aligned} U_{1\sin} &= \frac{2}{N_c}\sum_{n=1}^{N_c} U_o(n)\cdot\sin\left(\frac{2\pi}{N_c}n\right) \\ U_{1\cos} &= \frac{2}{N_c}\sum_{n=1}^{N_c} U_o(n)\cdot\cos\left(\frac{2\pi}{N_c}n\right) \\ U_1 &= \sqrt{{U_{1\sin}}^2 + {U_{1\cos}}^2} \end{aligned} \tag{8.6}$$

Then the ν-th order harmonic is

$$U_{\nu\sin} = \frac{2}{N_c}\sum_{n=1}^{N_c} U_o(n)\cdot\sin\left(\frac{2\pi}{N_c}\nu n\right)$$
$$U_{\nu\cos} = \frac{2}{N_c}\sum_{n=1}^{N_c} U_o(n)\cdot\cos\left(\frac{2\pi}{N_c}\nu n\right) \tag{8.7}$$

To avoid rough errors, the voltage harmonics can be evaluated up to the maximum order equal to $\nu < N_c/2$ (Shannon's theorem).

8.3 Computation of the Direct-Axis Inductance

In the computation of the direct-axis inductance, L_d, the magnetizing current is set to be zero and the stator windings are fed in such a way as to obtain a MMF distribution with the maximum value coincident with the polar axis. Since the magnetic field is synchronous with the rotor, a magnetostatic simulation is possible.

Thanks to the electrical and geometrical symmetry, only a portion of the machine is analyzed, that is, a portion equal to one pole or to half a pole.

The boundary conditions are assigned to constrain the flux lines to be tangential along the polar axis and normal to the interpolar axis. The boundary conditions are those assigned in the no-load simulation, shown in Figure 8.3.

In order to define the phase current, let us refer to the phasor diagram of Figure 8.7. With reference to the time instant t, it is posed

$$i_a(t) = I_M \sin(\omega t)$$
$$i_b(t) = I_M \sin\left(\omega t - \frac{2\pi}{3}\right)$$
$$i_c(t) = I_M \sin\left(\omega t - \frac{4\pi}{3}\right) \tag{8.8}$$

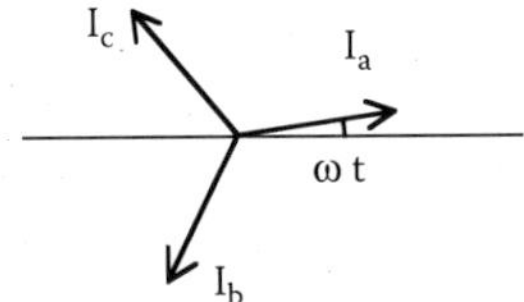

FIGURE 8.7
Reference phasor diagram of the stator winding currents.

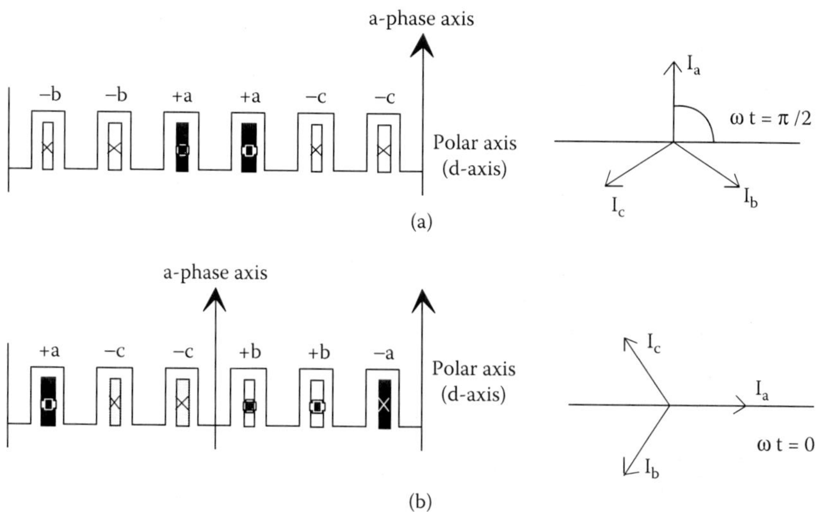

FIGURE 8.8
Combinations of the winding distribution and time instant, for computing the direct axis inductance.

Once the distribution of the winding is fixed, the three-phase currents are chosen so that the maximum of the MMF distribution coincides with the polar axis (d-axis). This corresponds to set $I_d = I_n$ and $I_q = 0$. Referring to one pole of the generator, Figure 8.8 shows two suitable combinations of winding distribution and time instant. For both of them, the boundary conditions of Figure 8.3(b) can be assigned.

With the configuration of Figure 8.8(a) the time instant with $\omega t = \pi/2$ has been chosen, so that

$$\begin{aligned} i_a &= I_M \\ i_b &= -\frac{I_M}{2} \\ i_c &= -\frac{I_M}{2} \end{aligned} \tag{8.9}$$

With the configuration of Figure 8.8(b) the time instant with $\omega t = 0$ has been chosen, so that

$$\begin{aligned} i_a &= 0 \\ i_b &= -\frac{\sqrt{3}}{2} I_M \\ i_c &= +\frac{\sqrt{3}}{2} I_M \end{aligned} \tag{8.10}$$

Once the three-phase currents i_a, i_b, and i_c, and the distribution of the windings have been fixed, i.e., the coefficients k_{aq}, k_{bq}, and k_{cq} for each slot, the current flowing in the q-th slot can be expressed as

$$
\begin{aligned}
i_q &= \frac{n_q}{n_{pp}} \sum_{j=a,b,c} k_{jq} i_q \\
&= \frac{n_q}{n_{pp}} \left(k_{aq} i_a + k_{bq} i_b + k_{cq} i_c \right)
\end{aligned}
\tag{8.11}
$$

8.3.1 Computation on the Solved Structure

Once the field solution is obtained, the direct axis inductance may be computed in several ways.

8.3.1.1 Computation of L_d by Means of the Magnetic Energy

Let W_{md} be the magnetic energy stored in the synchronous generator, due to the stator d-axis currents only. In linear conditions it is

$$
W_{md} = \frac{3}{2}\left(\frac{1}{2} L_d I_d^2 \right) \tag{8.12}
$$

With d-axis current only, it is $I_d = I_M$, and the synchronous d-axis inductance results in

$$
L_d = \frac{2}{3} \frac{2W_{md}}{I_M^2} \tag{8.13}
$$

Slightly different results are obtained with the distributions of Figure 8.8(a) and Figure 8.8(b) and the corresponding currents indicated in Equation (8.9) and Equation (8.10). This is due to the different harmonic contents of the two configurations. At last, we must bear in mind that this computation is correct only in case of a linear magnetic circuit.

8.3.1.2 Computation of L_d by Means of the Flux Linkage

The synchronous direct axis inductance is defined as the inductance of that phase whose axis coincides with the polar axis, when the three-phase windings are simultaneously fed. Let us refer to the combination of Figure 8.8(a). It is worth noticing that the polar axis coincides with the a-phase axis, so that $\vartheta = 0$. Since $I_d = I_M$, forcing d-axis currents only, and $I_a = I_M$, from Equation (8.9), thus $I_d = I_a$.

The flux linked with the a-phase winding, $\Lambda_a = \Lambda_d$, is computed from Equation (8.3). The inductance is given by

$$L_d = \frac{\Lambda_a}{I_a} = \frac{\Lambda_d}{I_M} \tag{8.14}$$

Similarly, the inductance can be computed from the winding distribution of Figure 8.8(b) and the currents given by Equation (8.10). In this case, $\vartheta = \pi/2$, $I_b = (\sqrt{3}/2)I_M$, and $\Lambda_d = (\sqrt{3}/2)\Lambda_b$ (forcing d-axis currents only, always $I_d = I_M$), then

$$L_d = \frac{\Lambda_b}{I_b} = \frac{2\Lambda_b}{\sqrt{3}I_M} \tag{8.15}$$

The difference between Equation (8.14) and Equation (8.15) indicates the different harmonic contributions of the two solutions.

8.3.1.3 Computation of L_d by Means of the Air-Gap Flux Density

From the field solution, it is possible to obtain the radial component of the flux density distribution in the air-gap, say $B_{gn}(\vartheta_r)$ where ϑ_r is the angular position referred to the polar axis (which is the rotor coordinate, as shown in Figure 8.1 expressed in electrical radiants). The fundamental harmonic of the magnetic flux density with respect to the direct axis is given by

$$B_{1dM} = \frac{2}{\pi}\int_0^{\pi} B_{gn}(\vartheta_r)\cos(\vartheta_r)d\vartheta_r \tag{8.16}$$

If the distribution of the radial component of the flux density $B_{gn}(n)$ is known in N_p points, the computation becomes

$$B_{1dM} = \frac{2}{N_p}\sum_{n=1}^{N_p} B_{gn}(n)\cos(n\Delta\vartheta)\Delta\vartheta \tag{8.17}$$

where $\Delta\vartheta = \pi/N_p$ (assuming a simulation of one pole of the generator).

From the flux density value (8.17), the flux is obtained as the product of the mean value of the sinusoidal distribution (of the fundamental harmonic) by the surface corresponding to one pole. Then, the peak value of the d-axis flux linkage is

$$\Lambda_d = B_{1dM}\frac{k_w N}{2}\frac{DL_{Fe}}{p} \tag{8.18}$$

where D is the stator bore diameter, N is the number of series conductors per phase, and k_w is the winding factor. Finally

$$L_{dm} = \frac{\Lambda_d}{I_M} \tag{8.19}$$

The last computation adopts the fundamental harmonic of the flux density distribution, then is almost independent of the harmonic content.

In addition, while the leakage flux in the air-gap and in the slots is considered in Equation (8.13) and Equation (8.14), it is not considered when Equation (8.19) is used. Thus, letting L_σ be the leakage inductance, the direct axis inductance is obtained as

$$L_d = L_\sigma + L_{dm} \tag{8.20}$$

It is worth noticing that Equation (8.14) and Equation (8.19) can be applied even without setting the magnetizing current equal to zero. In such a case, the flux linkage due to the magnetizing current must be previously computed and considered constant. It has to be subtracted from the total flux linkage. With reference to the a-phase, Equation (8.14) and Equation (8.19) become

$$L_d = \frac{\Lambda_d - \Lambda_{dm}}{I_M} = \frac{\Lambda_a - \Lambda_{ma}}{I_M} \tag{8.21}$$

where $\Lambda_{dm} = \Lambda_{ma}$ indicates the a-phase flux linkage due to the magnetizing current.

8.4 Computation of the Quadrature Axis Inductance

The computation of the quadrature axis inductance, L_q, is carried out in the same way as the computation of the L_d inductance. The phase currents and the winding distribution must be chosen so as to have an MMF distribution with the maximum value coincident to the interpolar axis, i.e., the q-axis. Referring to Figure 8.8, the two time instants (i.e., the angles ωt) that characterized the three-phase currents i_a, i_b, and i_c with respect the winding distribution have to be exchanged. The boundary conditions are suitably adjusted, as shown in Figure 8.9.

The q-axis inductance is then obtained as follows.

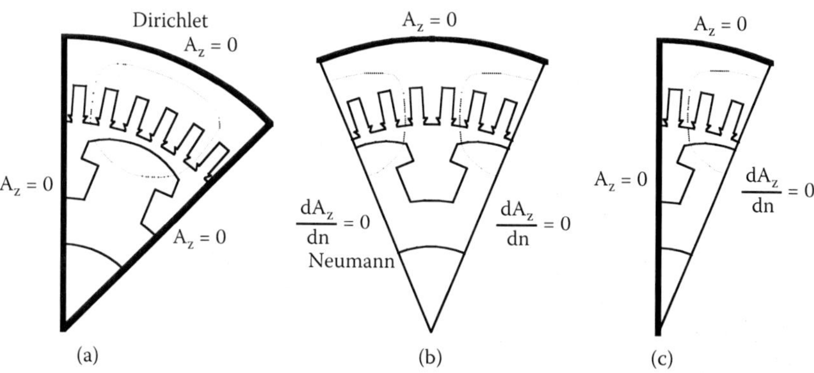

FIGURE 8.9
Boundary conditions for computing the quadrature axis inductance.

8.4.1 Computation of L_q by Means of the Magnetic Energy (Linear Case)

$$L_q = \frac{2}{3}\frac{2W_{mq}}{I_M^2} \tag{8.22}$$

8.4.2 Computation of L_q by Means of the Flux Linkage

$$L_q = \frac{\Lambda_q}{I_M} \tag{8.23}$$

8.4.3 Computation of L_q by Means of the Air-Gap Flux Density

$$B_{1qM} = \frac{2}{\pi}\int_{o}^{\pi} B_{gn}(\vartheta_r)\sin(\vartheta_r)d\vartheta_r \tag{8.24}$$

$$\Lambda_q = B_{1qM}\frac{k_w N}{2}\frac{DL_{Fe}}{p} \tag{8.25}$$

$$L_{qm} = \frac{\Lambda_q}{I_M} \tag{8.26}$$

8.5 Self- and Mutual Inductances

The self- and mutual inductances are obtained by feeding only one phase winding and computing the flux linked with the three-phase windings. For instance, only the a-phase winding is fed with a current $i_a = I_a$. The three-phase flux linkages Λ_a, Λ_b, and Λ_c are computed as given in Equation (8.3). Then

$$L_a(\vartheta) = \frac{\Lambda_a(\vartheta)}{I_a}$$

$$M_{ba}(\vartheta) = \frac{\Lambda_b(\vartheta)}{I_a} \tag{8.27}$$

$$M_{ca}(\vartheta) = \frac{\Lambda_c(\vartheta)}{I_a}$$

As highlighted, the self- and mutual inductances depend on the angular position ϑ. The most interesting values are in the position $\vartheta = 0$ (d-axis corresponding to the a-phase axis) and $\vartheta = -\pi/2$ (q-axis corresponding to the a-phase axis). The a-phase flux linkage Λ_a assumes the maximum value when $\vartheta = 0$, and minimum value when $\vartheta = \pm\pi/2$. It can be verified that $M_{ba} \approx M_{ca} \approx M \approx -L_a/2$. Then it is

at $\vartheta = 0$ $\quad \Lambda_a(0)$ maximum $\quad L_q \approx L_a(0) - M(0) \approx 3/2\ L_{a,max}$

at $\vartheta = \pm\pi/2$ $\quad \Lambda_a(\pi/2)$ minimum $\quad L_q \approx L_a(\pi/2) - M(\pi/2) \approx 3/2\ L_{a,min}$ (8.28)

8.6 Saturation Effect

The inductances computed above are influenced by the nonlinearity of the magnetic material of the rotor and stator. They depend on the values of the currents that are imposed in the analysis. In addition, Equation (8.14) and Equation (8.21) do not yield the same result, which is the computation with null or non-null magnetizing current.

The inductances computed as described above assume the meaning of apparent inductances. The differential inductances may be computed from two field solutions, as the ratio between the flux linkage variation and the corresponding current variation, as described in Chapter 5.

8.7 Computation of L_d and L_q with Any Current

Until now the analysis of the flux linkages and the d- and q-axis inductances has been carried out separately, feeding the windings so as to obtain the MMF distribution with the maximum value coincident to the d-axis or the q-axis. However, it is possible to carry out an analysis with stator currents independent of the particular rotor position ϑ.

In this case, each section of Figure 8.2(a), (b), or (c) can be used, assigning the periodic boundary conditions as shown in Figure 8.10. They are even periodic boundary conditions in Figure 8.10(a), and odd periodic boundary conditions in Figure 8.10(b) and Figure 8.10(c).

On the contrary, it is not possible to reduce the study to the section of Figure 8.2(d), since the periodic boundary conditions cannot be assigned to this section.

The d- and the q-axis flux linkage can be obtained from the three-phase flux linkages or from the air-gap flux density distribution.

Starting from the three-phase flux linkages Λ_a, Λ_b, and Λ_c, computed as given from Equation (8.3), the d- and the q-axis flux linkages Λ_d and Λ_q are achieved by means of the transformation $T_{abc/dq}$ (see the Appendix to this chapter):

$$\begin{aligned}\Lambda_d &= \frac{2}{3}\left[\Lambda_a \cos\vartheta + \Lambda_b \cos\left(\vartheta - \frac{2\pi}{3}\right) + \Lambda_c \cos\left(\vartheta - \frac{4\pi}{3}\right)\right] \\ \Lambda_q &= -\frac{2}{3}\left[\Lambda_a \sin\vartheta + \Lambda_b \sin\left(\vartheta - \frac{2\pi}{3}\right) + \Lambda_c \sin\left(\vartheta - \frac{4\pi}{3}\right)\right]\end{aligned} \tag{8.29}$$

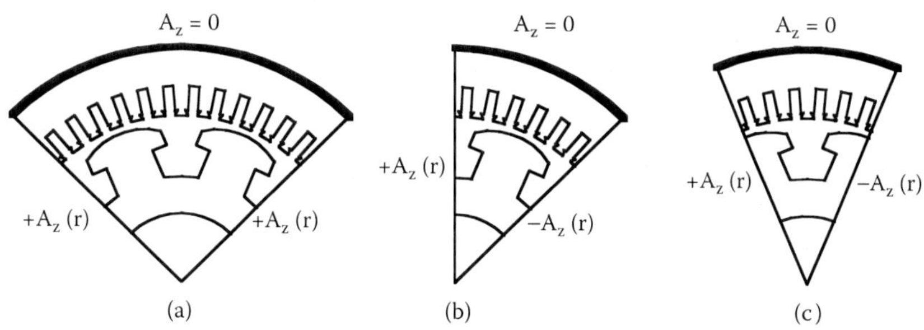

FIGURE 8.10
Periodic boundary conditions.

Alternatively, starting from the radial component of the air-gap flux density distribution $B_{gn}(\vartheta)$, the two-axis fundamental harmonic components are obtained from Equation (8.16) and Equation (8.24). The d- and q-axis flux linkages Λ_{1d} and Λ_{1q} are then obtained from Equation (8.18) and Equation (8.25). At this point, the d- and q-axis currents are computed from the three-phase currents I_a, I_b, I_c, always by means of the transformation $T_{abc/dq}$ (see the Appendix to this chapter), as

$$
\begin{aligned}
I_d &= \frac{2}{3}\left[I_a \cos\vartheta + I_b \cos\left(\vartheta - \frac{2\pi}{3}\right) + I_c \cos\left(\vartheta - \frac{4\pi}{3}\right)\right] \\
I_q &= -\frac{2}{3}\left[I_a \sin\vartheta + I_b \sin\left(\vartheta - \frac{2\pi}{3}\right) + I_c \sin\left(\vartheta - \frac{4\pi}{3}\right)\right]
\end{aligned}
\tag{8.30}
$$

Finally, neglecting the cross coupling between the d- and q-axis (that will be considered in Chapter 10), the d- and q-axis inductances result in

$$
\begin{aligned}
L_d &= \frac{\Lambda_d}{I_d} \\
L_q &= \frac{\Lambda_q}{I_q}
\end{aligned}
\tag{8.31}
$$

8.8 Computation of the Machine Characteristics

The finite element method can be used to simulate the steady-state operation of the synchronous generator. During steady state, the three-phase currents and voltages vary sinusoidal with the time; thus the complex symbolic notation can be used. According to the phasor diagram of Figure 8.11, the simulation is carried out as follows:

Input quantities:

- Thanks to the synchronous speed of the generator, a magnetostatic analysis is carried out, freezing a particular instant of its operation (this corresponds to analyzing the machine in the rotor reference frame). A constant synchronous speed ω_m is considered; then the operating electrical frequency is $\omega = p\omega_m$.
- The position of the rotor ϑ is fixed.
- The magnetizing current I_e is fixed.
- The maximum value of the stator current I_M and the phase angle ωt is chosen for the armature currents. From these values the three-phase currents I_a, I_b, and I_c are obtained.

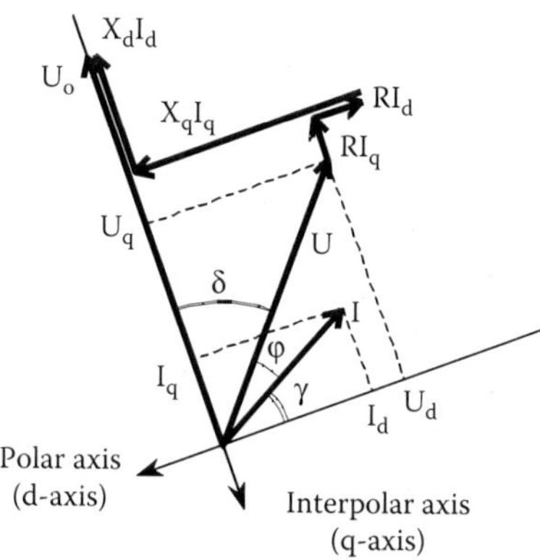

FIGURE 8.11
Phasor diagram of the synchronous generator.

Computed quantities:

- From the field solution, the three-phase flux linkages Λ_a, Λ_b, and Λ_c are computed.
- By means of the transformation $T_{abc/dq}$, the d- and q-axis components of current, i.e., I_d I_q, and flux linkages, i.e., Λ_d Λ_q, are computed.
- The d- and q-axis voltages are

$$\begin{aligned} U_d &= -RI_d + \omega\Lambda_q \\ U_q &= -RI_q - \omega\Lambda_d \end{aligned} \tag{8.32}$$

- The sign refers to the generating mode of the electrical machine, that is, the positive current direction is in phase with the voltage. The phasor U_o in Figure 8.11 is $\pi/2$ radiants lagging the polar axis (d-axis).
- The electrical angles are

$$\begin{aligned} \delta &= \tan\frac{U_d}{U_q} && \text{load angle} \\ \gamma &= \tan\frac{I_q}{I_d} && \text{current angle} \\ \varphi &= \frac{\pi}{2} - \gamma - \delta && \text{angle between voltage and current} \end{aligned} \tag{8.33}$$

In order to obtain the machine characteristics, various simulations should be carried out, changing one or more input quantities.

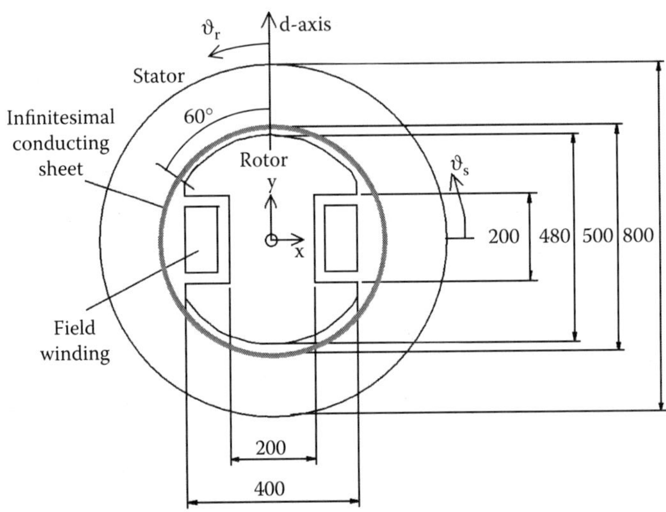

FIGURE 8.12
Ideal synchronous machine.

8.9 Example

The aim of the following example is to show the flux plots and the air-gap flux density distribution of the synchronous generator, during no-load and full-load operations. The effects of the armature reaction are highlighted, feeding the machine with d-axis currents only, q-axis currents only, and both of them together.

Rather than analyzing an effective figure of the synchronous generator, let us refer to the ideal structure drawn in Figure 8.12. This is a two-pole machine, formed by a rotor with salient poles through which the field windings are wound, and by a slotless stator. In the inner surface of the stator there is an infinitesimal conducting sheet carrying a linear current density distribution. Such a distribution is considered to be sinusoidal, corresponding to an ideally sinusoidal distribution of the winding.

In Figure 8.12, the dimensions of the generator are reported in mm. A linear magnetic material is used, with a constant relative permeability μ_r = 5000.

The first simulation deals with the no-load operation, without stator current and with a magnetizing current feeding the rotor winding. In the conducting bar, equivalent to the magnetizing winding, the total current is set N_eI_e = 13500 A. Figure 8.13 shows the corresponding flux plot and the air-gap flux density distribution. Since the pole shoe is not shaped — which is a constant air-gap under the pole — the flux density distribution looks like a quasi-square distribution. The maximum flux density corresponds to the

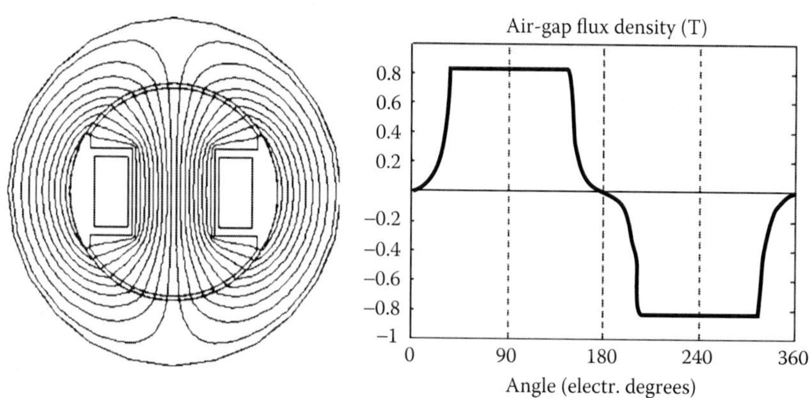

FIGURE 8.13
No-load flux plot (a) and air-gap flux density distribution (b).

predictable value by Ampere's law $B_o = \mu_0 N_e I_e/(2\,g)$, with the air-gap length g = 10 mm.

In the two following simulations, the machine is fed by a d-axis stator current only and by a q-axis stator current only, respectively. In the stator conducting sheet, a linear current density is imposed with the suitable distribution. In order to have a sinusoidal MMF distribution with the maximum value right on the d-axis, the following linear current density must be imposed:

$$J_{sd}\left(\vartheta_r\right) = J_{dM}\ \sin\left(\vartheta_r\right) \tag{8.34}$$

where ϑ_r is the rotor coordinate and the positive sign means a current direction equal to the z-axis direction.

Similarly, to have a sinusoidal MMF distribution with the maximum value centered right on the q-axis, the following linear current density must be imposed:

$$J_{sq}\left(\vartheta_r\right) = -J_{qM}\ \cos\left(\vartheta_r\right) \tag{8.35}$$

Figure 8.14 shows the simulation with d-axis current density distribution. It is useful to consider the stator coordinate $\vartheta_s = \vartheta_m + \vartheta_r$ (see Figure 8.12). Since the rotor position is $\vartheta_m = \pi/2$ (see Figure 8.1), a linear current density $J_{sd}(\vartheta_s) = J_{dM}\cos(\vartheta_s)$ has to be imposed. A negative current value has been fixed, which is a demagnetizing current, so that $J_{dM} = -20000$ A/m. Figure 8.14 shows the corresponding flux plot and the air-gap flux density distribution. The flux density distribution looks like a sinusoidal distribution in correspondence to the pole shoes, while it decreases elsewhere, due to the decrease of the permeance of the magnetic paths.

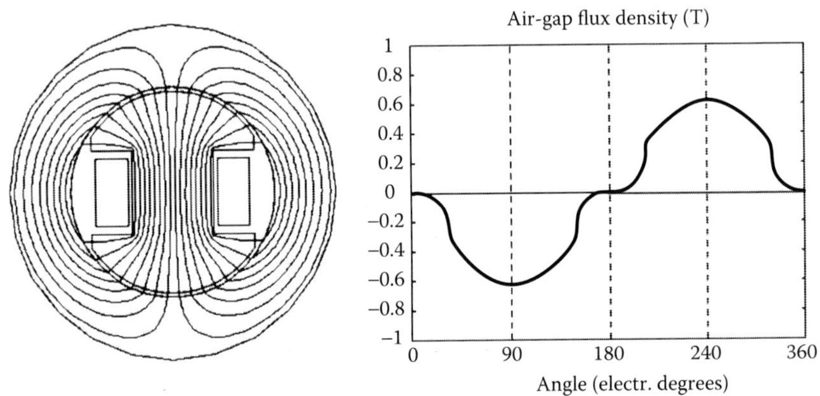

FIGURE 8.14
Flux plot (a) and the air-gap flux density distribution (b) with d-axis current only.

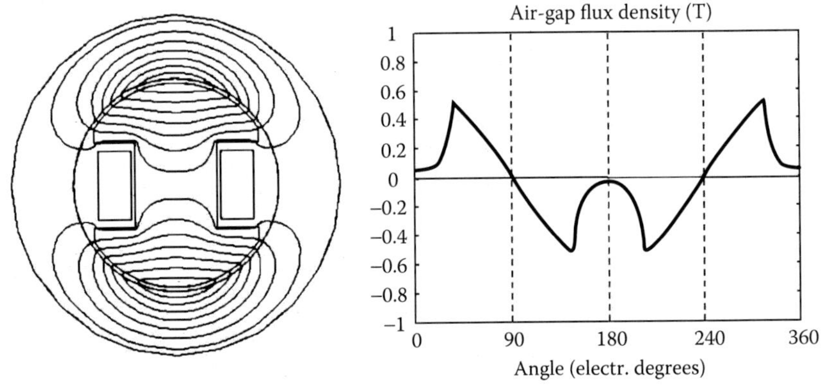

FIGURE 8.15
Flux plot (a) and the air-gap flux density distribution (b) with q-axis current only.

Similarly, Figure 8.15 shows the flux plot and the air-gap flux density distribution when the machine is fed by a q-axis current. A linear current density $J_{sq}(\vartheta_s) = -J_{qM}\sin(\vartheta_s)$ is assigned with J_{qM} = 20000 A/m (positive). Once again, the flux density distribution looks like a sinusoidal distribution in correspondence with the pole shoes, while it decreases elsewhere. Comparing the flux density distributions of Figure 8.14 and Figure 8.15, the reduction of the flux density corresponding to the symmetry axis is almost 10 times.

Figure 8.16 shows the flux plots and the air-gap flux density distributions, with magnetizing current in the rotor and d-axis current in the stator. Because of the negative d-axis current, the demagnetizing effect of the stator current is evident.

Figure 8.17 shows the flux plots and the air-gap flux density distributions, with magnetizing current in the rotor and q-axis current in the stator. In this

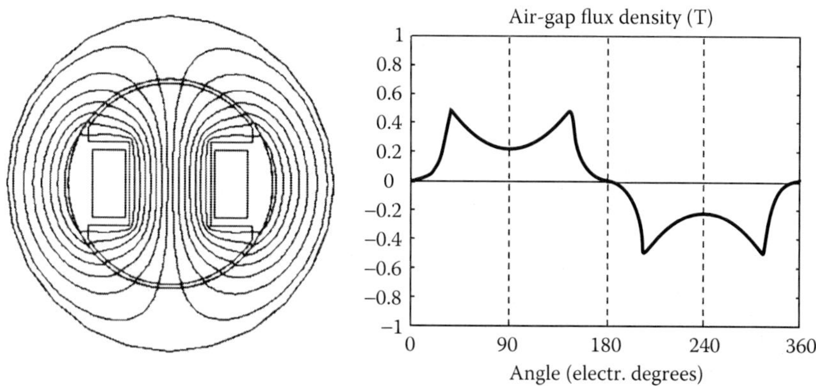

FIGURE 8.16
Flux plot (a) and air-gap flux density distribution (b) with magnetizing and d-axis currents.

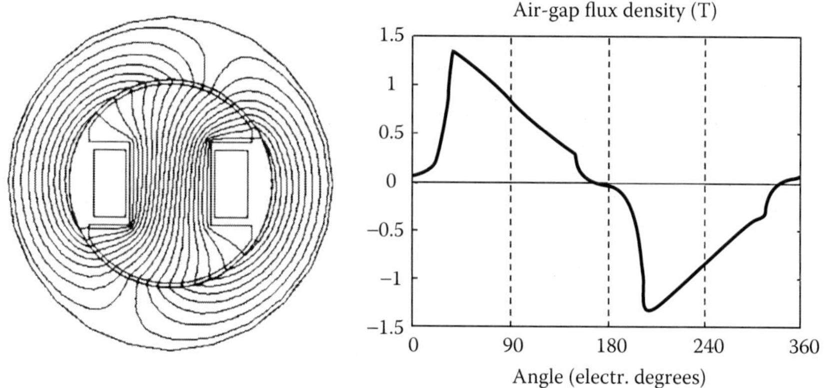

FIGURE 8.17
Flux plot (a) and air-gap flux density distribution (b) with magnetizing and q-axis currents.

case, the distorting effect of the stator current is evident. Since the material has a constant μ_r, the saturation does not occur.

Finally, Figure 8.18 shows the flux plot and the air-gap flux density distribution, when the rotor winding is fed by the magnetizing current, and the stator winding is fed by both d- and q-axis currents. Comparing Figure 8.18 and Figure 8.13, the effect of the armature reaction is evident.

8.10 Appendix: The Transformation abc-dq

The analysis of the three-phase machines is simplified using a change of variables for the electrical quantities, as currents, voltages, and flux linkages.

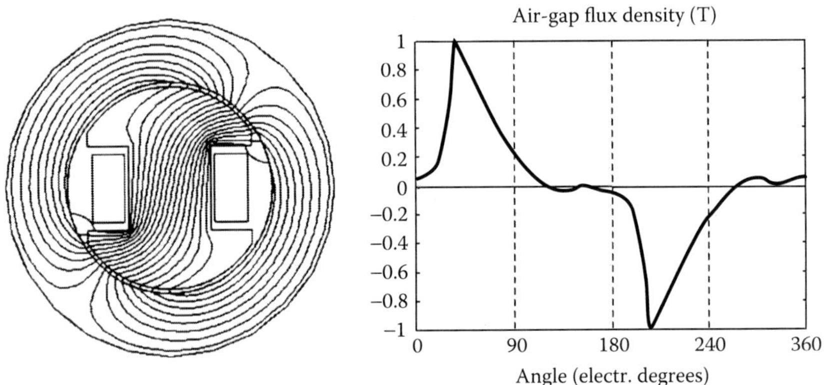

FIGURE 8.18
Flux plot (a) and air-gap flux density distribution (b) with magnetizing, d- and q-axis currents.

The physical meaning of the procedure is to replace the three windings, fixed to the stator and displaced of $2\pi/3$ electrical radiants (see Figure 8.1), with two windings, rotating at the same speed of the rotor. They are displaced of $\pi/2$ electrical radiants, and their axes correspond to the d- and the q-axis respectively, as shown in Figure 8.19. These two windings are equivalent to two stator windings connected to a collector, with the rotating brushes along the lines of the d- and the q-axis. The new reference frame (d, q) is called the *synchronous rotating reference frame*, while the initial reference frame (a, b, c) is called the *stationary reference frame*.

The transformation does not have to modify the magnetic quantities in the air-gap of the machine, i.e., the MMF and flux density distributions. In other words, an observer located in the rotor should not notice if the stator is fed by three stationary windings a, b, and c, or by two rotating windings d and q. The three currents flowing in the three-phase winding, i.e., i_a, i_b, i_c, yield to three MMFs f_a, f_b, f_c that are distributed in the air-gap with the maximum value along the axis of the phases a, b, and c, respectively (see Figure 8.1). The resulting MMF distribution may be conveniently split in two distributions, f_d, f_q, with the maximum value along the d- and the q-axis, respectively (see Figure 8.18).

Let ϑ be the electrical angle individuating the position of the d-axis with respect to the a-phase axis, i.e., $\vartheta = p\vartheta_m$, the transformation required to obtain the MMFs. f_d, f_q from the MMFs f_a, f_b, f_c is given by

$$
\begin{aligned}
f_d &= \frac{2}{3}\left[f_a \cos\vartheta + f_b \cos\left(\vartheta - \frac{2\pi}{3}\right) + f_c \cos\left(\vartheta - \frac{4\pi}{3}\right)\right] \\
f_q &= -\frac{2}{3}\left[f_a \sin\vartheta + f_b \sin\left(\vartheta - \frac{2\pi}{3}\right) + f_c \sin\left(\vartheta - \frac{4\pi}{3}\right)\right]
\end{aligned} \tag{8.36}
$$

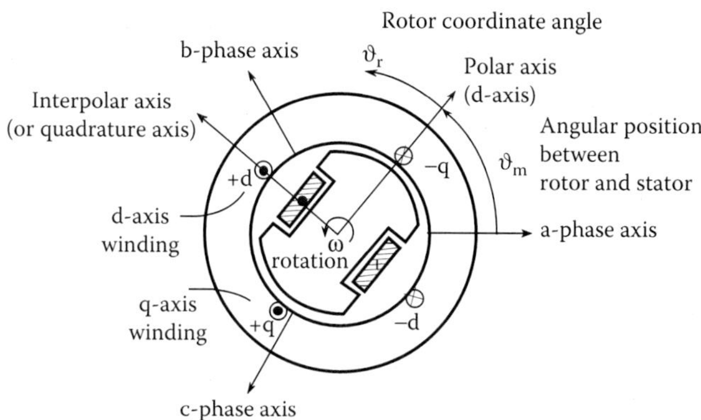

FIGURE 8.19
Synchronous rotating d- and q-axis windings.

or, using vector notation, written as

$$\begin{bmatrix} f_d \\ f_q \end{bmatrix} = \frac{2}{3} \begin{bmatrix} \cos\vartheta & \cos\left(\vartheta - \frac{2\pi}{3}\right) & \cos\left(\vartheta - \frac{4\pi}{3}\right) \\ -\sin\vartheta & -\sin\left(\vartheta - \frac{2\pi}{3}\right) & -\sin\left(\vartheta - \frac{4\pi}{3}\right) \end{bmatrix} \cdot \begin{bmatrix} f_a \\ f_b \\ f_c \end{bmatrix} \tag{8.37}$$

The transformation matrix is indicated by $T_{abc/dq}$. The phasor $f_d + jf_q$ is called the *space phasor*. The same transformation (8.37) is applied to the currents i_a, i_b, i_c and to the voltages v_a, v_b, v_c obtaining i_d, i_q and v_d, v_q, respectively. It is possible to verify that the voltage equations in the synchronous reference frame are given by

$$\begin{aligned} v_d &= Ri_d + \frac{d\lambda_d}{dt} - \omega\lambda_q \\ v_q &= Ri_q + \frac{d\lambda_q}{dt} + \omega\lambda_d \end{aligned} \tag{8.38}$$

where R is the winding resistance, ω is the electrical speed, given by $\omega = d\vartheta/dt$. The signs are reported with the notation of the motor. Finally, λ_d and λ_q are the d- and q-axis flux linkages. In the case of linear conditions, they are expressed as

$$\begin{aligned} \lambda_d &= L_d i_d + \lambda_{dm} \\ \lambda_q &= L_q i_q \end{aligned} \tag{8.39}$$

L_d is the d-axis synchronous inductance, and L_q is the q-axis synchronous inductance. The relationship between the L_d, L_q and the three-phase self-inductance is given in Equation (8.28). Then, λ_{dm} is the flux linked by the d-axis winding due to the magnetizing current only. Since the d-axis and the q-axis are orthogonal, there is no q-axis flux linkage due to the magnetizing current (see Figure 8.18).

At last, the electromagnetic torque developed by the 2p-pole synchronous machine can be expressed as

$$T = \frac{3}{2} p\left(\lambda_d i_q - \lambda_q i_d\right) \tag{8.40}$$

Some advantages of the transformation (8.36)–(8.37) are

1. Since the d- and q-axis are rotating with the rotor, and the stator reluctance variation is negligible (apart from almost negligible reluctance variation due to the stator slotting), the self-inductance of the two d- and q-axis windings is constant, unlike for the three-phase a, b, c windings in which it is a function of the rotor position ϑ_m.
2. Since the d- and q-axis are at $\pi/2$ electrical radiants and posed on two symmetry machine axes, they are not mutually coupled, as is clear by Equation (8.39). Conversely, the a-, b-, and c-phase windings are mutually coupled and this coupling is a function of the rotor position ϑ_m as well. In reality, with high saturation, the d- and q-axis windings may show a mutual coupling (the so-called *cross coupling*). This phenomenon will be analyzed in Chapter 10.
3. During steady-state operations, the electrical quantities in the stationary reference frame a, b, c are sinusoidal in time, while the electrical quantities in the rotating reference frame d, q are constant. For instance, observe the voltage components along the d- and q-axis, reported in Equation (8.32).
4. As a consequence of the points (1) and (2), the dynamic analysis of the machine in the synchronous reference frame is very simplified.

Finally, let us point out that the factor in Equation (8.36) and Equation (8.37) could be not only 2/3, even though such a choice is the most used in the literature. With this choice, it is

1. The rotating d- and q-axis windings are characterized by a number of series conductors equal to (3/2)N, where N is the number of series conductors of each a-, b-, or c-phase winding.
2. The maximum value of the electrical quantities (voltage, current, and flux linkage) in the two reference frames, a, b, c and d, q, is the same. In particular, during steady-state operations, the magnitude

of the space phasor $I_d + jI_q$ corresponds to the maximum value of the sinusoidal waveforms i_a, i_b, i_c.

3. The transformation is not conservative, as can be observed from the presence of the factor 3/2 in the torque equation (8.40).

References

1. R.H. Park, "Two Reaction Theory of Synchronous Machines, Part I," in *Trans. of American Institute of Electrical Engineers*, vol. 48, p. 716, 1929.
2. R.H. Park, "Two Reaction Theory of Synchronous Machines, Part II," in *Trans. of American Institute of Electrical Engineers*, vol. 52, p. 352, 1933.
3. M. Liwschitz-Garik and C.C. Whipple, *Electric Machinery*, Vol. I–II, D: Van Nostrand Co., New York, 1947.
4. C. Concordia, *Synchronous Machines*, John Wiley & Sons, New York, 1951.
5. G. Kron, *Equivalent Circuits of Electric Machinery*, John Wiley & Sons, New York, 1951.
6. W.V. Lyon, *Transient Analysis of Alternating Current Machinery*, John Wiley & Sons, New York, 1954.
7. A.S. Langsdorf, *Theory of Alternating Current Machinery*, McGraw-Hill, New York, 1955.
8. B.J. Chalmers, *Electric Motor Handbook*, Butterworth, London, 1988.
9. A.E. Fidzgerald, C. Kingsley, Jr., and S.D. Umans, *Electric Machinery*, McGraw Hill, New York, 1983.
10. S.A. Nasar (editor), *Handbook of Electric Machines*, McGraw-Hill, New York, 1987.

9

Surface-Mounted Permanent Magnet Motors

This chapter deals with the study of a three-phase surface-mounted permanent magnet motor. Both squarewave and sinewave current-fed motors are considered. In particular, the torque ripple as a function of the rotor position is investigated.

9.1 Introduction

The surface-mounted permanent magnet motor can be considered as a synchronous machine where the magnetizing winding is replaced with a permanent magnet. Thanks to the actual performance of the recent permanent magnet materials, such motors exhibit high efficiency and high torque-to-volume ratio.

In addition, the permanent magnet motor may be designed in different shapes. They may be built with high length-to-diameter ratio, when high speed and low inertia are required, e.g., for machine tools, or with low length-to-diameter ratio, when low speed and high torque are required, e.g., for direct drives and motor-wheel-in-traction applications. In recent years, permanent magnet generators and motors were developed; however, the following analysis deals with the permanent magnet motors.

The magnetic structure of two configurations of a four-pole surface-mounted permanent magnet motor is shown in Figure 9.1. The first configuration represents the classical solution with an inner rotor and an outer stator; the second configuration is with an outer rotor and an inner stator. In both configurations, the permanent magnets are radially or parallel magnetized, and produce an almost rectangular waveform of air-gap flux density. The d-axis corresponds to the polar axis in the middle of the rotor pole. The q-axis is leading the d-axis of $\pi/2$ electrical radians.

In order to highlight the motor versatility to be designed in various shapes, Figure 9.2 shows a radial-flux surface-mounted permanent magnet motor configuration. This is formed by two flat cylinders over which the permanent magnets, which are axially magnetized, are placed. Within the two permanent

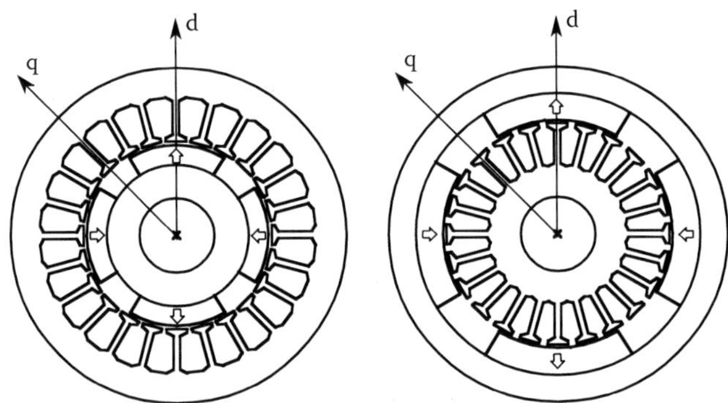

FIGURE 9.1
Surface-mounted permanent magnet motor configurations: (a) with inner rotor and (b) outer rotor.

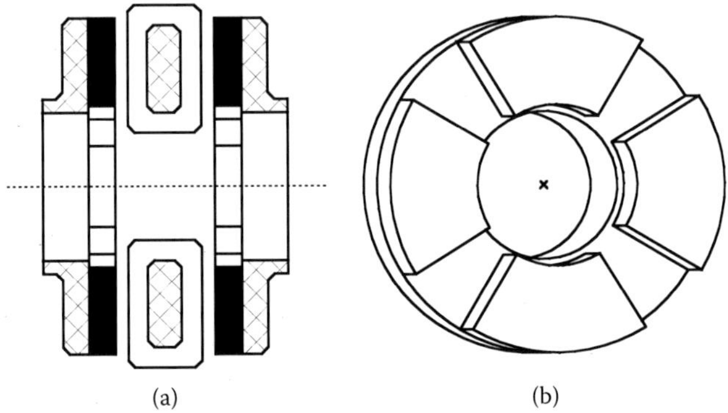

FIGURE 9.2
Axial flux surface-mounted permanent magnet motor configuration.

magnet cylinders, a stator cylinder has been inserted, holding the three-phase armature windings. Figure 9.2(b) shows a detail of the rotor cylinder.

The surface-mounted permanent magnet motors fed by an electrical drive are classified as: (1) squarewave current-fed motors (also called trapezoidal brushless or dc brushless motors), and (2) sinewave current-fed motors (also called sinusoidal brushless or ac brushless motors).

They are characterized by different winding distributions in such a way as to obtain a different induced EMF waveform. In the squarewave current-fed motors, the EMF waveform should be ideally trapezoidal and the windings are fed by squarewave currents, which are synchronized with the EMFs. The ideal waveforms of the three-phase EMFs and the corresponding forced currents are shown in Figure 9.3. In the sinewave current-fed motors, the induced EMF should be ideally sinusoidal and the windings are fed by sinewave currents, synchronized with the EMFs. In both cases, the instantaneous motor torque is ideally constant.

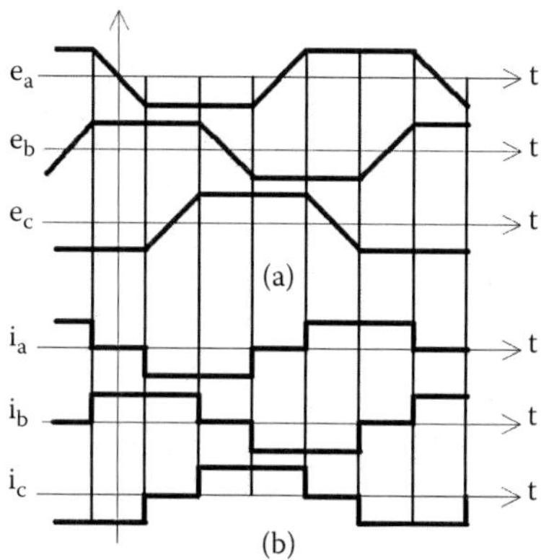

FIGURE 9.3
Ideal EMF (a) and current (b) waveforms in a trapezoidal brushless motor.

Neglecting the edge effects and supposing an identical behavior of the magnetic field along the whole machine length, a two-dimensional analysis is carried out in the (x, y) plane. As for the synchronous generators (see Chapter 8), the machine length is the net iron length L_{Fe}. The magnetic vector potential and the current density have only a z-axis component, which is $\mathbf{A} = (0, 0, A_z)$ and $\mathbf{J} = (0, 0, J_z)$; then the flux density and the field strength vectors have components only on the plane (x, y), which are $\mathbf{B} = (B_x, B_y, 0)$ and $\mathbf{H} = (H_x, H_y, 0)$.

Along the magnetization direction, the permanent magnets are characterized by a B-H curve like that shown in Figure 9.4. This shows the residual flux density B_{res}, the coercive force H_c, and the knee magnetic field strength H_{knee}. A generic working point is represented by the couple of values (H_o, B_o). In order to avoid an irreversible demagnetization, the magnetic field strength cannot go down below H_{knee}. Since the permanent magnet material works in the linear part of the characteristic, the curve is approximated with a straight line, as also shown in Figure 9.4. Then, for the permanent magnet

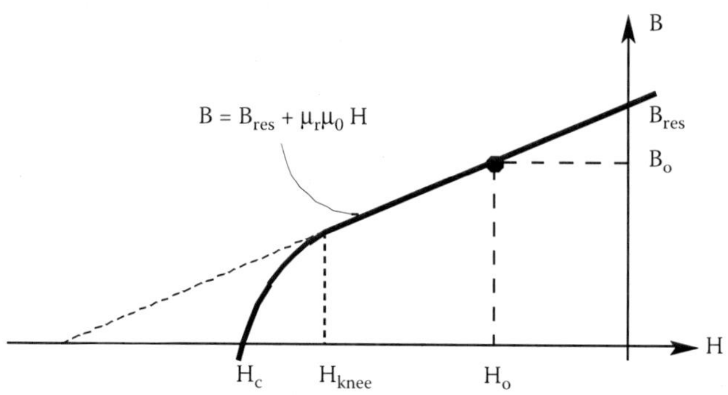

FIGURE 9.4
B-H curve of the permanent magnet and straight-line approximation.

a constant relative differential permeability μ_r is assumed. The flux density vector in the permanent magnets is then given by

$$\mathbf{B} = \mathbf{B}_{res} + \mu_r \mu_0 \mathbf{H} \tag{9.1}$$

where $\mathbf{B}_{res} = (B_{res,x}, B_{res,y}, 0)$. Then the two-dimensional magnetostatic problem is described by the differential equation

$$\frac{\partial}{\partial x}\left(\frac{1}{\mu_r}\frac{\partial A_z}{\partial x}\right) + \frac{\partial}{\partial y}\left(\frac{1}{\mu_r}\frac{\partial A_z}{\partial y}\right) = -\mu_0 J_z - \frac{\partial}{\partial x}\left(\frac{B_{res,y}}{\mu_r}\right) + \frac{\partial}{\partial y}\left(\frac{B_{res,x}}{\mu_r}\right) \tag{9.2}$$

The symmetry between the pole pairs and the corresponding periodic boundary conditions are the same of the synchronous generator; see Chapter 8.

9.2 Computation of the No-Load Characteristic

The permanent magnet rotor produces an air-gap magnetic flux that links the stator winding. When the rotor moves, an EMF is induced in the stator windings that corresponds to the rate of change of the flux linkage.

For the computation of the no-load flux linkage, the permanent magnet is magnetized (the B-H curve of Figure 9.4 is used), while the stator winding is open-circuited, which means a null conductivity is considered. Stator and rotor laminations are defined by the B-H curves of the used magnetic material.

As far as the boundary conditions are concerned, a null magnetic vector potential is fixed along the external stator circumference and along the d-axis (Dirichlet's condition), while a null normal derivative of the magnetic vector potential is assigned along the q-axis (Neumann's condition). Such boundary conditions are represented in Figure 9.5.

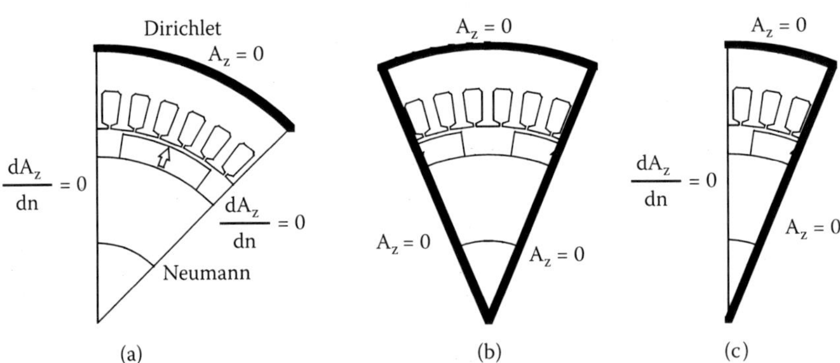

FIGURE 9.5
Boundary conditions in the no-load simulation.

The three-phase winding is placed within the stator slots, and may be single-layer, full-pitched winding or a double-layer, chorded winding. The first solution is mainly adopted with squarewave current-fed motor, in order to obtain EMFs with a trapezoidal waveform. The second solution is adopted in sinewave current-fed motor, in order to obtain EMFs with a sinusoidal waveform. Some examples of such windings are shown in Figure 8.4 and Figure 8.5 of Chapter 8.

9.2.1 Computation on the Solved Structure

9.2.1.1 *Flux Linkage*

Once the field solution is obtained, the magnetic vector potential $A_z(x, y)$ is known in each point of the domain. The magnetic flux due to the permanent magnet, which is linked with the j-th phase winding, is

$$\Lambda_{j,pm} = 2pL_{Fe}\frac{n_q}{n_{pp}}\sum_{q=1}^{Q/2p} k_{jq}\frac{1}{S_q}\int_{S_q} A_z dS \qquad j = a, b, c \tag{9.3}$$

where n_q is the number of the conductors in the slot, n_{pp} is the number of parallel paths of the winding, Q/2p is the number of slots per pole, and k_{jq} assumes the values 0, ±1, or ±0.5, depending on the orientation and whether the coils in the q-th slot belong to the j-th phase (see Chapter 8, Section 8.2.1.1).

9.2.1.2 *Induced EMF*

In case of sinewave current-fed motors, the EMF waveform can be assumed to be sinusoidal. Hence, the EMF may be easily estimated as follows. For the sake of convenience, the a-phase is chosen so as to link the maximum magnetic flux (the a-phase axis correspondent to the d-axis). Let $\Lambda_{a,pm}$ be the maximum flux linkage. Then, at constant rotor speed $\omega_m = \omega/p$, the RMS value of the a-phase EMF is

$$E_a = \frac{1}{\sqrt{2}}\omega\Lambda_{a,pm} \tag{9.4}$$

From the same field solution, supposing a one-slot step rotation, it is possible to find some points of the EMF waveform, as described in Chapter 8, Section 8.2.1.2.

In case of squarewave current-fed motors, the following procedure is required.

9.2.2 Computation for Various Rotor Positions

9.2.2.1 *Flux Linkage and Induced EMF*

For a rapid estimation of the flux linkages, one field solution is enough. The rotor is fixed in a convenient position with respect to the stator, usually with one winding linking the maximum magnetic flux, as shown in the previous section.

Conversely, a more accurate computation should make provision for an analysis at various rotor positions. In this way, it is possible to built the flux linkage waveform for each phase winding, as a function of the mechanical angle ϑ_m, i.e., $\lambda_{pm}(\vartheta_m)$. Therefore, the fundamental and the higher-order harmonics of the flux linkage can be obtained by means of the Fourier series expansion. It yields

$$\lambda_{pm}(\vartheta_m) = \sum_{k=1}^{N_k} \Lambda_{pm,k} \cos(k\vartheta_m) \tag{9.5}$$

where the higher order of the series is N_k. In the series expansion, only cosinusoidal waveforms have been adopted, thanks to a suitable choice of the reference angle $\vartheta_m = 0$. Really, this assumption is not valid in the case of fractional-slot winding without symmetry between the North and South poles.

The no-load EMF is obtained by deriving the flux linkage with respect to the time. Assuming a constant rotor speed, it is practical to derive the flux linkage with respect to the mechanical angle ϑ_m, and to multiply by the mechanical speed ω_m, which is

$$e(\vartheta_m) = \frac{d\lambda_{pm}(\vartheta_m)}{dt} = \frac{d\lambda_{pm}(\vartheta_m)}{d\vartheta_m}\frac{d\vartheta_m}{dt} = \frac{d\lambda_{pm}(\vartheta_m)}{d\vartheta_m}\omega_m \tag{9.6}$$

Since the flux linkage in Equation (9.6) is derived numerically, this operation may give rise to errors. It is convenient to express the flux linkage using the Fourier series expansion, as in Equation (9.5), and then to compute the EMF as the series of the derivatives of each flux linkage harmonic, as in

$$e(\vartheta_m) = \omega_m \sum_{k=1}^{N_k} -k\,\Lambda_{pm,k} \sin(k\vartheta_m) \tag{9.7}$$

Of course, the EMF can be expressed as a function of the electrical angle ϑ, replacing $\vartheta_m = \vartheta/p$, or as a function of the time, replacing $\vartheta_m = \omega_m t$.

9.2.2.2 *Cogging Torque*

In the various positions of the rotor, it is also possible to compute the cogging torque. This is due to the interaction between the permanent magnets and the stator slotting. Such torque may be computed by means of Maxwell's stress tensor or as the derivative of the magnetic energy with the position.

In the first case, the torque is obtained by integrating Maxwell's stress tensor along a surface containing the rotor. Because of the two-dimensional problem, this corresponds to integrating Maxwell's stress tensor along a line

l_g in the middle of the air-gap, and then multiplying the result by the active length of the rotor, L_{Fe}. Assuming a 2p-pole machine, it yields

$$T_{cog} = \frac{D-g}{2}\frac{L_{Fe}}{\mu_0} p \int_{l_g} B_r B_\vartheta dl \tag{9.8}$$

where B_r is the radial component of the flux density (normal to the line l_g), B_ϑ is the azimuthal component of the flux density (tangential to the line l_g), D is the inner stator diameter, and g is the air-gap length. The number of pole pairs p is used assuming that the simulation has been carried out on two pole-pieces of the machine only. A different factor has to be used with different simulation (e.g., 2p should be used instead of p when only one pole-piece is simulated).

Because of the numerical nature of the finite element method, the result may depend on both the position of the integration line and the number of points chosen for the numerical integration. Instead of Equation (9.8), it is better to compute the average value of the torque over the entire air-gap surface S_g. Then the torque is

$$T_{cog} = \frac{L_{Fe}}{g\mu_0} p \int_{S_g} r B_r B_\vartheta dS \tag{9.9}$$

where r is the dummy radius.

In the second case, the cogging torque may be computed from the variation of the magnetic energy W_m variation with respect to the rotor angular position ϑ_m. On open-circuit the stator currents are null, hence there is no electrical energy exchange between the motor and external sources. Corresponding to an elementary rotation $d\vartheta_m$, the sum of the variation of the magnetic energy stored in the system and the work done by the magnetic field is null: $dW_m + T_{cog} d\vartheta_m = 0$. Then, the cogging torque is equal to the negative derivative of the magnetic energy with respect to the rotating angle ϑ_m.

The computation of the magnetic energy in the air and in the iron is carried out in the classical way. The computation of the energy in permanent magnet materials is discussed later in this chapter. It is convenient to omit the work of magnetization (which is an irreversible process) and to refer the magnetic energy density to the point (0, B_{res}) of the B-H curve, as reported in Figure 9.6(a). Since the torque depends on a difference of the energy, this useful assumption does not cause any error. As a consequence, the energy in the permanent magnet is defined as

$$\begin{aligned} W_{m(pm)} &= \int_{\tau_{pm}} \int_0^B \mathbf{H} \cdot d\mathbf{B} \, d\tau \\ &= L_{Fe} \int_{S_{pm}} \int_0^B \mathbf{H} \cdot d\mathbf{B} \, dS \end{aligned} \tag{9.10}$$

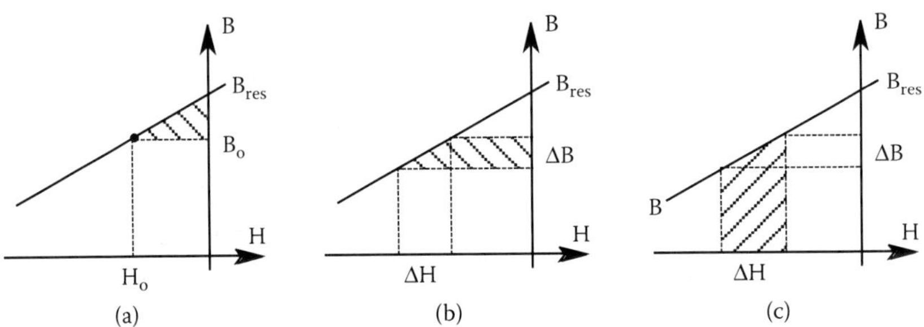

FIGURE 9.6
Magnetic energy and coenergy density in permanent magnets.

Bearing in mind the linear approximation of the B-H curve, given by Equation (9.1), supposing that **H** is parallel to $\mathbf{B}_{res}$, then Equation (9.10) becomes

$$\begin{aligned} W_{m(pm)} &= \frac{1}{2\mu_r\mu_0} L_{Fe} \int_{S_{pm}} \left(B^2 - B_{res}B\right) dS \\ &= \frac{\mu_r\mu_0}{2} L_{Fe} \int_{S_{pm}} H^2 \, dS \end{aligned} \tag{9.11}$$

In an analogous way, the magnetic coenergy is defined as

$$\begin{aligned} W'_{m(pm)} &= \int_{\tau_{pm}} \int_0^H \mathbf{B} \cdot d\mathbf{H} \, d\tau \\ &= \frac{1}{2\mu_r\mu_0} L_{Fe} \int_{S_{pm}} B^2 \, dS \\ &= \frac{\mu_r\mu_0}{2} L_{Fe} \int_{S_{pm}} \left(H + \frac{B_{res}}{\mu_r\mu_0}\right)^2 dS \end{aligned} \tag{9.12}$$

The variation of the magnetic energy density corresponding to any rotation of the rotor is shown in Figure 9.6(b). The variation of the magnetic coenergy density in the permanent magnet is shown in Figure 9.6(c).

Finally, from the magnetic energy W_m of the whole system, the cogging torque is given by

$$T_{cog} = -\frac{dW_m}{d\vartheta_m} \tag{9.13}$$

In the computation of the torque at various rotor positions the numerical errors caused by the mesh changes should be avoided. Such errors may be

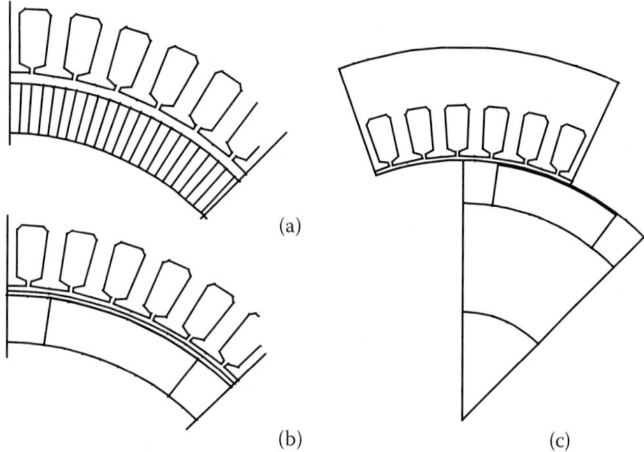

FIGURE 9.7
Tricks to rotate the structure without mesh change, so as to avoid numerical errors.

so high that they significantly alter the solution, especially when low torques are investigated (such as the cogging torque).

To avoid or to reduce such errors, two tricks are suggested.

The first is to divide the permanent magnet in various pieces, as shown in Figure 9.7(a). The length of each piece should correspond to the minimum rotation that has been fixed. According to the rotor position, the magnetic B-H curve of the air or of the permanent magnet is assigned to each piece. In this way, a fictitious rotation is obtained, except that it operates on the material definition and not on the drawing of the motor. As a consequence, the mesh of the domain remains always the same.

The second trick is to split the air-gap into two parts, as illustrated in Figure 9.7(b). Then the rotor and the corresponding air-gap part is rotated, keeping fixed the other air-gap part together with the stator, as shown in Figure 9.7(c). The mesh remains the same, fixed to the structure, both in the movable and in the fixed part. Of course, for each position, the boundary conditions have to be adapted to the exact domain contour. It is worth noticing that the boundary line between the two air-gap parts must be divided properly, in such a way as the stator mesh correctly matches the rotor mesh in each rotor position.

9.3 Computation of the Inductances

The computation of the inductances requires a magnetostatic field simulation. By geometric and electric symmetry, only one pole-piece of the machine is usually considered.

With the hypothesis of linear operating conditions (which is often verified in the surface-mounted permanent magnet motors), the permanent magnet is assumed to be demagnetized, i.e., with null residual flux density B_{res}. Thus, the magnetic flux due to the permanent magnet results in being null. The stator windings are fed so that the flux lines are normal or tangential to the boundary of the domain, allowing only Neumann's or Dirichlet's boundary conditions to be assigned.

The inductances can be computed from the magnetic energy, from the flux linkage, or from the air-gap flux density distribution, with the same procedures presented in Chapter 8, Sections 8.3 and 8.4.

9.4 Computation of the Torque

The motor torque is caused by the interaction between the permanent magnet and the stator currents. For the sake of generalization, the stator currents are considered to be independent of the rotor position ϑ. The analysis is reduced to a part of the machine, usually one pole or two poles according to the stator geometry. Of course, suitable boundary conditions are assigned.

The computation of the torque is carried out in different ways:

1. by means of Maxwell's stress tensor
2. by means of the virtual work principle, evaluating the changes of magnetic energy in two different rotor positions
3. by means of the interaction between flux linkage and current components in the synchronous rotating reference frame

The three ways for computing the motor torque are described next.

9.4.1 Computation by Means of Maxwell's Stress Tensor

As explained in the previous chapter, Maxwell's stress tensor is integrated on an arc in the middle of the air-gap, and the result is multiplied by the radius and the motor stack length L_{Fe}. However, as in Equation (9.9), to reduce the numerical error, the torque is computed as the average value on the whole air-gap surface S_g, as

$$T = \frac{L_{Fe}}{g\mu_o} p \int_{S_g} r B_r B_\vartheta dS \tag{9.14}$$

The pole pairs number p is used in Equation (9.14), assuming that the simulation is carried out again on two pole-pieces of the machine.

9.4.2 Computation by Means of the Virtual Work Principle

This method is based on the balance of the energy change corresponding to an elementary rotation of the rotor. The different energy terms are computed by integrating the volume magnetic energy density on the various components of the domain.

In nonlinear conditions, the magnetic coenergy has to be adopted (see Chapter 5). The torque is computed as

$$T = +\left.\frac{dW'_m}{d\vartheta_m}\right|_{i=const} \tag{9.15}$$

where the magnetic coenergy in the permanent magnet material is computed as indicated by Equation (9.12).

As highlighted previously, this method requires at least two field simulations, with the rotor in two different positions.

9.4.3 Computation by Means of Flux Linkages and Currents

At last, the motor torque can be computed from the interaction between the flux linkages and the current. It is advantageous to consider these quantities in the synchronous rotating reference frame, by means of the transformation $T_{abc/dq}$ (see the Appendix in Chapter 8). The torque is

$$T = \frac{3}{2} p\left(\Lambda_d I_q - \Lambda_q I_d\right) \tag{9.16}$$

where the d- and q-axis currents are given by Equation (8.30), and the d- and q-axis flux linkages are given by Equation (8.29).

The angle ϑ is the electrical angle between the reference a-phase axis and the d-axis. The flux linked with the j-th phase (j = a, b, c) is obtained from the average value of the magnetic vector potential A_z over the slot surfaces S_{Cu}.

It is worth bearing in mind that the torque computed by Equation (9.16) is obtained as the energy balance at the terminals of the motor windings. Such a balance is based on the assumption of sinusoidal distributed windings. As a consequence, the average torque is correct, but the torque ripple is lower. It could be verified that MMF harmonics are only partially included in Equation (9.16). In addition, such a computation method does not consider the cogging torque due to the interaction of the permanent magnet and the stator slotting. In fact, the torque computed from Equation (9.16) is always null when the stator currents are null (i.e., $I_d = I_q = 0$).

9.4.4 Dynamic Computation

Up to now, the torque computation has been carried out referring to the stator currents equal to their fixed ideal values. Because of the winding

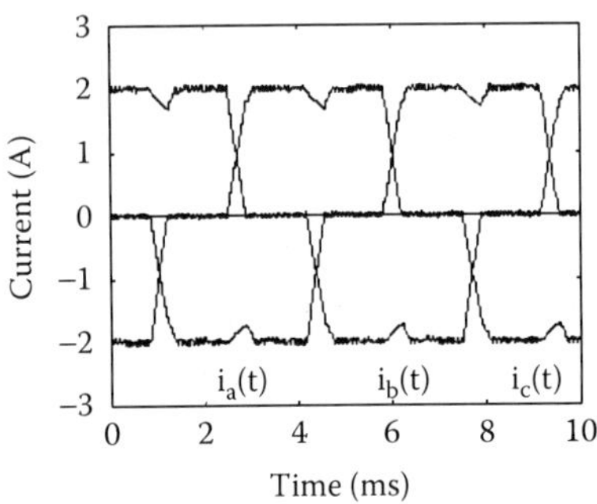

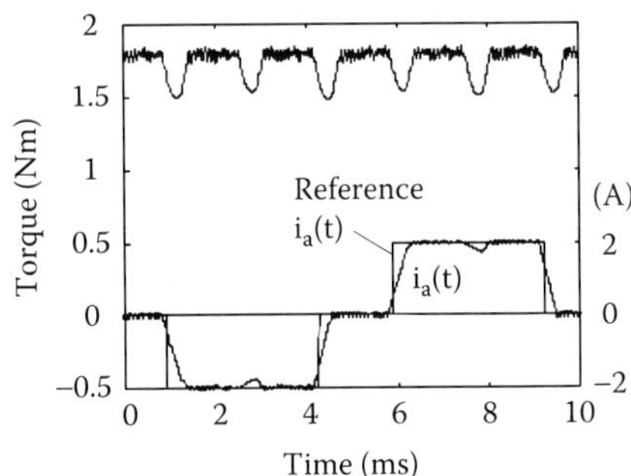

FIGURE 9.8
Currents and torque waveforms.

inductances and the induced EMFs, together with the limited value of the source voltage, the electrical drive is not always able to feed the motor with ideal currents. On the contrary, the currents often follow their references with a delay.

This delay is mainly appreciable in squarewave current-fed motors, since the currents cannot exhibit rapid changes as those required by the ideal currents shown in Figure 9.3(b). The current is as distorted as the rotor speed is high. Consequently, a constant torque is not obtained, but a torque ripple takes place, which is called the commutation effect. Figure 9.8(a) shows the typical three-phase current waveforms, obtained by a dynamic simulation at a supply frequency f = 100 Hz. In the figure, the rise and fall times of the current are clear. The delay of the phase current with respect to its reference, highlighted in Figure 9.8(b), causes the torque ripple, shown in Figure 9.8(b) as well.

For the computation of the instantaneous torque, a simple procedure is adopted. All the motor parameters are known, i.e., the resistance R, self- and mutual inductances L and M, the induced EMF waveform $e(\vartheta_m)$. With fixed motor speed, the latter may be expressed as a function of the time, as $\vartheta_m = \omega_m t$. Using a two-level inverter, the voltage forced at the motor terminals can assume the two values $v = -V_{dc}$ or $v = +V_{dc}$ ($v = 0$ using a three-level inverter), where V_{dc} is the dc bus voltage. The voltage level is chosen by the control comparing the current reference (ideal current) with the effective current.

The dynamic analysis is carried out, from the step integration of the differential equation

$$v_j(t) = Ri_j(t) + (L - M)\frac{di_j(t)}{dt} + e_j(t) + v_N(t) \tag{9.17}$$

for each j-th phase winding (j = a, b, c). In Equation (9.17), $v_N(t)$ indicates the star-center potential, which is computed at each step from the forced terminal voltages.

9.5 Example

This section discusses some results of finite element analysis of a surface-mounted permanent magnet motor. The motor data are reported in Table 9.1; the magnetic structure and the winding are sketched in Figure 9.9. It is a squarewave current-fed brushless motor with a low number of slots per pole and a chorded winding. Such a solution is adopted in low-power applications, with the aim of having minimum end-winding length and higher motor efficiency, in spite of a higher torque ripple.

With such a motor structure, a two-pole section of the machine is simulated, using the periodic conditions on the two boundaries. Different rotor positions are simulated splitting the permanent magnet in elementary pieces, as shown in Figure 9.7(a).

A rotation of 40 mechanical degrees (120 electrical degrees) is considered although a rotation of 30 mechanical degrees would be enough.

9.5.1 No-Load Simulation

Figure 9.10 shows the flux lines at no-load when the rotor is in the position $\vartheta_m = 10$ mechanical degrees.

TABLE 9.1

Brushless Motor Data

Stator Data		
$D_e = 125$	(mm)	stator external diameter
$D = 61$	(mm)	stator inner diameter
$L_{Fe} = 38$	(mm)	stack length
$2p = 6$		number of poles
$Q = 9$		number of slots
$N = 1458$		number of series conductors per phase
Stator Slot Data		
$w_{so} = 3.55$	(mm)	width of slot opening
$w_t = 7.5$	(mm)	width of stator tooth
$w_{so} = 4$	(mm)	height of slot opening
$h_s = 22$	(mm)	total height of slot
Rotor Data		
$D_r = 44.7$	(mm)	rotor external diameter
$D_{sh} = 18$	(mm)	shaft diameter
Permanent Magnet Data		
$l_m = 7.5$	(mm)	permanent magnet thickness
$2\alpha_{pm} = 55$	(deg)	mechanical permanent magnet angle
$B_{res} = 0.4$	(T)	residual flux density (Ferrite)
$\mu_r = 1.05$	(-)	differential relative permeability

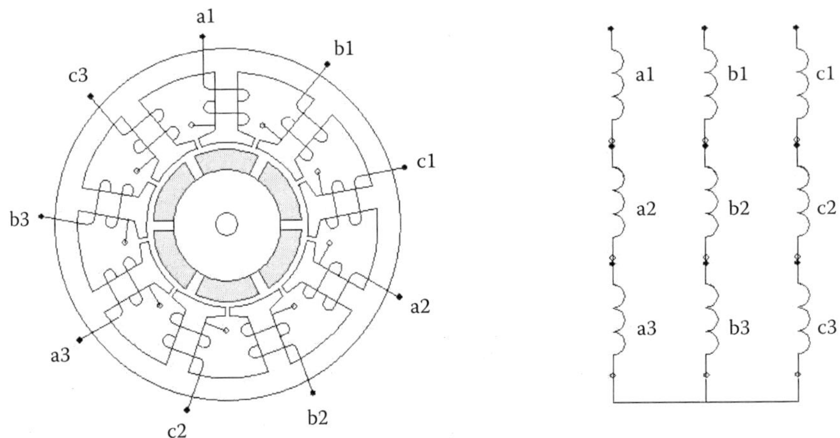

FIGURE 9.9
Brushless motor magnetic structure.

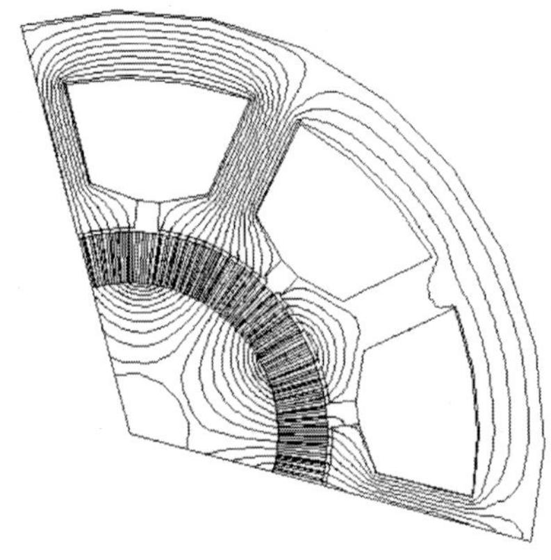

FIGURE 9.10
Flux lines at no-load operation; a rotation of $\vartheta_m = 10$ mechanical degrees is considered.

The three-phase flux linkages are shown in Figure 9.11(a). They are approximately flat for a portion of 40 mechanical degrees (120 electrical degrees). It follows that, with constant speed, the back EMFs exhibit a constant value for a corresponding period of time (one third of the electrical period), as required in a trapezoidal brushless drive; see Figure 9.3(a).

The EMF waveform is not exactly trapezoidal because of the chorded winding. Thus, a constant torque is not achieved even when ideal square-wave currents of Figure 9.3(b) feed the stator windings.

The magnetic energy is shown in Figure 9.11(b). The energy variation indicates that, during the rotation, there is a cogging torque due to the interaction of the permanent magnet with the stator slotting.

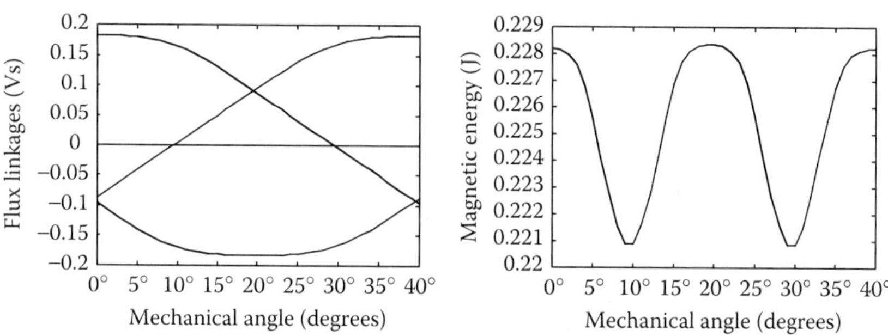

FIGURE 9.11
Flux linkages and magnetic energy during the no-load operation.

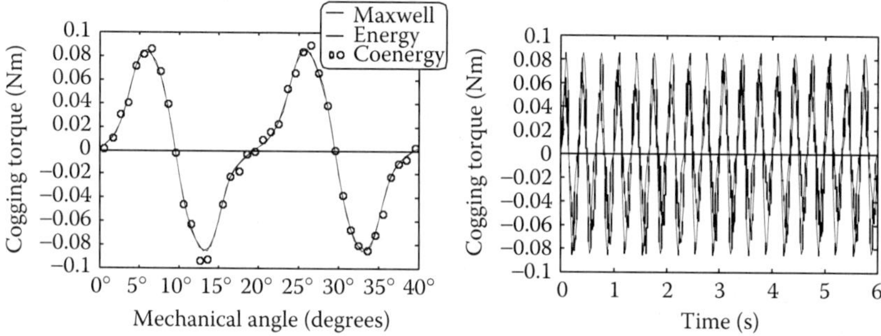

FIGURE 9.12
Cogging torque: simulated (a) and measured (b) values.

The cogging torque behavior is shown in Figure 9.12(a). It has been computed in different ways by means of Maxwell's stress tensor, using Equation (9.9), drawn in a thin line; by means of the derivative of the magnetic energy, using Equation (9.13), drawn in a bold line (almost indistinguishable from the previous one); and by means of the derivative of the magnetic coenergy, using Equation (9.15) with null currents, drawn using dots. It is worth noticing the good match among the three methods.

At last, Figure 9.12(b) shows the comparison between the simulated and the measured cogging torque. The measure system has been regulated to complete a whole rotation in 6 seconds. The figure shows a satisfactory agreement between simulated results and measurements. A small difference can be attributed to a nonperfect symmetry of the rotor poles and to a residual flux density slightly lower than that used in the simulations.

9.5.2 Simulations of Operations Under Load

The motor is fed by ideal squarewave currents, as depicted in Figure 9.3. They are centered on the constant value of the corresponding EMFs and their maximum value is 2 A.

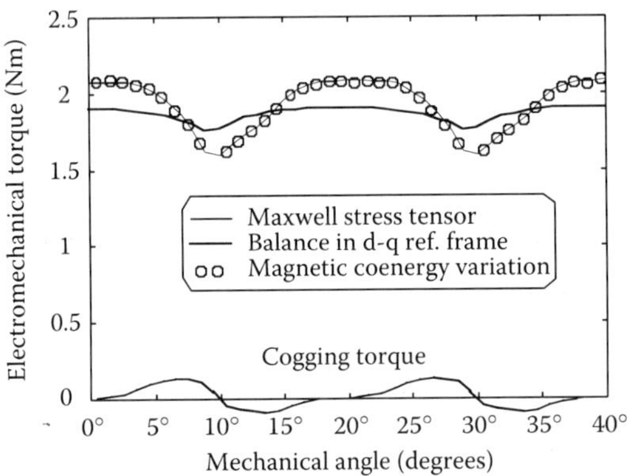

FIGURE 9.13
Electromechanic torque with ideal squarewave currents, I = 2 A.

The electromechanical torque behavior is shown in Figure 9.13. It is computed by means of the methods described above. The thin line refers to the integration of Maxwell's stress tensor, as given by Equation (9.14). The circles refer to the torque computed by the derivative of the magnetic coenergy with constant currents, as in Equation (9.15). With squarewave currents, the stator currents are constant for 20 mechanical degrees, so that this method is easily applied.

Finally, the bold line refers to the torque computed from the flux linkages and currents in the synchronous reference frame, as in Equation (9.16). As pointed out earlier, it is observable that the torque computed by the latter method shows the same average value, but a lower ripple. For a comparison, the same Figure 9.13 shows the cogging torque, which has been computed before.

References

1. N. Demerdash et al., "Comparison between features and performance characteristics of fifteen-hp samarium-cobalt and ferrite-based brushless d.c. motors," *IEEE Trans. on PAS*, vol. 102, pp. 104–112, 1983.
2. H. Weh, H. Mosebach, H. May, "Design concepts and force generation in inverter-fed synchronous machines with permanent magnet excitation," in *IEEE Trans on Magnetics*, vol. MAG-20, no. 5, pp. 1756–1761, 1984.
3. B.J. Chalmers, S.A. Hamed, G.D. Baines, "Parameters and performances of a high-field permanent-magnet synchronous motor for variable-frequency operation," *IEEE Proc.* Pt. B, vol. 132, no. 3, 1985.
4. N. Boules, "Prediction of No-Load Flux Density Distribution in Permanent Magnet Machines," *IEEE Trans. on IA*, vol. 21, no. 4, pp. 633–643, 1985.
5. K. Kobayashi and M. Goto, "A brushless DC motor of a new structure with reduced torque fluctuations," *Elec. Eng. Jpn*, vol. 105, no. 3, pp. 104–112, 1985.

6. T. Kenjo and S. Nagamori, *Permanent-Magnet and Brushless DC Motors*, Oxford University Press, 1985.
7. T. Nehl, N. Demerdash, and F. Fouad, "Impact of winding inductances and other parameters on the design and performance of brushless d.c. motors," *IEEE Trans. on PAS*, vol. 104, pp. 2206–2213, 1985.
8. T. Li and G. Slemon, "Reduction of cogging torque in permanent magnet motors," *IEEE Trans. on MAG*, vol. 24, no. 6, pp. 2901–2903, 1988.
9. B.J. Chalmers, "Influence of saturation in brushless permanent-magnet motor drives," in *IEEE Proc. Pt. B*, vol. 139, pp. 51–52, January 1992.
10. Z.Q. Zhu, D. Howe, E. Bolte, and B. Ackermann, "Istantaneous Magnetic Field Distribution in Brushless Permanent Magnet DC Motors, Part I: Open-Circuit Field," *IEEE Trans. on MAG*, vol. 29, no. 1, pp. 124–135, 1993.
11. N. Bianchi, S. Bolognani, "A Procedure for the Electromagnetic and Thermal Integrated Design of PM Brushless Motors," *SPEEDAM*, Taormina, 8–10 Giugno 1994, pp. 13–18.
12. N. Bianchi, S. Bolognani, "Optimal design of a PM brushless motor," in *Proc. of International Conference on Electrical Machine, ICEM '94*, Parigi, 5–8 September 1994, pp. 204--209.
13. N. Bianchi, S. Bolognani, "Analysis and design of a PM synchronous magnetic drive," in *Proc. of International Conference on Electrical Machine, ICEM '94*, Parigi, 5–8 September 1994, pp. 341–346.
14. D.C. Hanselman, *Brushless Permanent Magnet Motor Design*, McGraw-Hill, Inc., 1994.
15. J.M.D. Coey (editor), *Rare-Heart Iron Permanent Magnet*, Monographs on the Physics and Chemistry of Materials, Oxford Science Publications, Claredon Press, Oxford, 1996.
16. R.P. Deodhar, D.A. Staton, T.M. Jahns, and T.J.E. Miller, "Prediction of cogging torque using flux-mmf diagram technique," *IEEE Trans. on IA*, vol. 32, no. 3, pp. 569–575, 1996.
17. N. Bianchi, A. Lorenzoni, "Permanent Magnet Generators for Wind Power Industry: an Overall Comparison with Traditional Generators," in *Proceeding of IEEE International Conference on Opportunities and Advance in Electric Power Generation*, Durham (UK), 1996, pp. 49–54.
18. N. Bianchi, S. Bolognani, "Design Techniques for Reducing the Cogging Torque in Surface-Mounted PM Motors," accettata per presentazione e pubblicazione sui Proc. della conferenza internazionale *IEEE IAS Annual Meeting*, Rome, Italy, 8–12 October 2000.
19. R. Krishnan, *Electric Motor Drives*, Prentice-Hall, Englewood Cliffs, NJ, 2001.

Further References on AC Drives

1. W. Schuisky, *Berechnung Elektrischer Machinen*, Springer Verlag, Wien, 1960.
2. G.R. Slemon, A. Straughen, *Electric Machines*, Addison-Wesley, New York, 1980.
3. P. Sen, *Thyristor DC Drives*, John Wiley & Sons, New York, 1981.
4. E. Levi, *Polyphase Motors*, John Wiley & Sons, New York, 1984.
5. W. Leonard, *Control of Electrical Drives*, Springer-Verlag, Berlin, 1985 (first edition), 1996 (second edition).

6. P. Vas, *Vector Control of AC Machines*, Oxford Science Publications, Clarendon Press, Oxford, 1990.
7. I. Boldea and S.A. Nasar, *Vector Control of AC Drives*, CRC Press, Boca Raton, FL, 1992.
8. D.W. Novotny, T.A. Lipo, *Vector Control and Dynamics of AC Drives*, Oxford Science Publications, Clarendon Press, Oxford, 1996.
9. J.F. Gieras and M. Wing, *Permanent Magnet Motors Technology: Design and Applications*, Marcel Dekker, Inc., New York, 1996.
10. I. Boldea and S.A. Nasar, *Electric Drives*, CRC Press, Boca Raton, FL, 1999.

10

Interior Permanent Magnet and Reluctance Synchronous Motors

This chapter deals with the finite element analysis of an interior permanent magnet motor. The interior permanent magnet motor is used in traction applications or when high dynamic performance is required. It is a synchronous motor, fed by sinewave currents via an electrical drive; therefore the analysis of such a motor is carried out by means of the direct and the quadrature axis theory. Two magnetic models of the motor are presented, and the procedures for obtaining them are illustrated. An automatic algorithm is also described to process the data obtained from the finite element analysis. The synchronous reluctance motor is discussed at the end of the chapter, considering such a motor as a particular case of the interior permanent magnet motor.

10.1 Introduction

The key characteristic of the interior permanent magnet motor is that the permanent magnets are not on the surface of the rotor but are buried in the rotor laminations. Apposite holes are stamped in the rotor laminations to hold the permanent magnets and to obstruct the d-axis flux due to the stator current. For this reason, they are also called flux barriers.

Since the differential relative permeability of the permanent magnet is only slightly higher than the air permeability, the rotor exhibits magnetic paths with different magnetic permeance. Let the d-axis correspond to the polar axis (i.e., the permanent magnet axis). Thus the d-axis permeance is lower than the q-axis permeance, and the inductances are $L_d < L_q$. This is opposite of the usual situation of the salient pole synchronous machines with wound rotor (it is useful to compare with Chapter 8).

As a consequence of the rotor anisotropy, the motor takes advantage of the reluctance torque and the permanent magnet torque, showing a favorable torque-to-volume ratio. In addition, the interior permanent magnet motor is

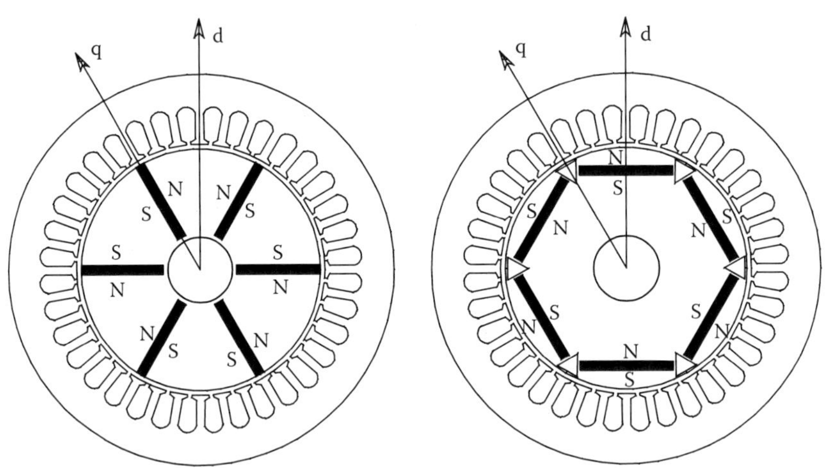

FIGURE 10.1
Different rotor structures of the interior permanent magnet motors.

appropriate for flux-weakening operations, when a constant power torque-speed characteristic is required, since it maintains a high power up to high operating speeds.

Although the stator is similar to that of an induction motor, the rotor may assume different topologies, according to the permanent magnet position. These configurations may be divided into two groups: rotor with tangential (or azimuthal) magnetized permanent magnets and with radial magnetized permanent magnets. They are shown for a 6-pole motor in Figure 10.1(a) and Figure 10.1(b), respectively.

In the first case, the magnetic flux of each pole is given by the sum of the flux of two permanent magnets. This configuration type is appropriate with permanent magnets of low residual flux density and with machines of high number of poles. The two permanent magnet surfaces are chosen so as their sum is higher than the air-gap pole surface, in order to obtain a higher air-gap flux density. In addition, the rotor shaft should be nonmagnetic to avoid flux loops.

In the second case, since the permanent magnet surface is lower than the pole surface, the air-gap flux density is lower than the permanent magnet flux density. Moreover, such a configuration shows a high q-axis permeance and a low d-axis permeance, which is a high rotor anisotropy.

In both configurations of Figure 10.1, there are magnetic bridges that are essential for support the rotor structure against the centrifugal forces. However, their thickness should be as narrow as admitted by mechanical considerations, since they are magnetic short-circuits of the PM flux.

Referring to a tangential (azimuthal) magnetization permanent magnet motor, some flux density waveforms are reported in Figure 10.2.

Figure 10.2(a) shows the air-gap flux density distribution due to the permanent magnet only. The average value is highlighted by the dashed line.

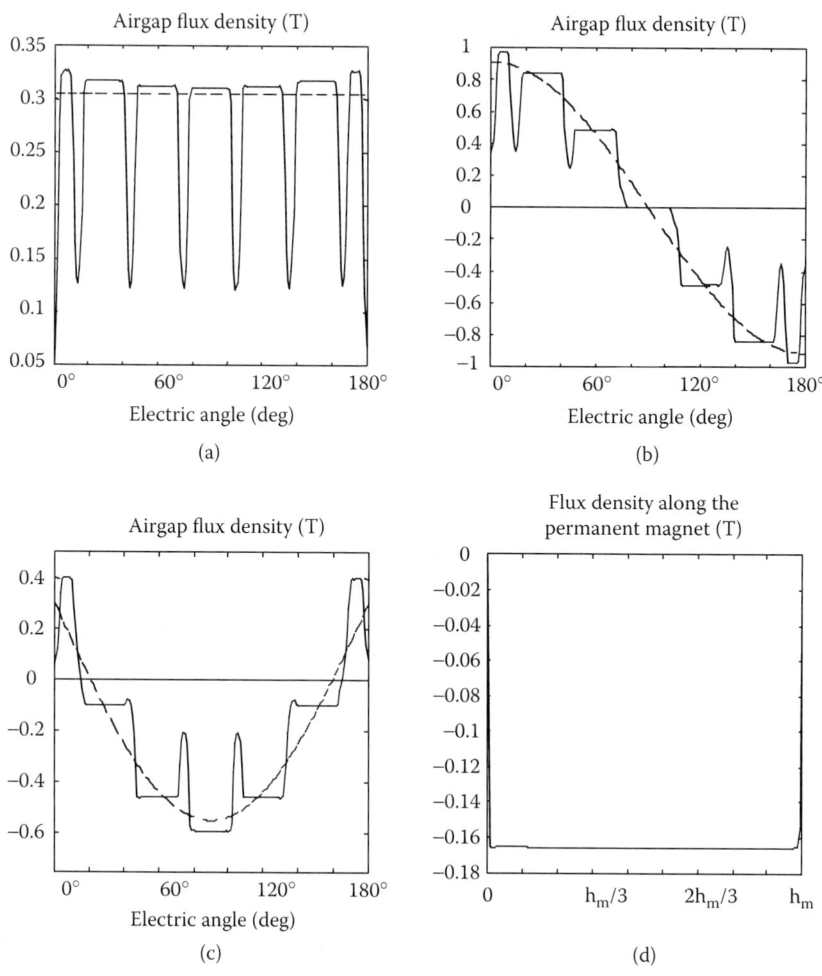

FIGURE 10.2
Air-gap flux density distributions due to the permanent magnet (a), d-axis current (b), q-axis current (c), and flux density distribution along the permanent magnet with negative d-axis current (d).

The effect of the slot opening is well manifested. Figure 10.2(b) and Figure 10.2(c) illustrate the air-gap flux density distribution due to the stator currents only (the permanent magnet has been demagnetized, i.e., $B_{res} = 0$), with only a d-axis current and only a q-axis current, respectively (the PM is unmagnetized). The fundamental harmonic of the distribution is highlighted by dashed lines. The slotting effect is again manifested, together with the effect of the winding distribution. Because of the flux barriers, when a d-axis current flows in the stator windings, some parts of the rotor assume a magnetic potential different from zero. These rotor parts are the iron "islands," bordered by the flux barriers and the air-gap. As a consequence, the maximum value of the flux density due to the d-axis current of Figure 10.2(c) is lower than that due to the q-axis current of Figure 10.2(b), although the stator is fed by the same current.

Finally, Figure 10.2(d) shows the flux density distribution on the surface of the permanent magnet, caused by a negative d-axis current only. The flux density is negative (demagnetizing) and is uniformly distributed on the whole surface. This check should be carried out at the maximum stator current and is necessary to ensure that the permanent magnets are not demagnetized irreversibly.

10.2 Characteristic Motor Parameters

The steady-state analysis of the interior permanent magnet motor is carried out in the synchronous rotating reference frame. The motor torque is expressed by Equation (9.16) and is reported as

$$T = \frac{3}{2} p\left(\Lambda_d I_q - \Lambda_q I_d\right) \tag{10.1}$$

The d- and q-axis steady-state voltages are

$$\begin{aligned} V_d &= RI_d - \omega\Lambda_q \\ V_q &= RI_q + \omega\Lambda_d \end{aligned} \tag{10.2}$$

Neglecting the resistive voltage drop, the voltage magnitude is

$$V^2 = \omega^2\left(\Lambda_d^{\;2} + \Lambda_q^{\;2}\right) \tag{10.3}$$

The flux linkages Λ_d and Λ_q are always expressed by using the flux linkage due to the permanent magnet Λ_m and the two axis inductances L_d and L_q. In general, they are dependent on the stator current, due to the saturation. These parameters represent the magnetic model of the motor and are easily obtained from the field solution.

10.2.1 Simplified Magnetic Model

The flux linkage due to the permanent magnet is obtained from the no-load simulation, as shown in Equation (9.3). Let Λ_m be its maximum value. It is considered constant and independent of the stator currents.

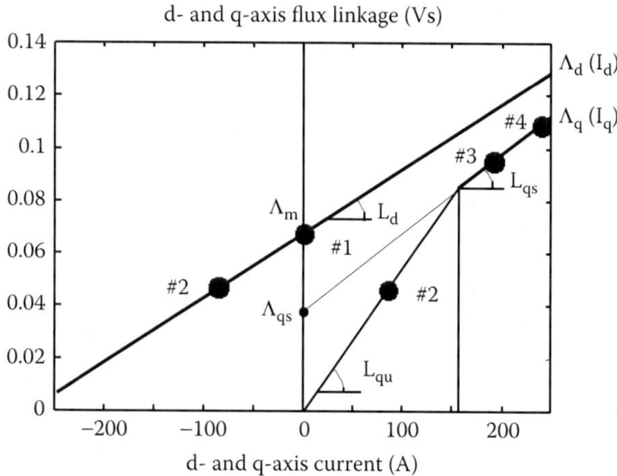

FIGURE 10.3
Simplified magnetic model of the interior permanent magnet motor.

For computing the d- and q-axis inductances, the same procedure described for the synchronous generator in Chapter 8 is adopted. They are indicated as L_d and L_q (eventually corrected to consider the three-dimensional effects, i.e., $L_d = L_\sigma + L_{dm}$ and $L_q = L_\sigma + L_{qm}$). As explained earlier, due to the particular rotor structure, it is $L_d < L_q$.

The d-axis inductance L_d is essentially constant, with the current. Then, the d-axis flux linkage becomes

$$\Lambda_d(I_d) = \Lambda_m + L_d I_d \tag{10.4}$$

On the contrary, the quadrature axis inductance L_q depends on the stator current, in particular on the q-axis current, which is $L_q(I_q)$. The q-axis flux linkage is $\Lambda_q = L_q(I_q) \cdot I_q$. For the sake of convenience, this relationship is approximated using piece-wise lines. Letting I_{qs} be the q-axis current at the beginning of the saturation, it is useful to define an unsaturated inductance L_{qu} with $I_q < I_{qs}$ and a saturated (differential) inductance L_{qs} with $I_q > I_{qs}$. The q-axis flux linkage becomes

$$\begin{aligned} \Lambda_q(I_q) &= L_{qu} I_q & I_q \le I_{qs} \\ &= \Lambda_{qs} + L_{qs} I_q & I_q \ge I_{qs} \end{aligned} \tag{10.5}$$

The magnetic characteristics that describe the dependence of the flux linkage on the currents are shown in Figure 10.3. An advantage of this magnetic model is that simple equations are employed. These are particularly advantageous in the design step of the interior permanent magnet motor or of the current control. Moreover, it is rapidly obtained from only four field solutions:

1. A first simulation with null stator currents allows the flux linkage due to the permanent magnet, Λ_m, to be obtained.
2. A second simulation, with low d- and q-axis current values (an indicative value could be $I_d = I_q$ equal to the 10% of the rated stator current), allows the two unsaturated inductances to be computed. Letting Λ_d and Λ_q be the flux linkages corresponding to the currents I_d and I_q, the two-axis inductances are

$$L_d = \frac{\Lambda_d - \Lambda_m}{I_d} \qquad L_{qu} = \frac{\Lambda_q}{I_q} \tag{10.6}$$

3. Two further simulations, with a high value of the q-axis currents only (for instance, equal to 80% and 100% of the rated current), allow the differential saturated inductance L_{qs} to be evaluated. If Λ'_q and Λ''_q are the q-axis flux linkages obtained with the two q-axis currents I'_q and I''_q, respectively, the inductance results

$$L_{qs} = \frac{\Lambda'_q - \Lambda''_q}{I'_q - I''_q} \tag{10.7}$$

Then, referring to Figure 10.3, the remaining parameters are given by

$$\Lambda_{qs} = \Lambda'_q - L_{qs} I'_q \tag{10.8}$$

$$I_{qs} = \frac{\Lambda_{qs}}{L_{qu} - L_{qs}}$$

Figure 10.3 refers to the simulation of an interior permanent magnet motor with radial magnetization, whose configuration is reported in Figure 10.1(b). The values obtained from the field solutions are highlighted by dots. The number close to each dot indicates the corresponding simulation step. The same values are reported in Table 10.1.

TABLE 10.1

Computed Values for the Interior Permanent Magnet Motor Model

	I_d (A)	I_q (A)	Λ_d (mVs)	Λ_q (mVs)	Motor Parameter
#1	0	0	67.5	—	Λ_m = 67.5 (mVs)
#2	–85	85	46.8	46.2	L_d = 0.243 (mH)
					L_{qu} = 0.543 (mH)
#3	0	192	—	95.05	
#4	0	240	—	108.9	L_{qs} = 0.288 (mH)

10.2.2 Cross-Coupling Effect

The cross-coupling is the magnetic interaction between the d- and the q-axis, which is essentially due to the saturation. This phenomenon may be interpreted as follows: the magnetic flux produced by the d-axis current saturates some zones of the machine, distorting the flux lines produced by the q-axis current, and vice versa.

The d- and q-axis flux linkages are considered to be functions of both the two axis currents, which is

$$\begin{aligned} \Lambda_d &= \Lambda_d(I_d, I_q) \\ \Lambda_q &= \Lambda_q(I_d, I_q) \end{aligned} \tag{10.9}$$

The flux linkages may be computed by integrating the magnetic vector potential or by processing the air-gap flux density distribution, as described in Chapter 8. An automatic procedure is also reported in Section 10.4.

In the case of linear magnetic circuit, the d- and q-axis flux linkages Λ_d and Λ_q vary linearly with the d- and q-axis currents, which is

$$\begin{aligned} \Lambda_d(I_d, I_q) &= \Lambda_d(I_d) = \Lambda_m + L_d I_d \\ \Lambda_q(I_d, I_q) &= \Lambda_q(I_q) = L_q I_q \end{aligned} \tag{10.10}$$

The dependence of the flux linkages on the currents, given in Equation (10.9) is reported in Figure 10.4. They refer always to the motor of Figure 10.1(b). It is worth noticing that the two flux linkages are not described by a single curve, but feel the effects of both the d- and q-axis currents. It is interesting to compare the result with that of Figure 10.3 (it refers to the same motor). The knowledge of the correct dependence of the flux linkages

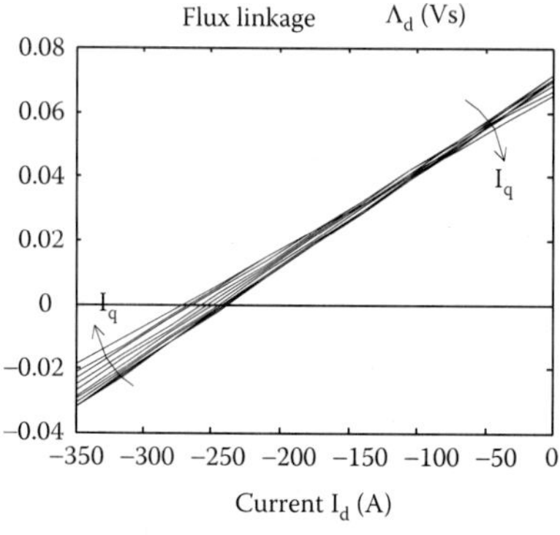

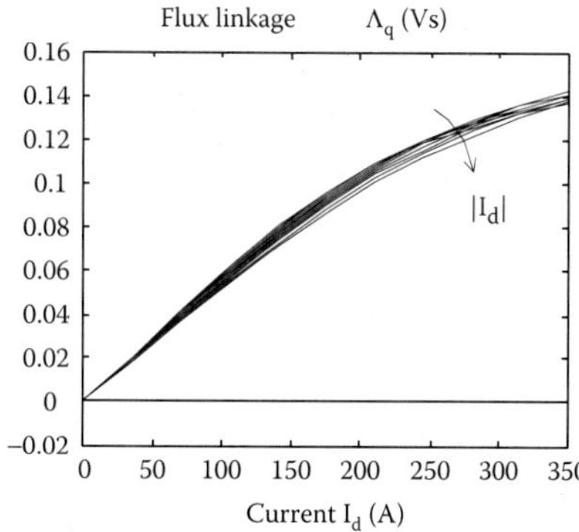

FIGURE 10.4
Flux linkages Λ_d and Λ_q vs. currents I_d and I_q.

on the currents is particularly useful for the design of electrical drive with very precise positioning and very high dynamic performance.

10.3 Torque-Speed Characteristic

Thanks to its anisotropic structure, the interior permanent magnet motor is well suited in applications requiring flux-weakening operations, that is, a torque-speed characteristic with a constant torque region at low speeds, and a constant power region at higher speeds. During steady-state operations, the limit operating region may be divided into two regions:

Constant torque region — At low speed, the motor operates at the rated current, say I_n, with I_d and I_q components chosen in such a way to obtain the maximum torque. It is the base torque T_b. This operating point can be maintained from null speed up to the base speed, say n_b, at which the terminal voltage reaches the rated value V_n.

Constant power region — The motor operates at speed higher than n_b with a demagnetizing (i.e., negative) d-axis current. Such a current weakens the flux linked by the stator winding, so that it is possible to increase the motor speed without exceeding the nominal voltage, as indicated by Equation (10.3) where $\omega = (2\pi/60)np$. On the other hand, a decrease of the torque is achieved, since the I_d and I_q currents move from the operating point of maximum torque. The operating region presents a torque decreasing with the speed and is so called the "constant power" region, even though it only approximates this behavior. Referring to the operating limit, the current phasor $I_d + jI_q$ maintains a constant magnitude and describes a circular trajectory, always increasing the negative value of the d-axis current. A wide operating region is obtained if the interior permanent magnet motor exhibits $\Lambda_m \approx L_d I_n$.

In order to individuate the operating regions by means of finite element analysis, the following procedure is suggested:

1. The value of the rated current I_n is fixed. Some simulations are carried out with different combinations of the base operating point defined by the current phasor $I_d + jI_q$. From the field solutions the flux linkages Λ_d and Λ_q are computed, and then the torque from Equation (10.1). The maximum torque T_b and the corresponding operating point are found. Because $L_d < L_q$, then $I_d < 0$ has to be imposed. In addition, it can be verified that the maximum torque is found with $I_q > |I_d|$.

2. Once the maximum torque operating point has been found, imposing the nominal voltage $V = V_n$ in Equation (10.3), the base speed is computed as $n_b = (60/2\pi)\omega/p$, where

$$\omega = \frac{V_n}{\sqrt{\Lambda_d^2 + \Lambda_q^2}} \tag{10.11}$$

   The constant torque region is then completely defined by T_b and n_b.

3. The d- and q-axis current components, I_d and I_q, are modified, increasing the negative value of I_d and decreasing the value of I_q, but always with a constant magnitude equal to I_n. From the corresponding field solution, the flux linkages Λ_d and Λ_q are computed, and thus the torque T from Equation (10.1) and the speed ω from Equation (10.11), and thus $n = (60/2\pi)\omega/p$. The values of torque T and speed n correspond to a point of the mechanical characteristic, in the decreasing torque region.
4. The procedure described in the step 3 is repeated until the required maximum speed is reached or, alternatively, when a null torque is reached.

10.3.1 Example

An interior permanent magnet motor is designed for an application requiring a constant torque region with base torque T_b = 10 Nm and base speed n_b = 2000 rpm, and a constant power region up to the maximum speed n_{fw} = 6000 rpm, at which the torque should be T_{fw} = 3.33 Nm.

The motor structure and the main data are reported in Figure 10.5. A single-layer, full-pitched winding is used. The rotor exhibits an anisotropy ratio (also called the saliency ratio) $\xi = L_q/L_d$ almost equal to 4, confirming the high anisotropy of this kind of rotor configuration.

Figure 10.6 shows the flux lines in one section of the motor pole. Figure 10.6(a) refers to the no-load operation. The flux leakage through the rotor

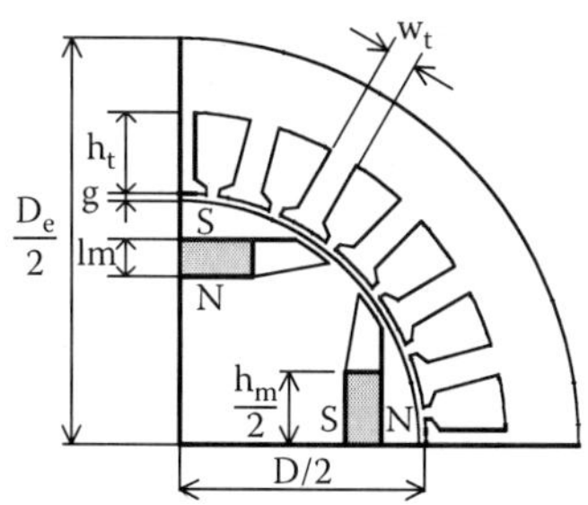

FIGURE 10.5
Motor structure and main data.

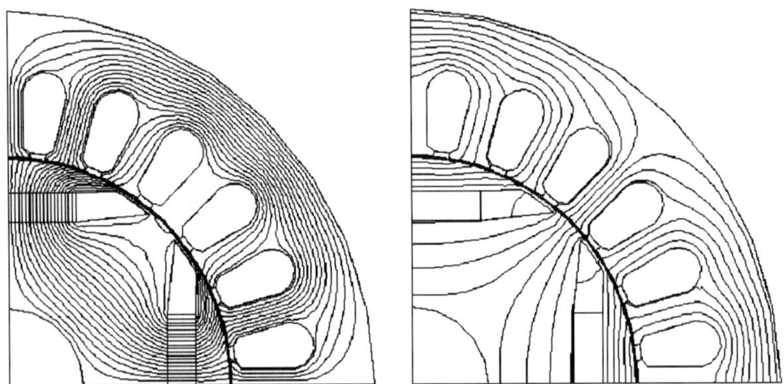

FIGURE 10.6
Flux lines with only permanent magnet (a) and only q-axis currents (b).

magnetic bridges is evident. The corresponding air-gap flux density distribution is shown in Figure 10.7(a1). Figure 10.7(a2) and Figure 10.7(a3) show the air-gap flux density distribution corresponding to the d-axis current equal to –30 A and –70 A, respectively, highlighting their demagnetizing effect.

Figure 10.6(b) shows the flux lines with only a q-axis current (the permanent magnet is assumed to be demagnetized, i.e., $B_{res} = 0$). The flux lines enter in the rotor, centered along the interpolar axis and flow by the side of the barriers. The air-gap flux density distribution with only a q-axis current $I_q = 30$ A is shown in Figure 10.7(b1). Figure 10.7(b2) shows the air-gap flux density distribution due to both permanent magnet and $I_q = 30$ A. Finally, Figure 10.7(b3) shows the flux density distribution increasing the q-axis current to $I_q = 70$ A.

The motor torque computed from the field solutions at different currents is shown in Figure 10.8. In the same figure the experimental data on a motor prototype are reported, confirming a good agreement between simulated and tested values.

Since the rated dc voltage is 48 V, reduced to 42 V for considering the voltage drop, a voltage limit $V_n = 24.2$ V (peak value of the sinusoidal waveform). According to the procedure illustrated above, the mechanical characteristics of Figure 10.9 are obtained, with different currents I_n and given voltage V_n.

The dashed line in Figure 10.9 confirms that the requirements are satisfied with a rated current of 70 A, peak value. The interior permanent magnet motor exhibits a constant power region in a speed range of 1:3.

10.4 Algorithm for an Automatic Computation

The procedures illustrated above are well suited to be implemented in a code for the automatic computation of the motor characteristics. These procedures interact with the finite element algorithms, leading each step of the simulation.

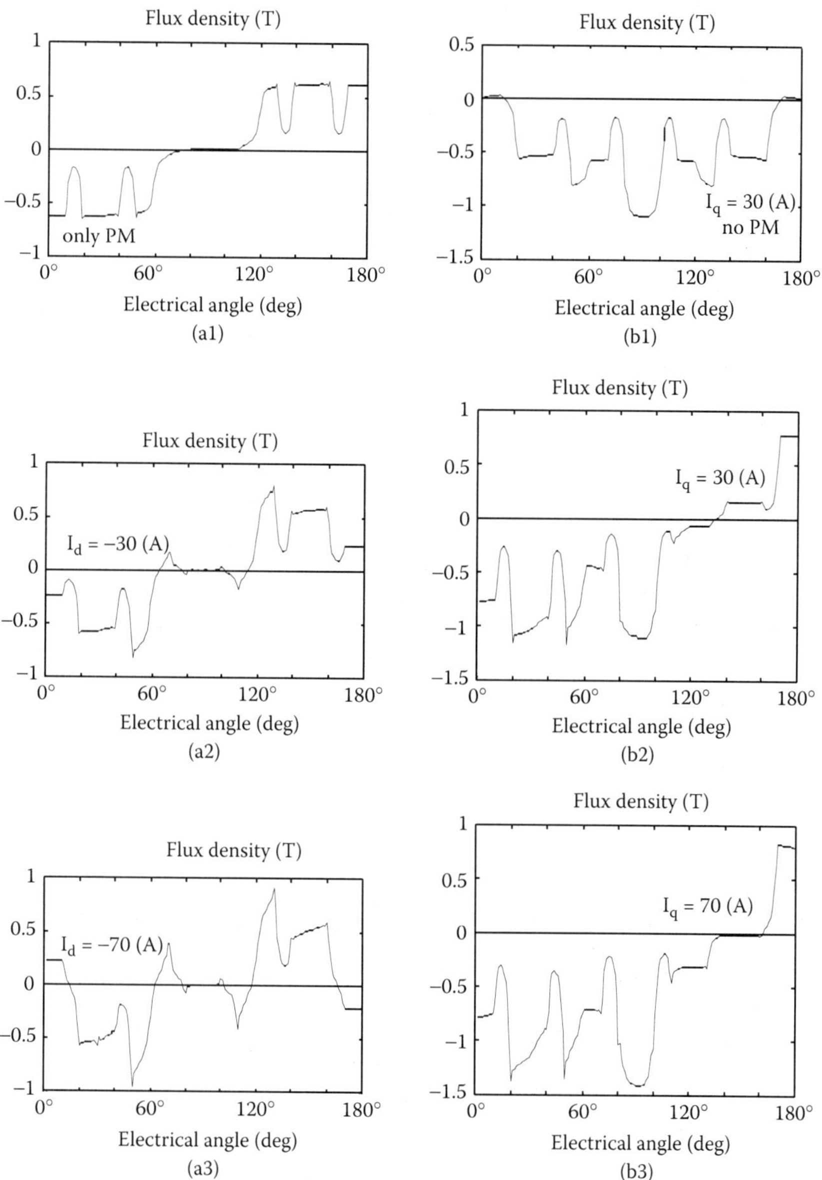

FIGURE 10.7
Effect of different currents on the air-gap flux density distribution.

At first, the code is reported for the automatic computation of the d- and q-axis flux linkages, corresponding to a generic current value.

In the example, the motor structure of Figure 10.10 is considered. A double-layer, one-slot chorded winding is adopted. The portion used in the simulation is that shown in Figure 8.5(b) in Chapter 8.

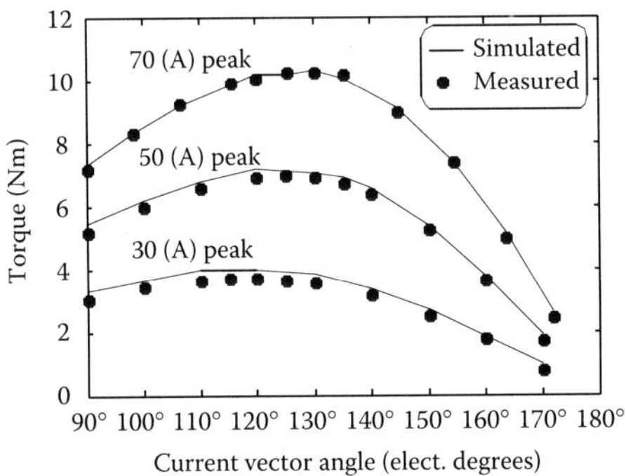

FIGURE 10.8
Motor torque versus the angle of the current vector (I_d, I_q).

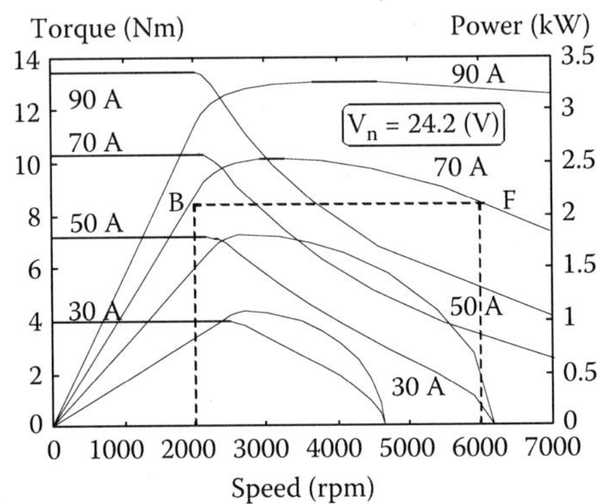

FIGURE 10.9
Torque and power versus speed, at different stator current values.

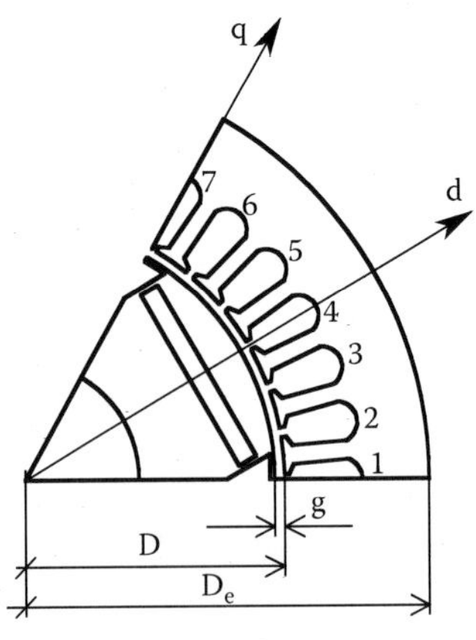

FIGURE 10.10
Structure of the interior permanent magnet motor and references.

Some preliminary variables are initially defined:

```
' Preliminary definitions
  DIM Bn1(180), Bn2(180), Bn3(180)
  pi = 3.1415926536
  dtheta = pi / 180
'-----------------------------------------
' Current values in the simulation
  Id = -113
  Iq = 185
```

Some FUNCTIONs are employed to assign the currents, the material characteristics, and the boundary conditions. Then the field problem is solved. From the field solution the magnetic quantities are processed. Let us assume that the air-gap flux density distribution is stored in the file `FileInduz$`, and the average magnetic vector potential within the stator slots are stored in files `FilePotential$`, one for each slot.

```
' Name of the file for flux density distribution
    FileInduz$ = "bgload"
' Name of the file for average magnetic vector potential
    FilePotential$ = "MagnVectPot"
' Assignment of the currents
    CALL AssignCurrent (Id, Iq)
' Assignment of the iron permeability
    CALL AssignMu
' Assignment of the permanent magnet parameters
    CALL AssignPM (mu, Br, ang)
' The field problem is solved
    CALL Solution
' the flux density distribution is stored in the file
    CALL Induction (FileInduz$)
' the flux linkage is computed by integrating Az
    CALL GlobalFlux
```

The FUNCTION that is used to assign the current in the stator slots is reported next. This FUNCTION is based on the distribution of the coils as shown in Figure 10.10(b). In order to impose the current components I_d and I_q, since the polar axis has been fixed corresponding to the a-phase axis, so that $\vartheta = 0$, the three-phase currents are given by

$$I_a = I_d \qquad I_b = -\frac{1}{2}\left(I_d - \sqrt{3}I_q\right) \qquad I_c = -\frac{1}{2}\left(I_d + \sqrt{3}I_q\right) \tag{10.12}$$

At this point, the current in the slot can be assigned as

$$\begin{aligned} I_1 &= \frac{n_{qs}}{4}\left(-I_a - I_a\right) & I_2 &= \frac{n_{qs}}{2}\left(-I_a + I_c\right) & I_3 &= \frac{n_{qs}}{2}\left(I_c + I_c\right) \\ I_4 &= \frac{n_{qs}}{2}\left(I_c - I_b\right) & I_5 &= \frac{n_{qs}}{2}\left(-I_b - I_b\right) & I_6 &= \frac{n_{qs}}{2}\left(-I_b + I_a\right) \\ I_7 &= \frac{n_{qs}}{4}\left(I_a + I_a\right) \end{aligned} \tag{10.13}$$

where n_{qs} is the number of series conductors per slot, given by n_q/n_{pp}, with n_{pp} the number of parallel paths (see Chapter 8). The current I_1 and I_7 are divided by 4, since only half a slot is simulated. Let N be the number of series conductors per phase, 2p the number of poles, and q_s the number of slots per pole per phase. Then, it results in

$$n_{qs} = \frac{N}{2pq_s} = \frac{3N}{Q} \tag{10.14}$$

```
SUB AssignCurrent (Id, Iq)
Ia = Id
Ib = -0.5 * (Id - SQR(3) * Iq)
Ic = -0.5 * (Id + SQR(3) * Iq)
Icava1 = (nqs / 4) * (-Ia - Ia)
Icava2 = (nqs / 2) * (-Ia + Ic)
Icava3 = (nqs / 2) * (Ic + Ic)
Icava4 = (nqs / 2) * (Ic - Ib)
Icava5 = (nqs / 2) * (-Ib - Ib)
Icava6 = (nqs / 2) * (-Ib + Ia)
Icava7 = (nqs / 4) * (Ia + Ia)
```

After the field solution has been obtained, the results stored in the files are processed. At first the air-gap flux density distribution is managed. The distribution of the air-gap flux density is given by 180 values; it is read from the file `FileInduz$`. These 180 values are stored in the variable `Bn1(nnn)`. The two fundamental harmonics are computed, corresponding to the direct axis ($\cos\vartheta_r$) and the quadrature axis ($\sin\vartheta_r$) as depicted by Equation (8.16) and Equation (8.24). The d- and q-axis flux linkages Λ_d and Λ_q are obtained as depicted by Equation (8.18) and Equation (8.25).

```
' Processing of the obtained results
'
' 1: Method of the airgap flux density
' Reading the file
    OPEN FileInduz$ FOR INPUT ACCESS READ AS #1
    FOR nnn = 1 TO 180
        INPUT #1, a, b, c
' The flux density value is in the third column
            Bn1(nnn) = c
    NEXT nnn
    CLOSE #1
' The two-axis fundamental components are computed
    Bid = 0
    Biq = 0
    FOR nnn = 1 TO 180
        theta = dtheta * nnn
        Bid = Bid + Bn1(nnn) * SIN(theta) * dtheta
        Biq = Biq - Bn1(nnn) * COS(theta) * dtheta
    NEXT nnn
    B1d = Bid * 2 / pi
    B1q = Biq * 2 / pi
    F1d = (B1d * D * L / p) * (kw * n / 2)
    F1q = (B1q * D * L / p) * (kw * n / 2)
```

The second method is based on the processing of the magnetic vector potential A_z on each coil section. As in Equation (8.2), the average magnetic vector potential in each q-th slot is computed as

$$A_{slot(q)} = \frac{1}{S_q} \int_{S_q} A_z \, dS \tag{10.15}$$

and it is stored in the files `FilePotential$`, one for each slot. Then these values are read from the files, for each slots of the domain (7 in the present example) and assigned to the variables `fslot(nnn)`.

From the distribution of the coil sections in the slots and from the symmetry conditions, the three-phase flux linkages Λ_a, Λ_b, and Λ_c are computed. In the specific example they are given by

$$\begin{aligned}
\Lambda_a &= 2p\frac{n_{qs}}{2}\left[-(\Lambda_1 + \Lambda_2) + (\Lambda_6 + \Lambda_7)\right] \\
\Lambda_b &= 2p\frac{n_{qs}}{2}\left[-(\Lambda_4 + 2\Lambda_5 + \Lambda_6)\right] \\
\Lambda_c &= 2p\frac{n_{qs}}{2}\left[(\Lambda_2 + 2\Lambda_3 + \Lambda_4)\right]
\end{aligned} \tag{10.16}$$

The two-axis flux linkages Λ_d and Λ_q are then computed, using the transformation $T_{abc/dq}$ (see the Appendix in Chapter 8). Since $\vartheta = 0$, the d- and q-axis flux linkage components are given by

$$\Lambda_d = \frac{2}{3}\left(\Lambda_a - \frac{\Lambda_b + \Lambda_c}{2}\right) \qquad \Lambda_q = \frac{1}{\sqrt{3}}\left(\Lambda_b - \Lambda_c\right) \tag{10.17}$$

Then it results:

```
' Processing of the obtained results
'
' 2: Method of the flux linkages
'  Reading the files corresponding to the 7 slots
    FOR nnn = 1 TO 7
        File$ = FilePotential$ +
                 + LTRIM$(STR$(nnn)) + ".dat"
        OPEN file$ FOR INPUT ACCESS READ AS #5
          INPUT #5, fcava(nnn)
          fcava(nnn) = fcava(nnn)
        CLOSE #5
    NEXT nnn
' The three-phase flux linkages are computed
    Fa0 = (fcava(1)+fcava(2))+(fcava(6)+fcava(7))
    Fb0 = -(fcava(4) + 2 * fcava(5) + fcava(6))
```

```
      Fc0 = (fcava(2) + 2 * fcava(3) + fcava(4))
      Fa = 2 * p * nqs / 2 * L * Fa0
      Fb = 2 * p * nqs / 2 * L * Fb0
      Fc = 2 * p * nqs / 2 * L * Fc0
' The two-axis flux linkages are computed
      Fd = 2 / 3 * (Fa - (Fb + Fc) / 2)
      Fq = 1 / SQR(3) * (Fb - Fc)
```

The flux lines corresponding to one pole of the machine are shown in Figure 10.11, referring to the no-load and the full-load operation, respectively. Some results obtained by applying the two proposed methods are reported in Figure 10.12. Figure 10.12 shows a good agreement between the two computation methods. In addition, the results are compared with experimental results obtained by feeding the motor at industry frequency 50 Hz and different loads.

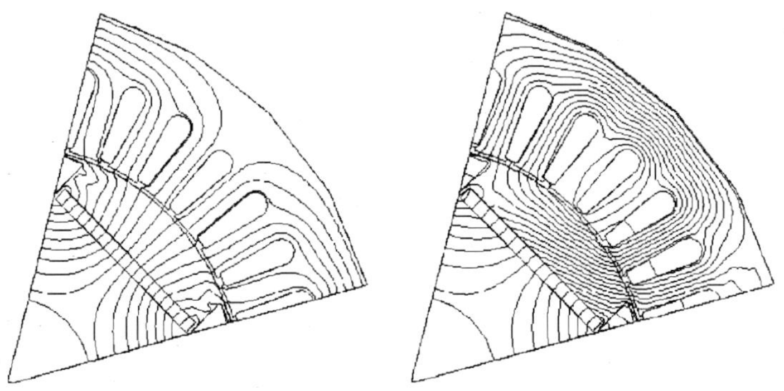

FIGURE 10.11
Flux lines during no-load (a) and the full-load (b) operations.

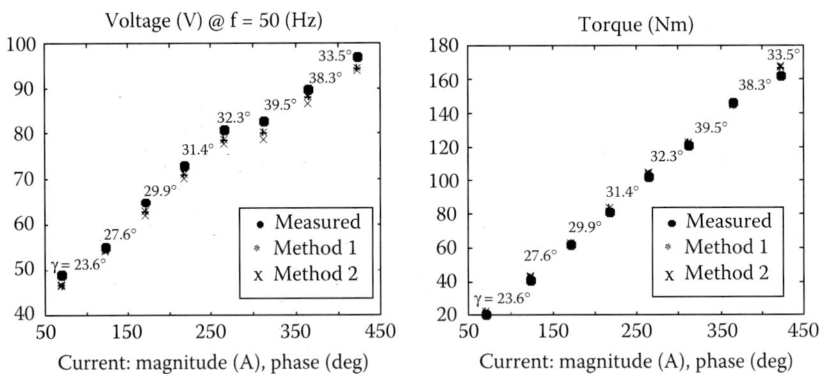

FIGURE 10.12
Comparison between simulated and measured results.

10.5 Synchronous Reluctance Motors

The synchronous reluctance motor may be considered as a particular case of the interior permanent magnet motor, obtained by taking the permanent magnet away from the rotor. The synchronous reluctance rotor is similar to the interior permanent magnet rotor, but generally it is designed with several barriers per pole. This is because the motor torque is only due to the rotor anisotropy, so that the saliency ratio has to be as high as possible.

Two reluctance motor configurations are shown in Figure 10.13. The first one refers to a transversally laminated rotor, and the second one refers to an axially laminated rotor. The commonly used notation for the d- and the q-axis of synchronous reluctance motor is adopted in Figure 10.13, that is, the d-axis is posed along the higher permeance path. This notation is opposite to the notation adopted above for the interior permanent magnet motor. In the reluctance motors, $L_d > L_q$ is obtained and the saliency ratio is defined by $\xi = L_d/L_q$.

In the case of the transversally laminated reluctance motor, shown in Figure 10.13(a), the rotor is formed by a stack of stamped laminations. The flux barriers are directly formed during the stamping process. There are a number of barriers per pole in the range between 2 and 5. The rotor structure is supported by thin magnetic bridges, whose thickness is designed by means of mechanical considerations. These magnetic bridges are saturated by the q-axis flux. This rotor configuration shows a saliency ratio up to 10.

The axially laminated reluctance rotor, illustrated in Figure 10.13(b), is formed by staking laminations and nonmagnetic insulating layers, which

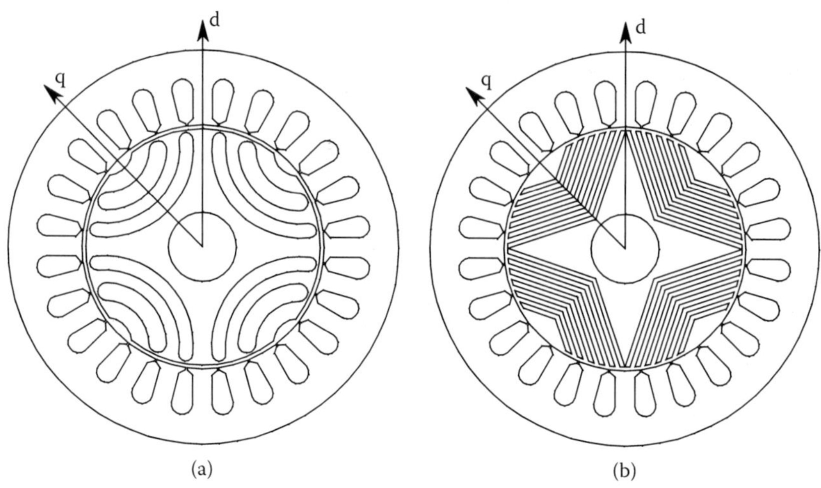

FIGURE 10.13
Synchronous reluctance motor configurations.

are bended and placed along the axial direction. The structure is supported by nonmagnetic spacings. The shaft is nonmagnetic as well. This rotor configuration shows a saliency ratio more than 20, in linear conditions. However, the rotor assembly is difficult and iron rotor losses occur, caused by the flux fluctuation in the rotor laminations due to the stator slotting.

Sometimes, for increasing the power factor or the performance in the constant power region, permanent magnets are inset within the flux barriers. They are plastic bonded low-energy permanent magnets, which are shaped easily.

10.5.1 The Transversally Laminated Reluctance Motor

In the simulation of the reluctance motor, particular care has to be taken with the rotor. The mesh of the domain should be mainly refined in the rotor and in the air-gap. In addition, attention should be given to the magnetic bridges, which work extremely saturated.

The motor may be fed by only a d-axis or only a q-axis current component, reducing the analysis to only half a pole-piece. Otherwise, the motor may be fed by whatever couple of current components I_d and I_q, so that the analysis is on one pole of the motor. The latter analysis allows a more accurate magnetic model to be developed, considering the cross-coupling effect, which greatly affects the reluctance motor performance.

As an example, Figure 10.14 shows the field lines of one pole of the synchronous reluctance motor with different combinations of the currents I_d and I_q. The a-phase axis has been chosen in correspondence with the direct axis.

Figure 10.14(a) shows the flux lines when the motor is fed by d-axis current only (I_d = 10 A, I_q = 0 A), imposing the three-phase currents I_a =10 A, I_b = –5 A, I_c = –5 A.

In Figure 10.14(b) the motor is fed by a q-axis current only (I_d = 0 A, I_q = 10 A), imposing the three-phase currents I_a = 0 A, I_b = 8.66 A, I_c = –8.66 A.

At last, Figure 10.14(c) refers to the motor fed by a current phasor with a 45 electrical degree ($I_d = I_q = 5\sqrt{2}$ A), which is I_a = 7.07 A, I_b = 2.59 A, I_c = –9.66 A.

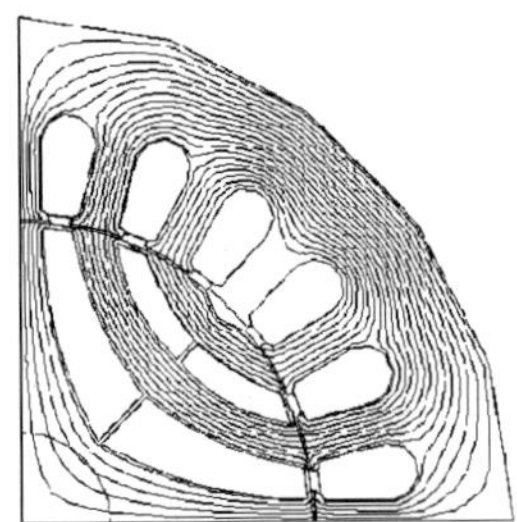
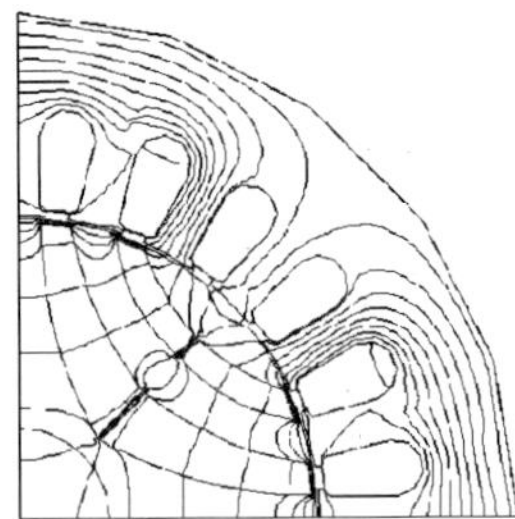
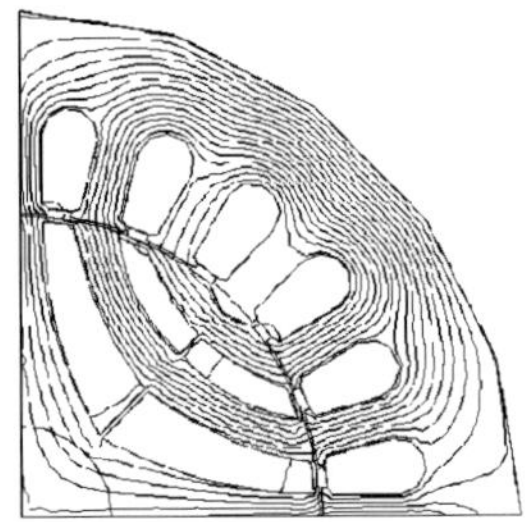

FIGURE 10.14
Flux lines in a synchronous reluctance motor fed by only d-axis current (a), only q-axis current (b), and two-axis current (c).

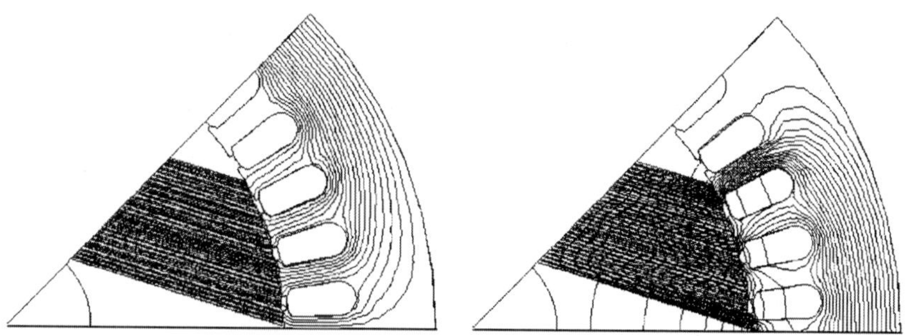

FIGURE 10.15
Flux plots with d-axis current only (a) and q-axis current only (b).

10.5.2 The Axially Laminated Reluctance Motor

Because of the several laminations that are thin and divided one to the other, the finite element simulation requires a mesh with a very large number of elements. For this reason, it is necessary to reduce the analyzed domain as much as possible, making use of any symmetry of the machine. In the following, only a quarter of pole is analyzed, feeding the stator with only d-axis or only q-axis current. As a consequence, the flux linkages Λ_d and Λ_q are computed separately.

With d-axis current only, the magnetic vector potential A_z is null along the direct axis. Homogeneous Dirichlet's boundary conditions are assigned along the d-axis, and homogeneous Neumann's boundary conditions along the q-axis, along which the flux density vector has only normal components.

In the same way, with q-axis current only, Dirichlet's and Neumann's boundary conditions are assigned along the q-axis and the d-axis, respectively.

An example is reported in Figure 10.15. A three-phase, four-pole 36-slot motor is considered with a single-layer, full-pitched winding. The rotor is composed by 58 magnetic laminations with thickness $t_l = 0.328$ mm spaced by an insulating sheet with thickness $t_i = 0.194$ mm.

Due to the heavy simulation of such a machine, the analysis may be limited to the rotor only. The stator is substituted by an infinitesimal conducting sheet placed along the periphery of the bore diameter D. In this sheet a surface current density is forced. It could be a d-axis surface current density as

$$J_{sd}\left(\vartheta_r\right) = \frac{mN\hat{I}}{\pi D}\sin(\vartheta_r) \tag{10.18}$$

or a q-axis surface current density as

$$J_{sq}\left(\vartheta_r\right) = -\frac{mN\hat{I}}{\pi D}\cos(\vartheta_r) \tag{10.19}$$

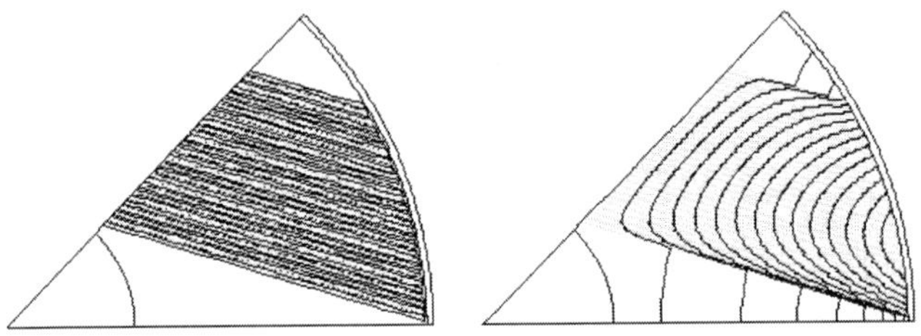

FIGURE 10.16
Flux lines with (a) d-axis and (b) q-axis surface current density distribution.

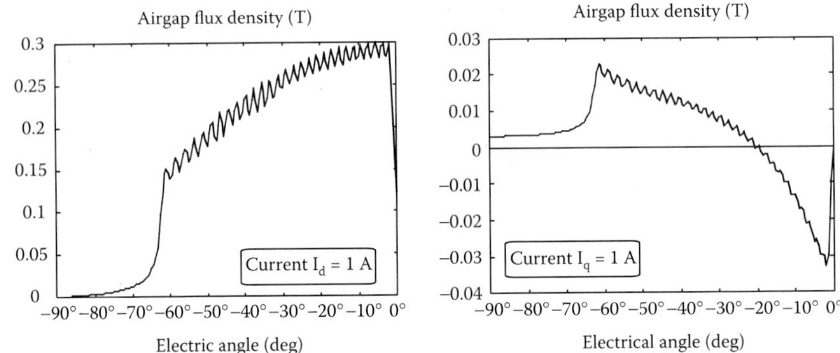

FIGURE 10.17
Air-gap flux density distributions in simulations of rotor only.

where m is the number of phases, N is the number of series conductors, and ϑ_r is the electrical angle referred to the rotor d-axis (as shown in Figure 8.19).

Figure 10.16 shows the flux lines with a d-axis and q-axis current distribution, respectively. In the first case, the flux lines are parallel to the laminations, while in the second case they are normal to the laminations. The corresponding air-gap flux density distributions are shown in Figure 10.17.

References

1. V.B. Honsinger, "The field and parameters of interior type AC permanent magnet machines," in *IEEE Trans. on PAS*, 101, 1981, pp. 867–876.
2. B. Sneyers, D.W. Novotny, T.A. Lipo, "Field weakening in buried permanent magnet AC motor drives," in *IEEE Trans. on Ind. Appl.*, vol. 21, no. 2, pp. 398–407, March/April 1985.
3. T.M. Jahns, G.B. Kliman, T.W. Neumann, "Interior permanent-magnet synchronous motors for adjustable-speed drives," in *IEEE Trans. on Ind. Appl.*, vol. 22, no. 4, pp. 738–747, July/August 1986.
4. R.F. Schiferl and T.A. Lipo, "Power capability of salient pole permanent magnet synchronous motors in variable speed drive applications," in *IEEE Trans. on Ind. Appl.*, vol. 26, no. 1, pp. 115–123, January/February 1990.

5. S. Morimoto, Y. Takeda, T. Hirasa, K. Taniguchi, "Expansion of operating limits for permanent magnet motor by current vector control considering inverter capability," in *IEEE Trans. on Ind. Appl.*, vol. 26, no. 5, pp. 866–871, September/October 1990.
6. A. Fratta, A. Vagati, F. Villata, "Design Criteria of an IPM Machine Suitable for Field-Weakened Operation," *Proc. of International Conference on Electrical Machines*, pp. 1059–1065, 1990.
7. A.K. Adnanes and T.M. Undeland, "Optimum torque performance in PMSM drives above rated speed," in *Ind. Appl. Society Annual Meeting Records*, pp. 169–175, 1991.
8. W.L. Soong and T.J.E. Miller, "Theoretical limitations to the field-weakening performance of the five classes of brushless synchronous AC motor drive," in *Proc. of Electrical Machines and Drives Conference*, pp. 127–132, Oxford (UK), 1993.
9. B.J. Chalmers, L. Musaba, D.F. Gosden, "Variable Frequency Synchronous Motor Drives for Electric Vehicles," *IEEE Trans. on Ind. Appl.*, vol. 32, no. 4, 1996, pp. 896–903.
10. E. Chiricozzi, F. Parasiliti, M. Villani, T. Higuchi, J. Oyama, E. Yamada, "Buried PM Synchronous Motor Design Optimization," in *2nd EPE Chapter Symposium on Electric Drive Design and Applications*, Nancy (F), June 1996, pp. 41–45.
11. N. Bianchi, S. Bolognani, "Parameters and Volt-Ampere Ratings of a Synchronous Motor Drive for Flux-Weakening Applications," in *IEEE Trans. on Power Electronics*, vol. 12, no. 5, September 1997, pp. 895–903.
12. N. Bianchi, S. Bolognani, B.J. Chalmers, "Comparison of Different Synchronous Motor Drives for Flux-Weakening Applications," in *Proc. of International Conference on Electrical Machine, ICEM '98*, Istanbul, Turkey, 2–4 September 1998, pp. 946–951.
13. N. Bianchi, S. Bolognani, "Magnetic Models of Saturated Interior Permanent Magnet Motors based on Finite Element Analysis," in *IEEE Industrial Applications Conference Records, IAS '98*, St. Louis, 12–15 October 1998, vol. 1, pp. 27–34.
14. B.J. Chalmers, R. Akmese, L. Musaba, "Design and field-weakening performance of permanent-magnet/reluctance motor with two-part rotor," in *IEEE Proc. Pt. B*, vol. 145, no. 2, 1998, pp. 133–139.
15. D.M. Ionel, J.F. Eastham, T.J.E. Miller, E. Demeter, "Design Considerations for Permanent Magnet Synchronous Motors for Flux-Weakening Applications," *IEEE Proc. Pt. B*, vol. 145, no. 5, 1998, pp. 435–440.
16. N. Bianchi, S. Bolognani, F. Parasiliti, M. Villani, "Prediction of Overload and Flux-Weakening Performance of an IPM Motor Drive: Analytical versus Finite Element Approach," in *Proc. of 8th European Conference on Power Electronics and Applications, EPE99*, Lausanne, 7–9 September 1999, DS 1.6, topic 5, pp. 1–8.
17. N. Bianchi, S. Bolognani, B.J. Chalmers, "Design Considerations for a PM Synchronous Motor with Rotor Saliency for High Speed Drives," in *IEEE Industrial Applications Conference Records, IAS '99*, Phoenix, Arizona, 3–7 October 1999, vol. 1, pp. 117–124.
18. N. Bianchi, S. Bolognani, M. Zigliotto, "Design and Development of a PM Synchronous Motor Drive for an Electrical Scooter," in *Proc. of ICEM 2000 International Conference*, Helsinki, Finland, 28–30 August 2000.
19. N. Bianchi, S. Bolognani, M. Zigliotto, "High Performance PM Synchronous Motor Drive for an Electrical Scooter," in *IEEE IAS Annual Meeting Records*, Rome, Italy, 8–12 October 2000.

20. A. Fratta and A. Vagati, "A Reluctance Motor Drive for High Dynamic Performance Applications," in *IEEE IAS Annual Meeting*, Atlanta, pp. 295–302, 1987.
21. R.E. Betz, "Control of Synchronous Reluctance Machines," in *Ind. Appl. Society Annual Meeting Records*, pp. 456–462, 1991.
22. R.E. Betz, M. Javanovic, R. Lagerquist, T.J.E. Miller, "Aspects of the control of synchronous reluctance machines including saturation and iron losses," in *Ind. Appl. Society Annual Meeting Records*, pp. 456–463, 1992.
23. A. Vagati, G. Franceschini, I. Marongiu, G.P. Troglia, "Design criteria of high performance synchronous reluctance motors," in *Ind. Appl. Society Annu. Meet.*, 1992, vol. I, pp. 66–73.
24. N. Bianchi, S. Bolognani, "Synchronous Reluctance Motor Drives: Overload and Flux-Weakening Performance Taking into Account Iron Saturation," in *Proc. of International Conference on Electrical Machines, ICEM '96*, Vigo (Spagna), 10–12 September 1996, vol. I, pp. 360–365.
25. N. Bianchi, S. Bolognani, "Parameters and Volt-Ampere Ratings of a Synchronous Reluctance Motor Drives for Flux-Weakening Applications Taking into Account Iron Saturation," in *Proc. of Conference on Power Electronics and Applications, EPE '97*, Trondheim (Norvey), 8–10 September 1997, vol. 3, pp. 631–636.
26. B.J. Chalmers, L. Musaba, "Design and Field-Weakening Performance of a Synchronous Reluctance Motor with Axially-Laminated Rotor," *IEEE Ind. Appl. Society Annual Meeting Conference Records*, 1997, pp. 271–278.
27. N. Bianchi, B.J. Chalmers, "Effect of the Distribution of the Laminations in an Axially Laminated Relactance Motor," in *Proc. of IEEE 9th International Conference of Electrical Machines and Drives, EMD '99*, Canterbury, UK, 1–3 September 1999, pp. 376–380.
28. N. Bianchi and T.M. Jahns, Eds., "Design, Analysis and Control of Interior PM Synchronous Machines," Tutorial Course Notes, IEEE IAS Meeting, Seattle, WA, Oct. 3, 2004, CLEUP, Padova.

11

Self-Starting Single-Phase Synchronous Motors

This chapter deals with the analysis of a self-starting single-phase synchronous motor. At first, finite element magnetostatic analysis is carried out to obtain the motor parameters as a function of rotor position and current. Then, these parameters are used in the analysis of dynamic performance of the motor.

In this particular case, a single-phase permanent magnet micro-motor is analyzed. It is characterized by an anisotropic stator that is necessary to develop a starting torque. Its stationary and dynamic operations are simulated. The algorithms for the prediction of the dynamic performance are reported.

11.1 Introduction

Electrical motors of very small dimensions are used in many devices, especially in civil applications. Since an alternative voltage is commonly available, the motors are often ac motors. Among them, the most utilized is the single-phase induction motor, with an auxiliary winding or a shaded pole to allow the rotor starting. In some applications, where a synchronous speed is required, hysteresis motors, reluctance motors, or permanent magnet motors are adopted. Additionally, when reduced dimensions are explicitly required, the use of permanent magnets allows miniaturized motors to be obtained.

Apart from the hysteresis motor, the main problem of the synchronous motors is the rotor starting. In order to obviate such a problem, a conducting cage may be inset in the rotor, so that an asynchronous starting is achieved. Also, an anisotropic stator may be designed, so as the rest positions of the rotor are different according to the stator current. The analysis and the design of such motors are laborious and require remarkable work.

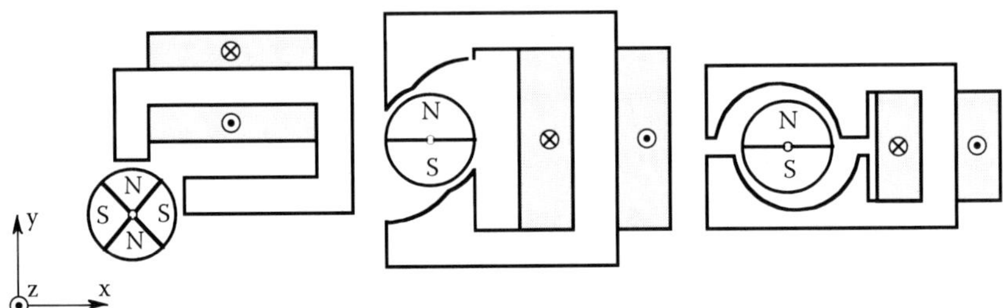

FIGURE 11.1
Structure of the single-phase permanent magnet motor.

11.2 Definition of the Motor Model

Some structures of the single-phase permanent magnet motor are shown in Figure 11.1. The stator core is obtained by magnetic laminations. The stator coil is wound around the stator core. A planar symmetry is recognized, and a two-dimensional analysis is carried out.

If the motor does not show any other symmetry axis, as easily observed from Figure 11.1, then there are further feasible simplifications of the domain. In Figure 11.1(a), the rotor axis is not aligned with the stator laminations, while in Figure 11.1(b) and Figure 11.1(c) there is an asymmetry between the rotor and stator poles.

Such a misalignment is required for achieving the starting torque. The rest position of the rotor with null currents is different from that position when a stator current feeds the coil. In other words, the magnetic field strength due to the stator current produces a force that tends to rotate the permanent magnet from its natural rest position.

The analysis of the stationary performance of the motor is carried out by means of magnetostatic field solutions. Neglecting the edge effects, the frontal section is drawn in the plane (x, y). The results of the magnetostatic field solution are then processed for studying the dynamic behavior of the motor.

Silicon-steel laminations are used for stator core. Particular care is taken with modeling the permanent magnet: the characteristic and the magnetization directions of the various rotor sectors have to be exactly defined.

Ferrite permanent magnet composites are usually employed. Plastic bonded permanent magnet materials reduce dimensions, keeping restrictive mechanical tolerances. In addition, the brittleness is lower than in sintered permanent magnet materials.

The B-H curve is essentially linear in the second quadrant, as seen in Figure 9.4. It is characterized by the residual flux density B_{res} and the coercive field strength H_c, or alternatively by the differential relative permeability μ_r. The rotor poles are defined as separate objects, with the form of a cylinder sector.

They are identical but characterized by the appropriate direction of magnetization. The resulting rotor is isotropic, since $\mu_r \approx 1$.

The boundary conditions are assigned assuming that the flux lines do not go out of the fixed domain. Hence, a null magnetic vector potential can be assigned along all the boundary of the domain.

The stator coil is formed by N_t turns, and is modeled by an equivalent conducting rectangular bar. The current flowing within the equivalent bar corresponds to N_t times the effective current of the winding.

According to the reference coordinate system adopted in Figure 11.1, the current is positive when its direction matches the z-axis direction (leaving the sheet), and negative when its direction is opposite to the z-axis direction (going into the sheet).

11.3 Computation of the Electrical Parameters

The stator flux linkage is a function of the stator current and of the rotor angular position, because of the permanent magnet, which is $\lambda = \lambda(\vartheta_m, i)$. Let us assume that the flux density is low so that the magnetic circuit is not saturated. Then, since the rotor is isotropic, the flux linkage can be expressed as

$$\lambda(\vartheta_m, i) = L_a\, i + \lambda_{pm}(\vartheta_m) \tag{11.1}$$

where L_a is the self-inductance of the winding. It is constant with the assumptions given above (linear magnetic circuit and isotropic rotor). Then, λ_{pm} is the flux linkage due to the permanent magnet; it is a function of the rotor angular position ϑ_m. With no hysteresis, Equation (11.1) defines a univocal correspondence among λ, i, and ϑ_m. Two of these quantities are chosen to be the state variables of the system (see Chapter 5).

The computation of the flux linkage, in a fixed rotor position and with a fixed stator current, is carried out by integrating the magnetic vector potential **A** over the equivalent bar. Since the magnetic vector potential has only the z-axis component, the flux linkage is given by

$$\lambda(\vartheta_m, i) = N_t L_{Fe} \left[\frac{1}{S_{Cu}} \int_{S_{Cu+}} A_z dS - \frac{1}{S_{Cu}} \int_{S_{Cu-}} A_z dS \right] \tag{11.2}$$

where L_{Fe} is the motor net length, S_{Cu+} is the equivalent conductive bar carrying a positive current (which is a current with direction corresponding to the z-axis), and S_{Cu-} is the equivalent conductive bar carrying a negative current (which is a current with direction opposite to the z-axis). The ratio between the integral of A_z and the bar surface gives the average value of the magnetic vector potential.

11.3.1 Flux Linkage due to the Permanent Magnet

The dependence of the flux linkage due to the permanent magnet on the rotor angle, which is $\lambda_{pm}(\vartheta_m)$, is analyzed by means of a series of simulations with different rotor positions ϑ_m and null stator current, as Equation (11.1) highlights.

The magnetization of the permanent magnet elements has to be defined from time to time, at each rotor position ϑ_m. The flux linkage has to be analyzed in an angular interval $2\pi/p$, where p is the pole pairs number. It is useful to choose the angular positions at regular intervals, e.g., fixing an angular interval $\Delta\vartheta_m$.

11.3.2 Self-Inductance Coefficient

In the computation of the self-inductance of the stator winding, only the flux linkage due to the stator current is considered. It is

$$L_a = \frac{\lambda(\vartheta_m, i) - \lambda_{pm}(\vartheta_m)}{i} \tag{11.3}$$

and can be computed from two field solutions at the same rotor angle ϑ_m, the former with null current, to obtain the flux linkage $\lambda_{pm}(\vartheta_m)$, and the latter with a stator current i, to obtain the flux linkage $\lambda(\vartheta_m, i)$.

Alternatively, the permanent magnet is demagnetized, so that $\lambda_{pm}(\vartheta_m) = 0$. This is achieved by assigning a null residual flux density, i.e., $B_{res} = 0$ to the permanent magnet elements, but without varying the relative permeability μ_r. It results in

$$L_a = \left.\frac{\lambda(\vartheta_m, i)}{i}\right|_{\lambda_{pm}=0} \tag{11.4}$$

With the hypothesis of linear magnetic circuit, L_a is constant and one computation is enough to determine it.

11.4 Computation of the Torque

The motor torque is a state function of the rotor angular position ϑ_m and the stator current i, i.e., $\tau_m = \tau_m(\vartheta_m, i)$.

A complete study of the torque should be carried out varying ϑ_m and i. The torque is evaluated in a number N_ϑ of angular positions placed at intervals $\Delta\vartheta_m$ for a complete angle $2\pi/p$, corresponding to two motor poles. Similarly, a number of N_i currents with fixed increment is chosen. Positive currents only may be studied. It can be assumed that the motor torque for positive current i and angle ϑ_m is the same torque developed with a negative current and a rotation of π/p, i.e., $\tau_m(\vartheta_m + \pi/p, -i) = \tau_m(\vartheta_m, i)$. Such an observation allows the number of magnetostatic field analysis to be reduced.

Computing the motor torque in all the N_ϑ rotor positions and with all the N_i stator currents, a matrix with N_i column and N_ϑ rows is built, containing the numerical evaluation of the state function $\tau_m = \tau_m(\vartheta_m, i)$. It is called the *torque matrix*.

11.5 Analysis of the Dynamic Performance

The stationary behavior of the synchronous permanent magnet motor is analyzed, by means of the field solutions. The flux linkage and the torque are expressed as functions of the current and the rotor position. These results are processed to carry out the analysis of the dynamic performance of the motor.

To do that, the equations describing the dynamic performance of the electrical and mechanical quantities are used:

$$v(t) = Ri(t) + L_a \frac{di(t)}{dt} + \frac{d\lambda_{pm}(t)}{dt} \tag{11.5}$$

$$\tau_m(t) = \tau_L + k_B \omega_m(t) + J \frac{d\omega_m(t)}{dt} \tag{11.6}$$

where v(t) is the forcing voltage source and R is the coil resistance. In Equation (11.6), $\tau_m(t)$ is the electromechanical torque developed by the motor, τ_L is the load torque, k_B is the friction coefficient, and $\omega_m(t)$ is the mechanical speed.

The dynamic simulation requires that the motor performance be known with any rotor position and with any stator current. Since in the previous field analyses, the flux linkage and the torque have been obtained in a discrete number of points, these values have been interpolated.

The flux linkage vector has been computed at N_ϑ different rotor positions. A one-dimensional interpolation is required to obtain the value corresponding at each angular position.

Similarly, the torque matrix contains the torque computed at N_ϑ rotor position with N_i current values. A 2D interpolation is needed to obtain the motor torque at any position and current.

In Section 11.7, a simple algorithm for the linear interpolation is presented. In Section 11.8, the algorithm for the dynamic analysis of the motor is described.

11.5.1 Simulation at Constant Speed

The first dynamic analysis is carried out at a constant rotor speed ω_m, supposing a rotor inertia so high as any speed variation is neglected. Hence, in the mechanical differential equation (11.6) the derivative term may be omitted. The problem is reduced to the integration of the electrical differential equation (11.5).

Let us assume a voltage supply with sinusoidal waveform of amplitude V_M and frequency f, while the constant rotor speed is fixed to $\omega_m = \omega/p = (2\pi f)/p$.

A simple procedure for the integration of the electrical differential equation (11.5) is described in the following. Eulero's method is used for the numerical integration. The algorithm is reported in Section 11.8. Because of the simplicity of the integration method, a short integration time has to be chosen to obtain a good result.

The time derivative of the flux linkage due to the permanent magnet is given by

$$\frac{d\lambda_{pm}}{dt} = \frac{d\lambda_{pm}}{d\vartheta_m}\frac{d\vartheta_m}{dt} = \frac{d\lambda_{pm}}{d\vartheta_m}\omega_m \tag{11.7}$$

since the variation of λ_{pm} with ϑ_m has been computed by the magnetostatic field analysis.

Assuming that the voltage v, and the derivative $d\lambda_{pm}/dt$ are given in the n-th instant, the time variation of the current is computed from Equation (11.6) as

$$\left(\frac{di}{dt}\right) = \frac{1}{L_a}\left[v - Ri(n-1) - \frac{d\lambda_{pm}}{dt}\right] \tag{11.8}$$

In Equation (11.8), i(n–1) is the current value at the (n–1)-th instant, and the current variation refers to the n-th instant. Then the current at the n-th instant is given by

$$i(n) = i(n-1) + \left(\frac{di}{dt}\right)dt \tag{11.9}$$

The corresponding torque is obtained by means of the 2D interpolation (see Section 11.7), considering an average value of current given by

$$i_m(n-1)=\frac{i(n)+i(n-1)}{2} \tag{11.10}$$

in each interval dt.

Since the mechanical speed ω_m is constant, the rotor position varies linearly:

$$\vartheta_m(n)=\vartheta_m(n-1)+\omega_m dt \tag{11.11}$$

11.5.2 Simulation of the Full Dynamic Performance

The full dynamic performance of the motor is obtained by integrating at the same time both the electrical and mechanical differential equations (11.5) and (11.6). This is particularly convenient for evaluating the starting performance of the rotor. Once again Eulero's integration method is used. The angular acceleration of the rotor is computed from the torque at the n-th instant and the angular speed at the (n–1)-th instant, which is

$$\left(\frac{d\omega_m}{dt}\right)=\frac{1}{J}\left[\tau_m(n)-\tau_L-k_B\omega_m(n-1)\right] \tag{11.12}$$

then the angular speed at the n-th instant is

$$\omega_m(n)=\omega_m(n-1)+\left(\frac{d\omega_m}{dt}\right)dt \tag{11.13}$$

The angular position at the end of each integration time is

$$\vartheta_m(n)=\vartheta_m(n-1)+\omega_m(n-1)dt+\frac{d\omega_m}{dt}\frac{dt^2}{2} \tag{11.14}$$

The rotor starting is simulated by imposing a sinusoidal voltage with amplitude V_M and frequency f, with different values of the initial voltage phase φ and with different initial angular rotor positions $\vartheta_m(0)$. In particular, the rotor starting has to be verified, assuming the rotor to be in its rest positions at null current, which is in the positions where $\tau_m(\vartheta_m,0) = 0$.

By means of this study one can verify if the rotor is able to start and synchronize itself at the fixed supply frequency. Section 11.8 deals with the algorithm of such an integration.

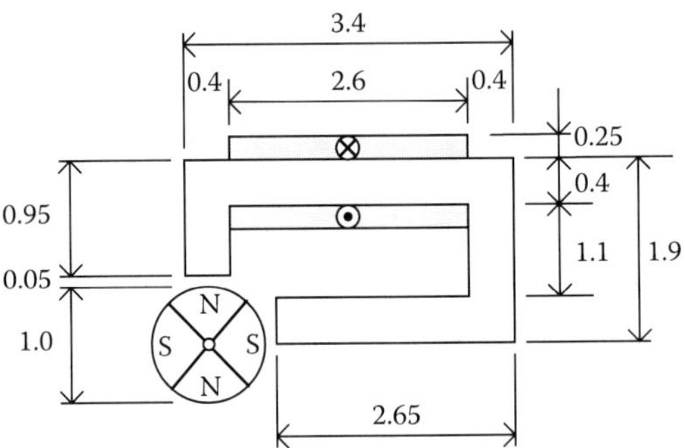

FIGURE 11.2
Sketch and main dimensions of the motor (measured in mm).

11.6 Example

The motor used in this example is a micro-motor, whose main dimensions are shown in Figure 11.2. The rotor is formed by an isotropic permanent magnet, which is a ceramic Barium-Ferrite magnet, characterized by a residual flux density B_{res} = 80 mT and a coercive field strength H_c = –60 kA/m. The external diameter is 1 mm and the inner diameter is 0.15 mm. Then the permanent magnet thickness is 0.425 mm.

The stator is formed by magnetic steel laminations with a thickness of 0.4 mm, and a stack length L_{Fe} = 5 mm. The coil is placed on one side of the lamination and is formed by N_t = 1000 turns with d_c = 25 μm copper wire diameter.

The four poles of the rotor are modeled by four different objects; the shape of each object is a circular sector. Each is characterized by a suitable magnetization direction. The flux lines due to the permanent magnet only are shown in Figure 11.3.

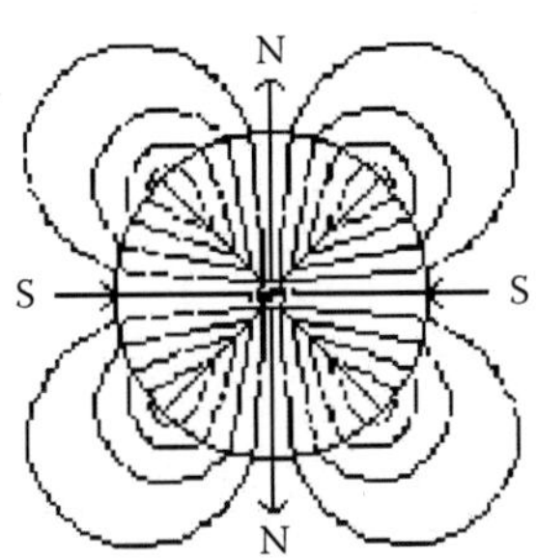

FIGURE 11.3
Flux lines due to the permanent magnet only.

The reference initial position $\vartheta_m = 0$ is chosen when the North pole is in front of the upper expansion of the stator expansion, as shown in Figure 11.2.

Figure 11.4 and Figure 11.5 show the flux lines changing the rotor position ϑ_m, starting from the initial position $\vartheta_m = 0$ up to the position $\vartheta_m = \pi/3$, with a counter-clockwise rotation.

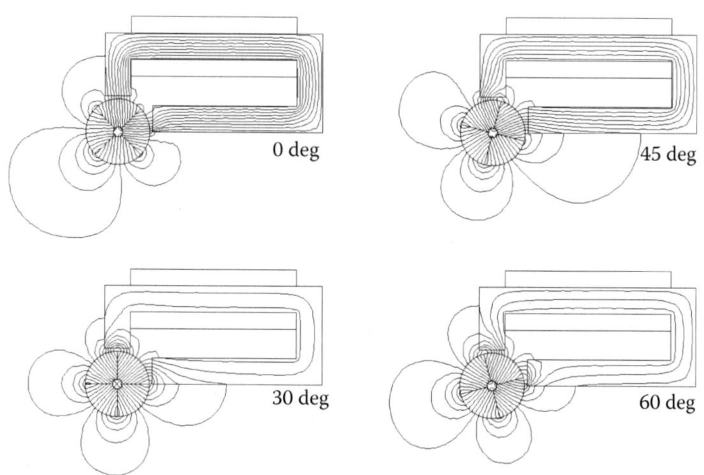

FIGURE 11.4
Flux lines at different rotor positions ϑ_m (at no-load).

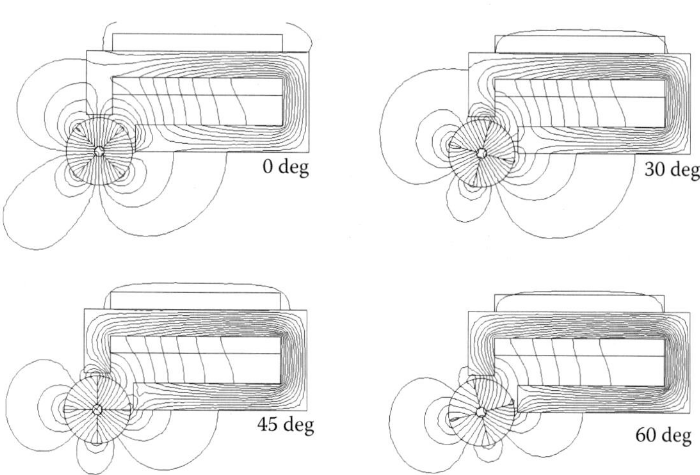

FIGURE 11.5
Flux lines at different rotor positions ϑ_m (under load).

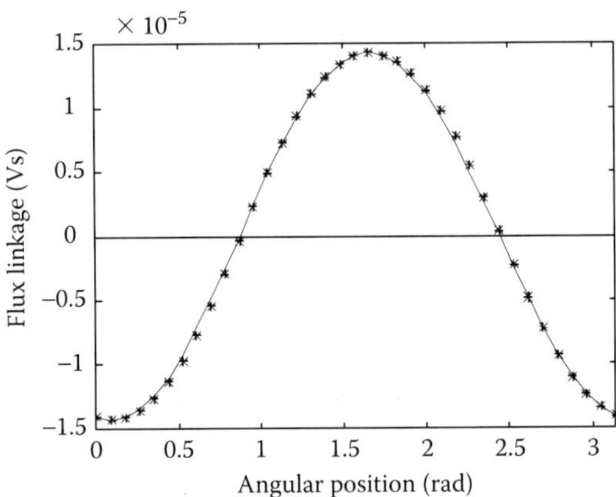

FIGURE 11.6
Flux linkage due to the permanent magnet only, simulated values, and interpolating function.

11.6.1 Computation of the Flux Linkage due to the Permanent Magnet

The flux linkage due to the permanent magnet has been computed at different rotor positions at a constant angular interval $\Delta\vartheta_m = \pi/36$ from $\vartheta_m = 0$ to $\vartheta_m = \pi$. The value of λ_{pm} is obtained from Equation (11.2), with $S_{Cu+} = S_{Cu-} = 65 \cdot 10^{-8}\ m^2$, by means of simulations with null stator current.

Figure 11.6. shows the flux linkage λ_{pm} as a function of the rotor position ϑ_m. Since the behavior is essentially sinusoidal, it is possible to approximate it by a sinusoidal waveform, given by

$$\lambda_{pm}(\vartheta_m) = -1.431 \cdot \cos\left[2\left(\vartheta_m - \frac{\pi}{36}\right)\right] \tag{11.15}$$

In Figure 11.6, stars indicate the simulated values while the solid line represents the interpolating function.

11.6.2 Computation of the Inductance

The inductance is computed by Equation (11.4), from a field analysis with a demagnetized permanent magnet ($B_{res} = 0$ T) and with a stator current 20 mA. The flux linkage is computed by Equation (11.2). The integration of magnetic vector potential A_z on the section S_{Cu+} is equal to $2.79 \cdot 10^{-11}$ Vsm, while the integration on the section S_{Cu-} is equal to $-0.0771 \cdot 10^{-11}$ Vsm. The flux linkage is then $\lambda = 2.2055 \cdot 10^{-5}$ Vs, so that the inductance results in $L_a = 1.6541$ mH.

11.6.3 The Torque Matrix

The torque motor contains the values of the motor torque $\tau_m = \tau_m(\vartheta_m, i)$ at different rotor positions ϑ_m and stator currents i. The angular position ranges

TABLE 11.1

Torque Matrix (Torque as a Function of the Rotor Position and Current)

Torque · 10^{-7} (Nm)

ϑ_m (rad)	Current (mA)								
	0	2.5	5	7.5	10	12.5	15	17.5	20
0	–0.020	–0.118	–0.199	–0.282	–0.354	–0.435	–0.521	–0.602	–0.689
0.131	–0.370	–0.340	–0.313	–0.283	–0.253	–0.218	–0.189	–0.183	–0.164
0.262	–0.595	–0.453	–0.315	–0.172	–0.035	0.113	0.247	0.301	0.425
0.396	–0.659	–0.411	–0.166	0.085	0.335	0.585	0.750	0.978	1.207
0.524	–0.558	–0.217	0.130	0.478	0.826	1.172	1.529	1.764	2.089
0.655	0.354	0.088	0.519	0.952	1.385	1.818	2.235	2.684	3.118
0.785	0.027	0.524	1.020	1.507	2.005	2.498	2.982	3.427	3.907
0.916	0.378	0.884	1.390	1.884	2.390	2.895	3.392	3.899	4.405
1.047	0.589	1.060	1.530	1.997	2.464	2.930	3.401	3.892	4.368
1.178	0.695	1.092	1.489	1.891	2.286	2.676	3.076	3.472	3.864
1.309	0.590	0.890	1.189	1.488	1.789	2.091	2.393	2.701	2.994
1.439	0.323	0.513	0.699	0.891	1.088	1.288	1.480	1.672	1.862
1.571	–0.020	0.060	0.145	0.226	0.311	0.390	0.469	0.543	0.622
1.70	0.365	–0.396	–0.426	–0.464	–0.491	–0.519	–0.546	–0.576	–0.6032
1.836	–0.609	–0.757	–0.897	–1.027	–1.171	–1.310	–1.449	–1.553	–1.705
1.964	–0.669	–0.916	–1.164	–1.411	–1.663	–1.912	–2.158	–2.406	–2.645
2.094	–0.557	–0.909	–1.266	–1.614	–1.962	–2.311	–2.661	–3.006	–3.355
2.225	0.356	–0.781	–1.164	–1.589	–2.082	–2.495	–2.941	–3.372	–3.816
2.356	0.021	–0.472	–0.965	–1.453	–1.952	–2.446	–2.943	–3.439	–3.937
2.487	0.376	–0.131	–0.640	–1.148	–1.657	–2.161	–2.664	–3.171	–3.678
2.618	0.595	0.127	–0.349	–0.820	–1.291	–1.765	–2.198	–2.668	–3.137
2.749	0.679	0.294	–0.101	–0.500	–0.898	–1.297	–1.696	–2.068	–2.462
2.879	0.589	0.292	–0.011	–0.309	–0.609	–0.909	–1.215	–1.513	–1.846
3.010	0.318	0.121	–0.071	–0.264	–0.456	–0.651	–0.851	–1.041	–1.236
3.142	–0.020	–0.118	–0.199	–0.282	–0.354	–0.435	–0.521	–0.602	–0.689

from $\vartheta_m = 0$ up to $\vartheta_m = \pi$, with an angular interval $\Delta\vartheta_m = \pi/24$. The current ranges from i = 0 A to i = 20 mA, with a current interval Δi = 2.5 mA. The positive current direction is reported in Figure 11.2.

The torque matrix contains the torque corresponding to $N_i = 9$ different currents and $N_\vartheta = 25$ angular positions, which is a $N_i \times N_\vartheta = 225$ elements matrix. The values are reported in Table 11.1. Figure 11.7 shows the behavior of the torque, as a function of the rotor positions with different stator currents. It is worth noticing that the current increase yields to an increase of the maximum torque, but also to a change of the angular positions at which the torque is zero.

The positions at null torque coincide neither with $\vartheta_m = 0$, nor with $\vartheta_m = \pi/2$. This is because the rotor axis does not coincide with the stator pole axes.

With a current 20 mA, the maximum torque is about $4.5 \cdot 10^{-7}$ Nm, at a rotor position almost equal to 1.047 rad (60 degrees). The torque becomes null two times during a rotation π, the first one at a rotor position of almost 0.2 rad (11 degrees), and the second one at a rotor position of almost 1.7 rad (97 degrees).

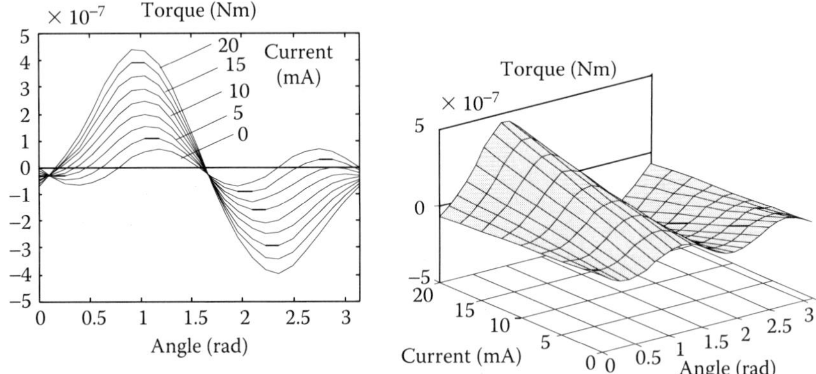

FIGURE 11.7
Motor torque as a function of the rotor position and the current.

11.6.4 Simulation of the Dynamic Performance

The stator coil resistance is R = 110 Ω, considering N_t = 1000 turns, with an average length l_m = 3 mm, of copper wire diameter d_c = 25 μm, and a resistivity ρ=0.0178 Ωmm²/m, at the working temperature 20°C. The system inertia is the sum of the rotor inertia and the load inertia. A total inertia, including rotor and load, is $J = 1.211 \cdot 10^{-13}$ kg·m². At last the friction coefficient has been set equal to $k_B = 1.8 \cdot 10^{-10}$ Nms.

11.6.4.1 Constant Speed Simulation

Let the rotor speed be $\omega_m = \omega/p$ = 314.16 rad/s and the electrical speed ω = 628.32 rad/s.

A sinusoidal voltage is applied, with maximum value V_M = 2V and frequency 100 Hz. The current and the torque waveforms are shown in Figure 11.8. As expected by a single-phase motor, the torque exhibits an alternate

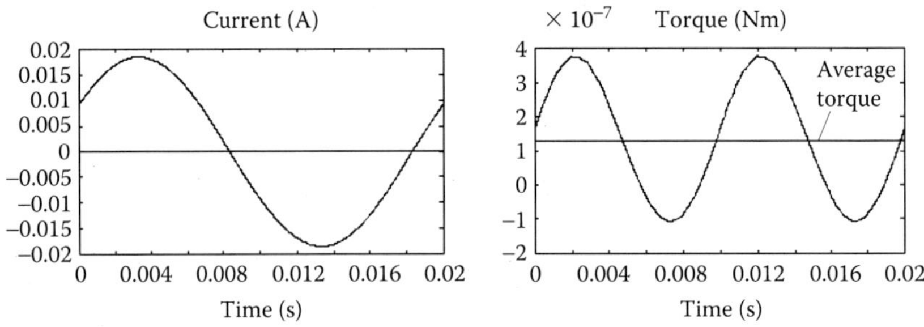

FIGURE 11.8
Current and the torque waveforms, assuming a constant rotor speed.

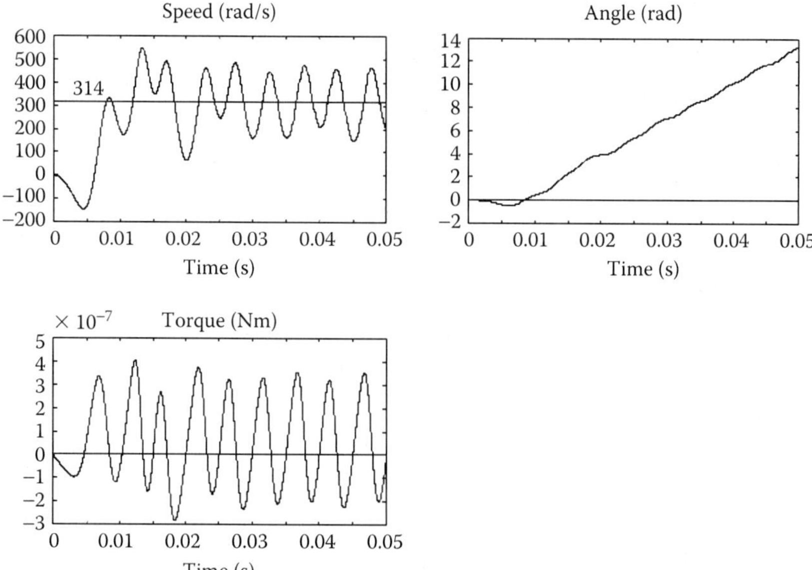

FIGURE 11.9
Rotor starting at $\varphi = 0$, $\vartheta(0) = 0$, f = 100 Hz.

waveform, with an average value different from zero, and an almost sinusoidal variation at a frequency double of the source frequency.

11.6.4.2 Simulation of the Rotor Starting

The rotor starting is analyzed forcing a sinusoidal voltage with $V_M = 2$ V and f = 100 Hz, with different values of the initial voltage phase φ and of the initial rotor position $\vartheta_m(0)$.

Figure 11.9 shows the rotor position, torque, and current waveforms, using an initial phase $\varphi = 0$ rad, and initial rotor position $\vartheta_m(0) = 0$ rad. A satisfactory starting is obtained: the rotor reaches the synchronous speed ω_m = 314 rad/s in less than 0.01 s. One can also observe that the steady-state speed is not constant (as assumed in the previous analysis), but oscillates around the synchronous speed. This is caused by the oscillating torque developed by the single-phase motor.

Figure 11.10 and Figure 11.11 show the same waveforms, using $\vartheta_m(0) = 0$, f = 100 Hz, and $\varphi = \pi/4$ rad and $\varphi = \pi/2$ rad, respectively. In the latter case, the synchronization of the rotor is obtained, but the speed direction is opposite to that of the previous cases. This is due to the nature of the single-phase motor, which does not produce a rotating field but a pulsating field. It follows that there is no preferential direction of rotation.

At last, Figure 11.12 shows a case with supply frequency 50 Hz, at which there is no synchronization of the rotor.

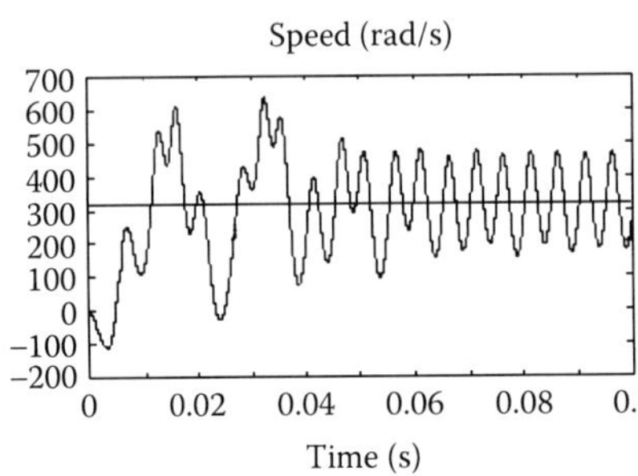

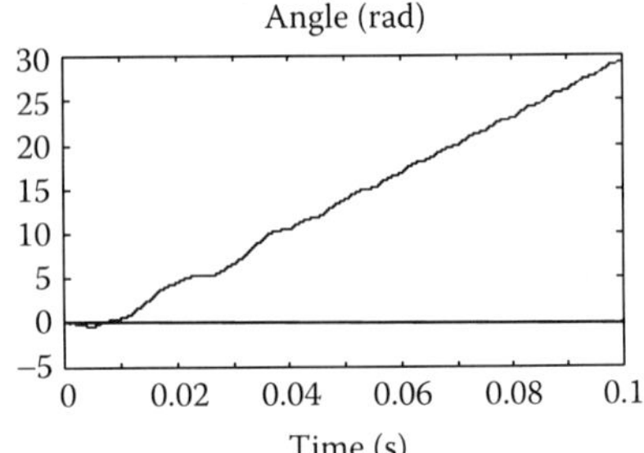

FIGURE 11.10
Rotor starting at $\varphi = \pi/4$, $\vartheta(0) = 0$, f = 100 Hz.

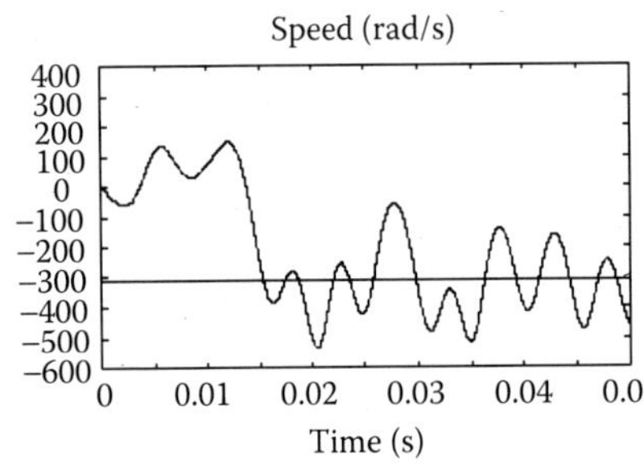

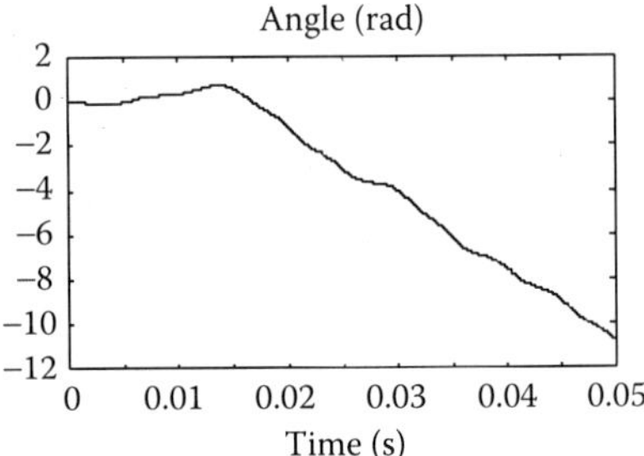

FIGURE 11.11
Rotor starting at $\varphi = \pi/2$, $\vartheta(0) = 0$, f = 100 Hz.

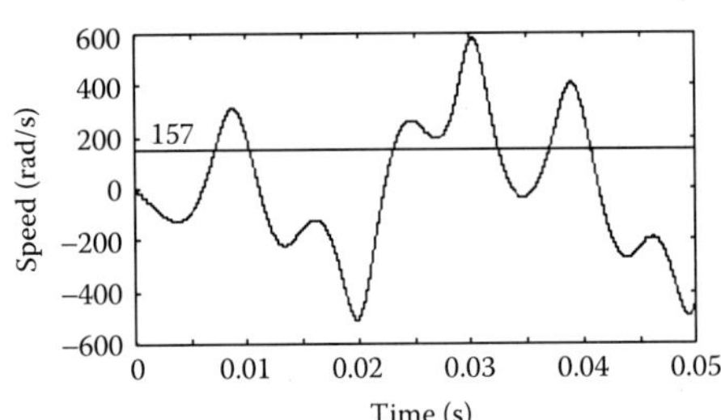

FIGURE 11.12
Rotor starting at $\varphi = \pi/2$, $\vartheta(0) = 0$, f = 50 Hz.

11.7 Two-Dimensional Linear Interpolation

Let $y(x_1, x_2)$ be a function of the two variables x_1 and x_2. It is assumed that the function $y(x_1, x_2)$ is known only in correspondence of a discrete number of values of x_1 and x_2. In other words, with a fixed number M of values of x_1, and a number N of values of x_2, the function $y(x_1, x_2)$ is known only in

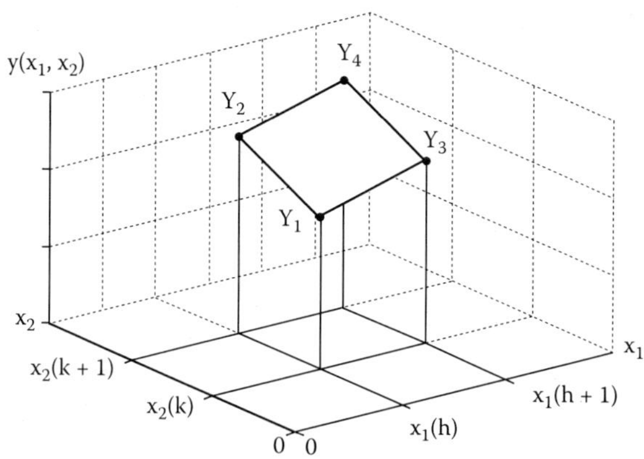

FIGURE 11.13
Interpolating plane.

M × N values. From the knowledge of such values, the value of the function $y(x_1, x_2)$ corresponding to a couple of variables (x_1, x_2) can be interpolated, with the constraint that $x_1(1) < x_1 < x_1(M)$ and $x_2(1) < x_2 < x_2(N)$.

The first step of the interpolation is to individuate the interval of the vector $[x_1(1), \ldots, x_1(M)]$ in which x_1 is contained. This involves individuating the number h so that $x_1 > x_1(h)$ and $x_1 < x_1(h + 1)$. Analogously, with the second variable, the number k is individuate, so that $x_2 > x_2(k)$ and $x_2 < x_2(k + 1)$.

The second step is to compute the four values of the function $y(x_1, x_2)$ corresponding to the combinations two by two of the variables, which is

Y_1 corresponding to the couple of variables $x_1(h)$, $x_2(k)$
Y_2 corresponding to the couple of variables $x_1(h)$, $x_2(k + 1)$
Y_3 corresponding to the couple of variables $x_1(h + 1)$, $x_2(k)$
Y_4 corresponding to the couple of variables $x_1(h + 1)$, $x_2(k + 1)$

as sketched in Figure 11.13.

Finally, the interpolation is carried out by means of the function

$$y(x_1,x_2) = y_1(x_1,x_2) + y_2(x_1,x_2) + y_3(x_1,x_2) + y_4(x_1,x_2) \tag{11.16}$$

where

$$y_1(x_1,x_2) = Y_1 \frac{x_1(h+1) - x_1}{x_1(h+1) - x_1(h)} \frac{x_2(k+1) - x_2}{x_2(k+1) - x_2(k)} \tag{11.17}$$

$$y_2(x_1,x_2) = Y_2 \frac{x_1(h+1) - x_1}{x_1(h+1) - x_1(h)} \frac{x_2 - x_2(k)}{x_2(k+1) - x_2(k)} \tag{11.18}$$

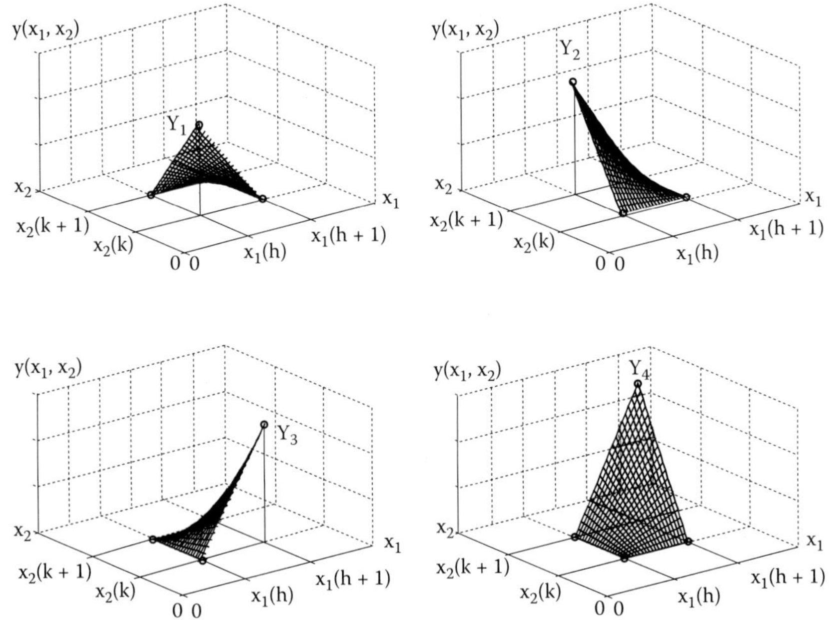

FIGURE 11.14
Shape of the interpolating functions.

$$y_3(x_1,x_2) = Y_3 \frac{x_1 - x_1(h)}{x_1(h+1) - x_1(h)} \frac{x_2(k+1) - x_2}{x_2(k+1) - x_2(k)} \tag{11.19}$$

$$y_4(x_1,x_2) = Y_4 \frac{x_1 - x_1(h)}{x_1(h+1) - x_1(h)} \frac{x_2 - x_2(k)}{x_2(k+1) - x_2(k)} \tag{11.20}$$

Each of these functions assumes a value different from zero in only one couple of variables, while it assumes a null value in the other three couples of variables (x_1, x_2). For instance, the first function, say $y_1(x_1, x_2)$, assumes the value Y_1 in the point of coordinates $x_1(h)$, $x_2(k)$; while a null value is assumed in the points of coordinates $x_1(h)$, $x_2(k + 1)$; $x_1(h + 1)$, $x_2(k)$; $x_1(h + 1)$, $x_2(k + 1)$.

The shapes of the interpolating functions are reported in Figure 11.14. Once the value that the four interpolating functions assume corresponds to the point (x_1, x_2), the final value is given by their sum.

11.7.1 Description of the Interpolating Algorithm

Some algorithms are reported in the following, using MatLab software. The simplicity of the functions has been preferred to the aim of an easier reading.

In the example, the interpolation algorithm has been written in the FUNCTION `torque.m`. The inputs of such a FUNCTION are the actual current *iact* and angular position *tetamec*, and the output is the corresponding torque. The matrix *Torq* contains the values of the motor torque, computed at the currents reported in the vector *VectI* and at the angular positions reported in the vector *VectA*.

From the input current *iact*, within the vector *VectI*, the two currents (lower and higher than *iact*) are individuated. Consequently, two columns are individuated within the torque matrix. The same research is repeated for the actual angular position *tetamec* within the vector *VectA*, individuating two rows within the torque matrix. From the intersection of the two rows and the two columns, the four torque values needed for the interpolation are obtained: the torque to be determined is in the interval of these four values.

Concerning the current sign, it has been assumed that the torque corresponding to a fixed current and angular position is the same of that computed with an opposite current and a rotation of π/p, i.e., $\tau_m(\vartheta + \pi/p, -i) = \tau_m(\vartheta, i)$. This allows the matrix dimensions to be significantly reduced.

The FUNCTION `torque.m` is reported in the following.

```
torque.m

function value = torque(iact, tetamec);

%Interpolation of the motor torque
%In the matrix "Torq", there are the computed torque values
%corresponding to the currents reported in vector "VECTI",
%and to the angular positions, in radians, reported in vector "VECTA"

Torq =[

 -0.0199  -0.1181  -0.1991  -0.2824  -0.3540  -0.4353  -0.5209  -0.6018  -0.6887
 -0.3703  -0.3401  -0.3130  -0.2826  -0.2529  -0.2176  -0.1886  -0.1830  -0.1638
 -0.5952  -0.4534  -0.3149  -0.1718  -0.0353   0.1128   0.2469   0.3013   0.4250
 -0.6588  -0.4114  -0.1657   0.0852   0.3348   0.5853   0.7502   0.9783   1.2065
 -0.5578  -0.2168   0.1297   0.4777   0.8259   1.1717   1.5294   1.7640   2.0893
 -0.3540   0.0876   0.5192   0.9523   1.3854   1.8184   2.2353   2.6842   3.1178
  0.0272   0.5236   1.0197   1.5073   2.0051   2.4983   2.9818   3.4268   3.9067
  0.3777   0.8839   1.3895   1.8838   2.3901   2.8951   3.3917   3.8985   4.4046
  0.5891   1.0601   1.5304   1.9973   2.4642   2.9299   3.4005   3.8920   4.3682
  0.6945   1.0918   1.4894   1.8905   2.2862   2.6760   3.0759   3.4718   3.8642
  0.5901   0.8897   1.1887   1.4882   1.7891   2.0906   2.3927   2.7005   2.9935
  0.3231   0.5133   0.6991   0.8913   1.0878   1.2881   1.4801   1.6721   1.8619
 -0.0203   0.0604   0.1451   0.2260   0.3109   0.3902   0.4690   0.5427   0.6223
 -0.3645  -0.3963  -0.4255  -0.4644  -0.4906  -0.5186  -0.5459  -0.5756  -0.6033
 -0.6089  -0.7573  -0.8969  -1.0267  -1.1709  -1.3098  -1.4490  -1.5531  -1.7047
 -0.6686  -0.9161  -1.1638  -1.4114  -1.6627  -1.9120  -2.1584  -2.4063  -2.6453
 -0.5572  -0.9090  -1.2660  -1.6137  -1.9622  -2.3111  -2.6610  -3.0055  -3.3552
 -0.3561  -0.7805  -1.1635  -1.5888  -2.0822  -2.4954  -2.9406  -3.3722  -3.8163
  0.0212  -0.4722  -0.9652  -1.4525  -1.9518  -2.4460  -2.9427  -3.4392  -3.9367
  0.3763  -0.1310  -0.6395  -1.1479  -1.6571  -2.1612  -2.6639  -3.1714  -3.6780
  0.5953   0.1270  -0.3495  -0.8196  -1.2911  -1.7654  -2.1984  -2.6678  -3.1371
  0.6792   0.2940  -0.1014  -0.4999  -0.8977  -1.2966  -1.6961  -2.0675  -2.4615
  0.5894   0.2921  -0.0111  -0.3093  -0.6091  -0.9091  -1.2146  -1.5130  -1.8462
  0.3177   0.1208  -0.0708  -0.2639  -0.4556  -0.6505  -0.8508  -1.0405  -1.2357
 -0.0199  -0.1178  -0.1988  -0.2821  -0.3540  -0.4353  -0.5210  -0.6018  -0.6887
]*(1e-7);
```

```
VectI=[0 2.5 5 7.5 10 12.5 15 17.5 20]*1e-3;
VectA=[0 0.1309 0.2618 0.3962 0.5236 0.6545 0.7854 0.9163 1.0472  1.1781 1.3090
1.4390 1.5708 1.7017 1.8362 1.9635 2.0940 2.2250 2.3561 2.4870 2.6180 2.7490
2.8790 3.0100 pi];

% It is forced that a negative current produces the same torque of a
% corresponding positive current with identical magnitude
% with an angular displacement of π/p radians (here p=2)

k=1;
z=1;

% Research of the interval for the current

% Control of the sign
if iact<0,
   iact =- iact;
   tetamec=tetamec-pi/2;
end;

% Control of the maximum current value

if iact < max(VectI),
    while iact >= VectI(k)
       k=k+1;
    end;
    else k=max(size(VectI));
 end;
 if iact > max(VectI),
    fprintf('The actual current is higher than maximum current')
    fprintf('the results of the torque computation are not correct!')
    k=max(size(VectI));
    val=0;
  else

% Research of the interval of the angular position

% Control of the angular position value

while tetamec < 0
   tetamec=tetamec+pi;
end;
while tetamec > pi
   tetamec=tetamec-pi;
end;
   if tetamec==pi,
      z=max(size(VectA));
   else
      while tetamec >= VectA(z)
        z=z+1;
     end;
   end;

% Research of the four values of the torque corresponding
% to the currents and angular position individuated above

   t1= Torq(z-1,k-1);
   t2= Torq(z-1,k);
   t3= Torq(z,k-1);
   t4= Torq(z,k);
```

```
% Interpolation of the torque corresponding to
% the current iact and the angular position tetamec

    com=(VectI(k)-VectI(k-1))*(VectA(z)-VectA(z-1));
    val1=t1*(VectI(k)- iact)*(VectA(z)-tetamec);
    val2=t2*( iact -VectI(k-1))*(VectA(z)-tetamec);
    val3=t3*(VectI(k)- iact)*(tetamec-VectA(z-1));
    val4=t4*( iact -VectI(k-1))*(tetamec-VectA(z-1));

% Value of the torque obtained by the interpolation

    value=(val1+val2+val3+val4)/com;

end;
```

11.8 Numerical Codes for the Motor Analysis

Some FUNCTIONs used for the analysis of the dynamic performance of the single-phase motor are now reported.

11.8.1 Code of the Computation of the Torque at Constant Speed

```
% Simulation of the motor performance at given constant speed
% with torque interpolation (the function torque.m is called)
% N.B.: The load torque is posed to zero

clear;
close all;

% Constant declaration

 dt=1e-5;              % time increment
 j=1.5e-12;            % rotor inertia
 vm=2;                 % maximum voltage
 we=2*pi*100;          % electrical speed 628.3185(100Hz)
 wm=we/2;              % mechanical speed wm=we/p; p=pole pair
 res=110;              % resistance of the stator coil at 20°C
 la=1.6541e-3;         % winding inductance

% Initial values of the variables

 fi=pi/6;              % voltage phase
 t=0;                  % initial istant
 dfdz=0;               % PM flux linkage derivative
 tetam(1)=0.738;       % initial rotor angular position
 tavrg=0;              % average torque
 tmot(1)=0;            % actual torque
 ip(1)=0;              % actual current
 im(1)=0;              % average current in the interval
 flux(1)=-1.4057e-5;   % magnetic flux of the PM
```

```
% Computation loop of the torque, speed, position and current

   for n=2:5000
      v=vm*sin(we*t+fi);                          % voltage in n-th time
      didt=(v-res*ip(n-1)-dfdz*wm)/la;            % current increment
      ip(n)=ip(n-1)+didt*dt;                      % current in n-th time
      im(n-1)=(ip(n)+ip(n-1))*0.5;                % c. average in interval
      tmot(n-1)=coppia(im(n-1),tetam(n-1)); % torque in the interval
      tetam(n)=tetam(n-1)+wm*dt;                  % position in n-th time
      tavrg=tavrg+tmot(n-1);                      % average torque
      flux(n)=(-1.431e-5)*cos(2*(tetam(n)-0.0277778*pi));  % PM flux
      dfdz=flux(n)-flux(n-1);                     % flux variation
      t=t+dt;                                     % time increment
   end;

   tavrg=tavrg/n                            % average torque

   end
```

11.8.2 Code of the Computation of the Dynamic Performance

```
% Simulation of the dynamic performance of a single-phase permanent magnet
motor
% with torque interpolation (the function torque.m is called)
% N.B.: The load torque is posed to zero

clear;
close all;

% Constant definition

dt=1e-5;          % time increment
j=1.5e-12;        % rotor inertia
vm=2;             % maximum voltage
we=2*pi*100;      % electrical speed 628.3185=100Hz
wm=we/2;          % mechanical speed wm=we/p;  p=pole pair
res=110;          % resistance of the stator coil at 20°C
la=1.6541e-3;     % winding inductance
kB=1.8e-10;       % friction coefficient

% Initial values of the variables

fi=pi/6;          % voltage phase
t=0;              % initial istant
dwmdt=0;          % speed derivative at t=0
dfdz=0;           % PM flux linkage derivative at t=0
tetam(1)=0.738;   % initial rotor angular position
tmot(1)=0;        % actual torque
ip(1)=0;          % actual current
im(1)=0;          % average current in the interval
flux(1)=-1.4057e-5;% magnetic flux of the permanent magnet
wm(1)=0;          % angular speed (starting)

% Computation loop of the torque, speed, position and current

for n=2:5000
v=vm*sin(we*t+fi);                       % voltage in n-th time
```

```
didt=(v-res*ip(n-1)-dfdz*wm(n-1))/la;     % current derivative
ip(n)=ip(n-1)+didt*dt;                    % current in n-th time
im(n-1)=(ip(n)+ip(n-1))*0.5;              % c. average in interval
tmot(n-1)=coppia(im(n-1),tetam(n-1));     % torque in the interval
dwmdt=(tmot(n-1)-kB*wm(n-1))/j;           % angular acceleration
wm(n)=wm(n-1)+dwmdt*dt;                   % angular speed
tetam(n)=tetam(n-1)+wm(n-1)*dt+dwmdt*(dt^2)*0.5;      % position
flux(n)=(-1.431e-5)*cos(2*(tetam(n)-0.0277778*pi));  % PM flux
dfdz= flux(n)- flux(n-1);                 % flux linkage derivative
t=t+dt;                                   % time increment
end;

end
```

References

1. P.C. Krause, *Analysis of Electrical Machinery*, McGraw-Hill, New York, 1986.
2. K. Deng, M. Mehregany, and A.S. Dewa, "A Simple Fabrication Process for Polysilicon Side-Drive Micromotors," in *IEEE Journal of Micromechanical System*, vol. 3, n. 4, December 1994, pp. 126–133.
3. A. Teshigahara, M. Watanabe, N. Kawahara, Y. Ohtsuka, and T. Hattori, "Performance of a 7 mm Microfabbricated Car," in *IEEE Journal of Micromechanical System*, vol. 4, n. 2, June 1995, pp. 76–80.

12

Switched Reluctance Motors

This chapter deals with the analysis of the switched reluctance motor. Since such a motor works in highly saturated, the chapter gives full details of how to choose the number of finite elements and how to compute the torque developed by the motor. We underline the effect of the number of finite elements on the local components of the magnetic fields, from which the torque is computed.

After we compute the electromagnetic quantities by the finite element method, they are used for the prediction of the dynamic performance of the motor, in a similar way as done in the previous chapter.

12.1 Introduction

The switched reluctance motor originated in 1842. However, it was not employed until several decades later, and then thanks to the development of the power electronic devices. The first variable-speed applications were proposed in the 1970s by Harris and Lawrenson.

The motor has a double saliency, i.e., salient poles in both stator and rotor. The phase currents are switched depending on the rotor position, by means of a very simple control strategy. The motor differs from the stepper motor for the following four reasons:

1. The number of rotor and stator poles is lower.
2. The rated torque and power are higher.
3. A position sensor is used to detect the rotor position (with the exception of some recent sensorless control applications), while in the stepper motor there is no position feedback.
4. A current control loop is normally applied, while the stepper motor is driven by voltage pulses, without a current control.

12.2 Operating Principle

Figure 12.1(a) shows a motor with $Q_s = 8$ stator poles and $Q_r = 6$ rotor poles. It is also called an 8/6 switched reluctance motor. Typical values of the pole number ratio Q_s/Q_r are 6/4, 8/6, 10/4, 10/6. Both stator and rotor are obtained by stacking laminations, since both of them carry variable magnetic fluxes. The winding is simply formed by coils wound around stator poles. Then the motor shows a very simple and robust structure, with very short end-winding lengths. The stator coils are placed, with the same orientation, around diametrically opposed stator poles, so that the number of phases is equal to the half of the number of stator poles, i.e., $m = Q_s/2$. Due to the motor geometry and the winding displacement, there is no mutual coupling among the phases.

The motor torque is essentially due to the reluctance of the magnetic circuit. The rotor moves so as to align its poles with the stator poles, whose coils are excited, maximizing the flux linkage.

The instantaneous electromechanical torque is a state function of the rotor position and the phase currents. It may be computed as the partial derivative of the magnetic coenergy with respect to the rotor position, as

$$T(\vartheta_m, i) = \frac{\partial W'_m(\vartheta_m, i)}{\partial \vartheta_m} \tag{12.1}$$

where ϑ_m is the mechanical angle between the rotor pole and the stator pole, as shown in Figure 12.1(a), and i is the current flowing through the stator coil. The analysis has to be carried out considering the dependence on both the variables ϑ_m and i.

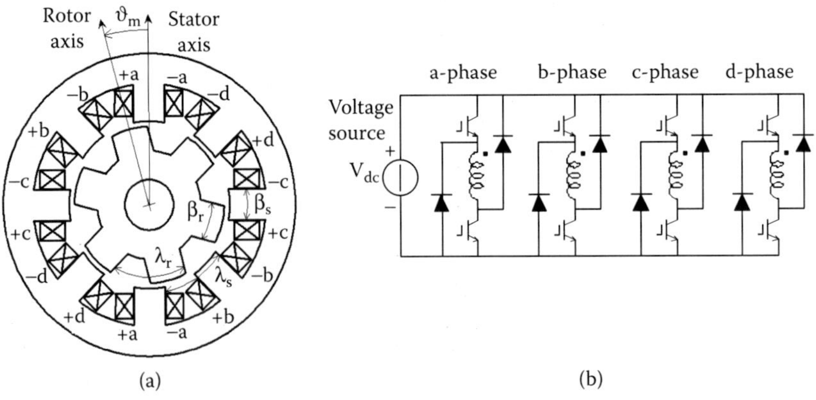

FIGURE 12.1
Structure of a switched reluctance motor with $Q_s = 8$ and $Q_r = 6$ (a); sketch of the static power electronic converter (b).

12.2.1 Linear Model

Although the switched reluctance motor works in a highly saturated way, it is useful to describe first its operating principles with the assumption of a linear magnetic circuit. In this case, magnetic energy and coenergy coincide and both can be described using the inductance coefficient $L(\vartheta_m)$. The latter is a function of the angular position ϑ_m but not of the current. It is

$$W'_m = W_m = \frac{1}{2}L(\vartheta_m)i^2 \tag{12.2}$$

hence the electromechanical torque (12.1) becomes

$$T(\vartheta_m, i) = \frac{1}{2}\frac{dL(\vartheta_m)}{d\vartheta_m}i^2 \tag{12.3}$$

The torque is proportional to the current squared, no matter its direction. This is advantageous for the power electronic converter, since only a unidirectional current is required. A possible configuration of the power electronic converter is shown in Figure 12.1(b). Moreover, again thanks to the dependence on the current squared, the starting torque is very high, similar to that of a direct-current series-excited motor.

As seen in Equation (12.3), both motor and generator operations are allowed. The sign of the torque comes from the derivative of the inductance. The torque is positive if the coils are fed when the inductance is increasing, and vice versa if the coils are fed when the inductance is decreasing. The rotation direction is determined by the chosen cyclic sequence of the phases.

According to the motor in Figure 12.1(a) with $\beta_r > \beta_s$, Figure 12.2 shows the ideal behavior of the a-phase inductance versus the angular position ϑ_m. The aligned position corresponds to $\vartheta_m = 0$, while the unaligned position corresponds to $\vartheta_m = \vartheta_0$. Four regions may be distinguished:

1. From ϑ_0 to ϑ_1 there is no overlap between the rotor and the stator pole pairs, so that the inductance is minimum.
2. From ϑ_1 to ϑ_2 the stator and rotor pole pairs are partially overlapped, starting from ϑ_1 where the two pole pairs start to overlap, up to ϑ_2 where they are completely overlapped. The phase inductance changes linearly from the minimum to the maximum value.
3. From ϑ_2 to ϑ_3 the two pole pairs are completely overlapped around the aligned position, due to the different width of the poles. The inductance remains constant at its maximum value.
4. From ϑ_3 to ϑ_4 the rotor pole pairs move forward reducing the overlap with the stator pole pairs, so that the inductance decreases linearly down to its minimum value.

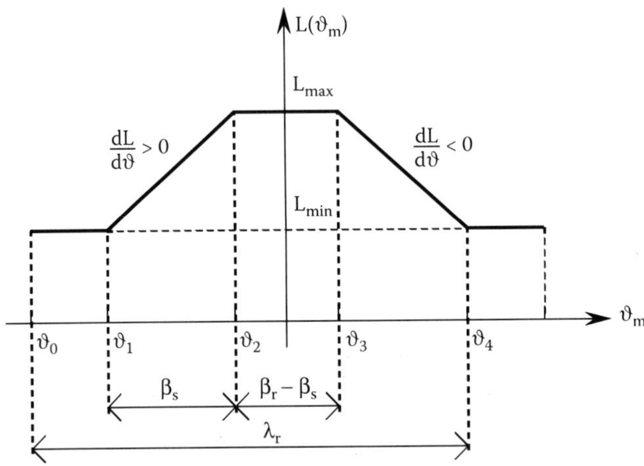

FIGURE 12.2
Phase inductance versus the rotor position.

According to Equation (12.3), it is observed that if the coil carries a current when the inductance is increasing (between ϑ_1 and ϑ_2), a positive torque is obtained. On the contrary, if the coil carries a current when the inductance is decreasing (between ϑ_3 and ϑ_4), a negative torque is obtained. In the two regions in which the inductance is constant, the current does not produce any torque. However, these regions are needed for the current commutations. In the period where the inductance is constant, the current is forced to increase from zero to the desired value, or to decrease from its non-null value down to zero.

12.2.2 Dynamic Performance

The equivalent electric circuit of each phase may be modeled as an RL circuit. A voltage equal to $+V_{dc}$ or $-V_{dc}$ is applied to its terminals, by means of the switches of the converter of Figure 12.1(b). The voltage equation is

$$v(t) = Ri(t) + \frac{d\lambda(t)}{dt} \tag{12.4}$$

Since the flux linkage λ is univocally determined by the current i and the rotor position ϑ_m, that is, $\lambda = \lambda(i,\vartheta_m)$, as explained in Chapter 5, Equation (12.4) becomes

$$v(t) = Ri(t) + \frac{\partial\lambda(i,\vartheta_m)}{\partial\vartheta_m}\frac{d\vartheta_m}{dt} + \frac{\partial\lambda(i,\vartheta_m)}{\partial i}\frac{di(t)}{dt} \tag{12.5}$$

where $\omega_m = d\vartheta_m/dt$ is the rotor speed. In linear conditions, Equation (12.5) may be rewritten as

$$v(t) = Ri(t) + \omega_m(t)i(t)\frac{dL(\vartheta_m)}{d\vartheta_m} + L(\vartheta_m)\frac{di(t)}{dt} \tag{12.6}$$

In the motoring operations, the phase coils are fed during the inductance increasing, fixing a specific conducting period. Let ϑ_{on} be the turn-on angle and ϑ_{off} be the turn-off angle. According to Figure 12.1(b), the switches are turned on when the rotor reaches the angular position ϑ_{on}. The voltage $v = +V_{dc}$ is applied to the coil terminal forcing the current to increase. Then the switches are turned off when the rotor reaches the angular position ϑ_{off}, the current flows through the two diodes, so that the voltage $v = -V_{dc}$ is automatically applied to the coil terminal, forcing the current down to zero.

12.3 Field Problem Statement

Since the phases are not mutually coupled, they can be considered separately. Only one phase is analyzed at a time and the results are extended to the other phases, simply taking into account the angular displacement. In reality, the saturation of the yokes may cause a mutual interaction among the phases. However, the currents of the phase are fed independently, with reduced overlap; therefore the mutual coupling can be often neglected.

The analysis of the motor consists of the computation of the flux linkage, the magnetic coenergy, and the torque as functions of the rotor position and the current.

Due to absence of the symmetry between rotor and stator, the analysis is carried out over the whole motor structure, as shown in Figure 12.3(a). However, the study can be reduced to only half a structure, as shown in Figure 12.3(b), with the advantageous reduction of the analysis domain.

Regarding the mesh of the domain, the higher gradient of the magnetic field strength is found in correspondence of the air-gap. In such a region, the number of elements is properly increased. Since the magnetic circuit of the motor is extremely saturated, it is necessary to have a high number of elements also in the stator and rotor poles, especially when they are partially overlapped, to the aim of guaranteeing the required accuracy of the field solution. The number of the elements might be established at first glance; however, it is better to carry out a preliminary analysis of the influence of the number of elements of the mesh on the field solution, as will be shown in the following example.

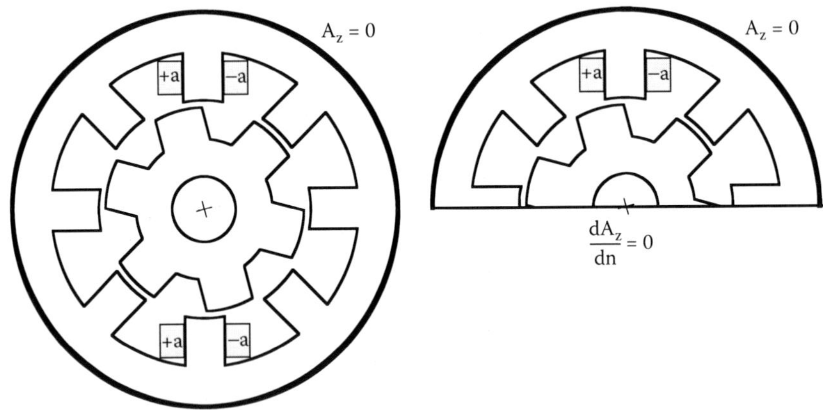

FIGURE 12.3
Structure of the switched reluctance motor for the finite element analysis.

12.4 Computation on the Solved Structure

The flux linkage $\lambda(\vartheta_m, i)$, the magnetic coenergy $W'_m(\vartheta_m, i)$, and the static torque $T(\vartheta_m, i)$ are computed from the field solutions. These quantities have to be computed as a function of the angular position ϑ_m and of the phase current i. Their computation was described in previous chapters, and is omitted here.

The torque can be evaluated from Equation (12.1) or, alternatively, directly from the field solution, by means of Maxwell's stress tensor. In the first case, it is important to take care with the angular displacement $\Delta\vartheta_m$. In the second case, it is possible to verify that the computation depends greatly on the finite element density in the air-gap and in the surrounding regions. The main reason for this dependence is the irregular and sharp behavior of the tangential component of the flux density in the air-gap, computed using the finite element method. Conversely, the radial component maintains a quite uniform behavior and is almost independent of the adopted number of mesh elements. Consequently, the choice of the integration line influences the computation of Maxwell's stress tensor. Such a computation improves the subdividing of the air-gap into two or three parts.

12.4.1 Dynamic Analysis

The magnetic parameters (which are state functions of the state variables ϑ_m and i) are determined from magnetostatic analyses at different values of ϑ_m and i. In particular the flux linkage λ and the torque T are found as state variables. They are necessary for the following prediction of the steady-state and dynamic performance of the motor.

A dynamic analysis of the switched reluctance motor is practically compulsory, due to the phase commutations and the nonlinearity of the magnetic circuit. The aim of the dynamic analysis is to evaluate the waveform of the motor torque for given turn-on ϑ_{on} and turn-off ϑ_{off} angles. It is also possible to determine these two angles so as to maximize the average torque and minimize the torque ripple.

A finite element step-to-step analysis is possible, but it is not convenient. In fact, for any change in the control strategy, a new finite element simulation is required, with a consequent long-time consumption. Conversely, if the magnetic model of the motor is built, so that the state functions are known, the integration of the differential equations is carried out with no further finite element analysis. An example is reported at the end of this chapter.

The rotor speed ω_m is assumed to be fixed, neglecting the variation of the mechanical quantities. Conversely, we focus on the electrical quantity dynamics, which generally is faster.

Starting from Equation (12.5), it results that

$$\frac{di(t)}{dt} = \frac{1}{\left[\dfrac{\partial \lambda(i,\vartheta_m)}{\partial i}\right]} \left[v(t) - Ri(t) - \omega_m \frac{\partial \lambda(i,\vartheta_m)}{\partial \vartheta_m} \right] \tag{12.7}$$

where

$$L_{app} = \frac{\partial \lambda(i,\vartheta_m)}{\partial i} \tag{12.8}$$

represents the apparent inductance of the coil, and

$$E_m = \omega_m \frac{\partial \lambda(i,\vartheta_m)}{\partial \vartheta_m} \tag{12.9}$$

represents the motional EMF. Both of them are functions of the state variables current i and angular position ϑ_m. Equation (12.7) has to be integrated numerically, considering the angular position $\vartheta_m = \omega_m t$ as the integration variable. The values of L_{app} and E_m are updated at each integration step $\Delta\vartheta_m$. At last, the current is computed by integrating Equation (12.7), and the flux linkage $\lambda(\vartheta_m,i)$ and the motor torque $T(\vartheta_m,i)$ are obtained from the model built from the magnetostatic analysis.

Interpolation methods are used to limit the number of field analyses. Some values of the state variables are chosen; then the tables of the corresponding flux linkage and torque are built, and the other values are obtained by means of the interpolation. An example of a two-dimensional interpolation is described in Section 11.7 of Chapter 11.

12.5 Example

As an example, a switched reluctance motor with $Q_s/Q_r = 8/6$ is analyzed. Its main dimensions are reported in Table 12.1.

Figure 12.4 shows the flux lines when the a-phase coil is supplied. Figure 12.4(a) shows the complete section of the motor, Figure 12.4(b) shows only half a section, and Figure 12.4(c) displays a detail of the flux lines in the region of the stator and rotor poles when they are partially aligned. It is important to check this region, since the two poles operate in a very saturated

TABLE 12.1

Main Dimension of the Switched Reluctance Motor

	Quantity	Value		Quantity	Value
Q_s	Stator pole number	8	g	Air-gap	0.625 mm
Q_r	Rotor pole number	6	N_t	Number of turns	93
β_s	Stator polar arc	20.1 deg	R	Resistance	3 Ω
β_r	Rotor polar arc	21.6 deg	P_n	Rated power	4 kW
L_{Fe}	Axial length	152 mm	I_{max}	Maximum current	18 A

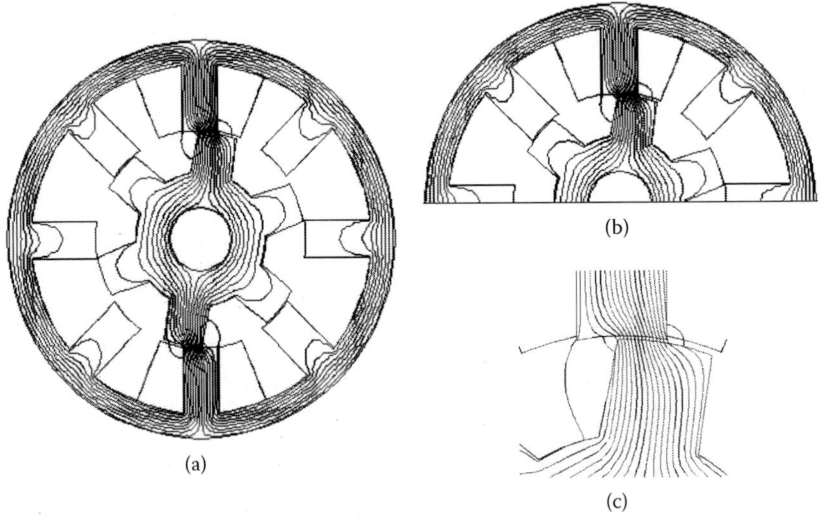

FIGURE 12.4
Flux lines in a switched reluctance motor.

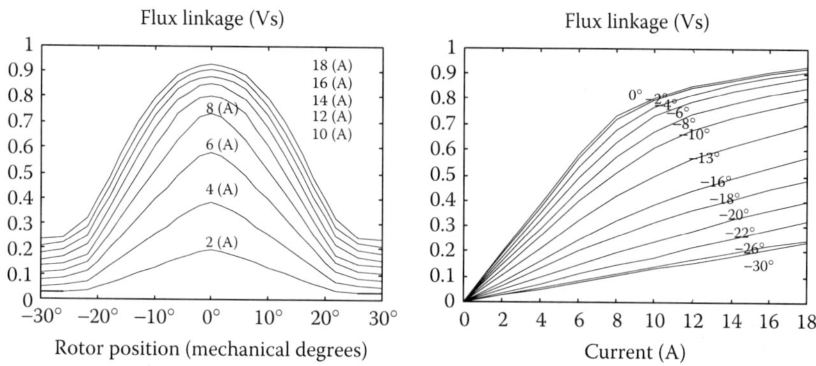

FIGURE 12.5
Flux linkage as a function of the rotor position (a) and of the current (b).

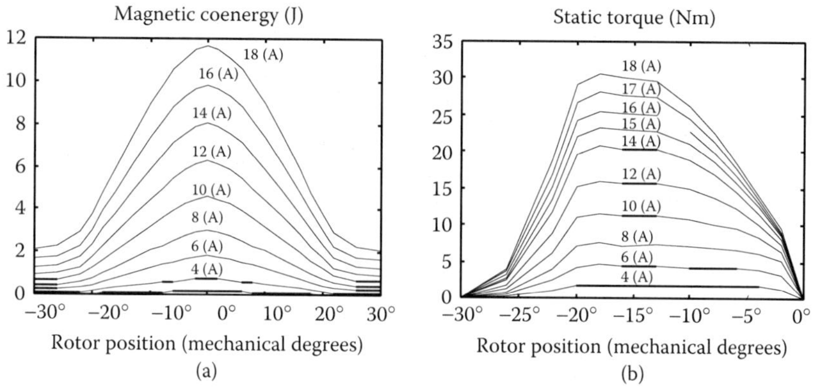

FIGURE 12.6
Magnetic coenergy (a) and static torque (b) versus rotor position and current.

manner. At the higher currents, they reach such a high saturation level that the differential magnetic permeability approaches unity.

Figure 12.5 shows the flux linkage as a function of the angular position and of the current, obtained from magnetostatic analysis. It is observed that when the pole pairs are not overlapped or when the current is low, the flux linkage is low. The motor works in linear conditions and the flux linkage is proportional to the current. On the contrary, when the current is high and when the pole pairs are overlapped, the effect of the magnetic saturation is manifest. The flux linkage is not more proportional to the current, and the slope of the curves decreases.

Figure 12.6(a) shows the magnetic coenergy as a function of the rotor angular position and the phase current. At given ϑ_m, the magnetic coenergy is proportional to the squared current only when the current is low and the poles are not aligned. When the current increases, due to the saturation, it becomes proportional to the current only. It follows that the torque-to-current ratio becomes almost constant.

Figure 12.6(b) shows the static torque as a function of the rotor angular position and the phase current. It is obtained in each position ϑ_m as the derivative of the magnetic coenergy with respect to ϑ_m, with fixed current. It is worth noticing that the torque is almost constant when the rotor position is in the range between $\vartheta_m = -5$ deg and $\vartheta_m = -20$ deg.

12.5.1 Effect of the Number of Elements of the Mesh

Figure 12.7(a) shows the behavior of the radial component B_r of the flux density in the middle of the air-gap, with two different numbers of finite elements in which the domain has been divided. The two behaviors are essentially uniform and independent of the number of elements of the mesh. In contrast, Figure 12.7(b) shows the behavior of the tangential component B_t of the flux density in the middle of the air-gap, in the same conditions. It is worth noticing that such a behavior is very sharp near the edge of the poles. In addition there are considerable variations of its shape with the number of elements. This yields a strong dependence of the torque computed by means of Maxwell's stress tensor on the chosen mesh. As a result, the number of elements of the mesh has to be carefully selected.

The two methods used for the torque computation are compared next. Figure 12.8(a) shows the static torque corresponding to the currents of 12 A and 18 A, at a fixed rotor position ϑ_m and different numbers of elements of the mesh. The torque has been computed as the variation of the magnetic coenergy and by means of Maxwell's stress tensor. It is worth noticing the strong dependence of the computed torque on the numbers of finite elements.

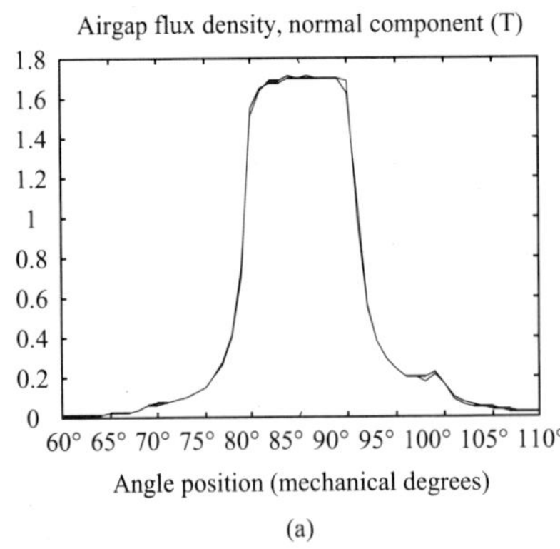

(a)

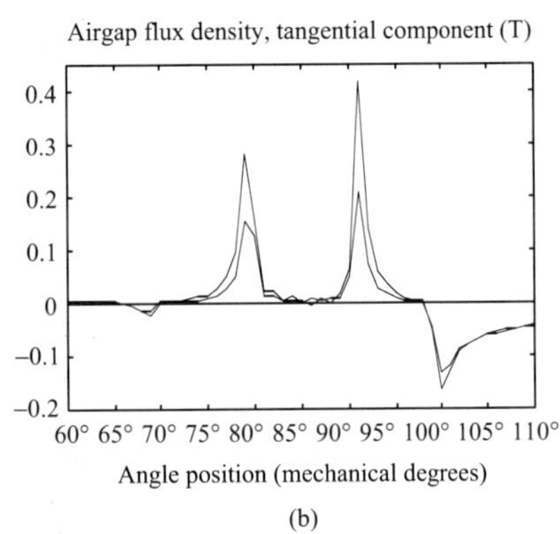

(b)

FIGURE 12.7
Normal component (a) and tangential component (b) of the air-gap flux density with different numbers of elements of the mesh.

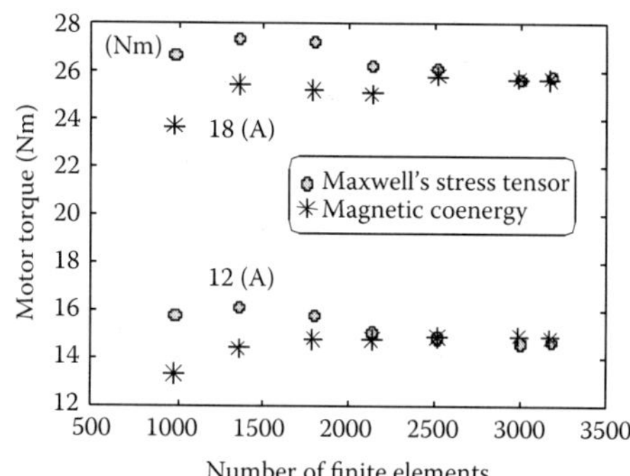

FIGURE 12.8
Dependence of the motor torque on the number of finite elements of the mesh.

The torque converges to an asymptotic value only with an increasing number of elements. It is interesting to note that the computation by means of the magnetic coenergy variation (an integral quantity) exhibits a convergence to the asymptotic value with a lower number of elements, compared to the computation by means of Maxwell's stress tensor (which process differential quantities).

12.5.2 Simulation of the Dynamic Performance

This section reports some simulation results of the dynamic performance of the motor. The waveforms of phase current, flux linkage, and torque are computed considering a dc voltage supply equal to $V_{dc} = 400V$.

Figure 12.9 shows the waveforms of the applied voltage, the flux linkage, the current, and the motor torque, referring to some fixed values of the rotor speed ω_m (considered constant in the dynamic analysis) and the commutation angles ϑ_{on} and ϑ_{off}. Such values are summarized in Table 12.2. In the simulations (b) and (c) the current is limited: a hysteresis current control has been introduced, in order to limit the maximum current value to 10 A.

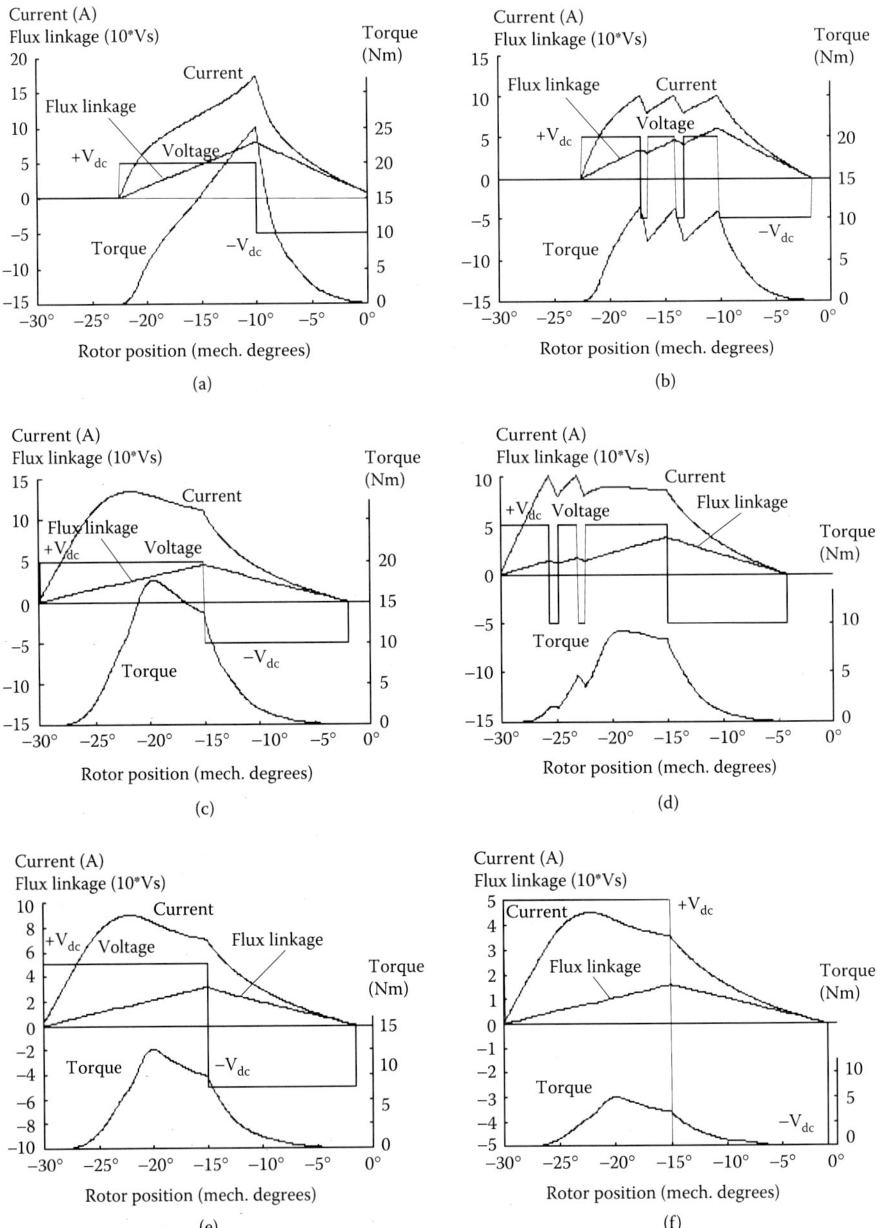

FIGURE 12.9
Simulations of the switched reluctance motor at constant rotor speed.

TABLE 12.2

Speed and Control Angles for the Dynamic Simulations

	n(rpm)	ϑ_{on} (deg)	ϑ_{off} (deg)	Current Limitation
(a)	1000	–22.5	–10.0	No
(b)	1000	–22.5	–10.0	Yes
(c)	2000	–30.0	–15.0	No
(d)	2000	–30.0	–15.0	Yes
(e)	3000	–30.0	–15.0	No
(f)	6000	–30.0	–15.0	No

References

1. P.J. Lawrenson et al., "Variable speed switched reluctance motors," in *IEE Proc., Electr. Power Appl.*, pt. B, vol. 127, no. 4, pp. 253–65, 1980.
2. J.F. Lindsay, R. Arumugam, and R. Krishnam, "Finite-Element Analysis Characterization of a Switched Reluctance Motor with Multitooth per Stator Pole," *IEE Proc., Electr. Power Appl.*, vol. 133, pp. 347–353, Nov. 1986.
3. P. Materu, R. Krishnan, "Estimation of switched reluctance motor losses," in *Conf. Rec. IEEE Ind. Applicat. Soc. Annu. Meeting*, Pittsburgh, pp. 79–90, October 1988.
4. I.D. Mayergoyz, "Dynamic Preisach Models of Hysteresis," in *IEEE Transactions on Magnetics*, vol. 24, pp. 2925–2977, 1988.
5. T.J.E. Miller, *Switched Reluctance Motor Drive*, Claredon Press, Oxford, 1989.
6. T.J.E. Miller, *Brushless Permanent-Magnet and Reluctance Motor Drive*, Claredon Press, Oxford, 1989.
7. I. Husain and M. Ehsani, "Torque Ripple Minimization in Switched Reluctance Motor Drives by PWM Current Control," *Proc. Appl. Power Electr. Conf.*, Orlando, FL, 1996, pp. 72–77.
8. P.C. Kjaer, P. Nielson, L. Anderson, and F. Blaabjerg, "A New Energy Optimizing Control Strategy for Switched Reluctance Motor," *IEEE Trans. Industry Appl.*, vol. 31, no. 5, pp. 1088–1095, 1995.
9. N. Bianchi, S. Bolognani, M. Zigliotto, "Prediction of Iron Losses in Switched Reluctance Motors," in *Proc. of 7th International Power Electronics and Motion Control Conference, PEMC '96*, Budapest, Hungary, 2–4 September 1996, vol. 3, pp. 223–228.
10. R. Krishnan, P. Vijayraghavan, "State of the Art: Acustic Noise in Switched Reluctance Motor Drives," *Proc. of IEEE-IECON Conf.*, Aachen (D), pp. 929–934, 1998.
11. P.G. Barrass and B.C. Mecrow, "Flux and Torque Control of Switched Reluctance Machines," *IEE Proc., Electr. Power Appl.*, vol. 145, no. 6, pp. 519–527, Nov. 1998.

13

Three-Phase Induction Motors

This chapter deals with the steady-state analysis of a three-phase induction motor, supplied by sinusoidal voltages. Two different approaches of analysis are described. The first one is based on the combination of the field solution and the equivalent circuit, while the second one is based exclusively on the field solution. Both methods are applied to a two-dimensional domain; thus the three-dimensional effects are computed analytically.

The two methods are distinguished by their analysis approaches. The first method reproduces the indirect motor tests, that is, the no-load test and the locked rotor (or short-circuit) test. From the field solutions characterized by forced current sources, the motor parameters are determined so that the equivalent circuit is built. From an equivalent circuit, the main characteristics of the motor under load are studied with given voltage.

Conversely, the second method combines the field solution directly with the electrical circuit analysis, forcing the voltage source on the motor terminals. Thus, the motor operations under load are reproduced.

13.1 Introduction

Each component of the field quantities is assumed to vary sinusoidally with the time. The symbolic notation is adopted. A dot is placed over each field quantity to represent the complex phasor. The value at the instant t assumed by the components G_x, G_y, and G_z of the generic field quantity G(P,t), that depends on the point P = (x,y,z) and the time t, is given by

$$\begin{aligned} G_x(P,t) &= \text{Im}\left[\dot{G}_x(P,t)\right] = \text{Im}\left[\hat{G}_x(P)e^{j(\omega t+\vartheta_x(P))}\right] \\ G_y(P,t) &= \text{Im}\left[\dot{G}_y(P,t)\right] = \text{Im}\left[\hat{G}_y(P)e^{j(\omega t+\vartheta_y(P))}\right] \\ G_z(P,t) &= \text{Im}\left[\dot{G}_z(P,t)\right] = \text{Im}\left[\hat{G}_z(P)e^{j(\omega t+\vartheta_z(P))}\right] \end{aligned} \tag{13.1}$$

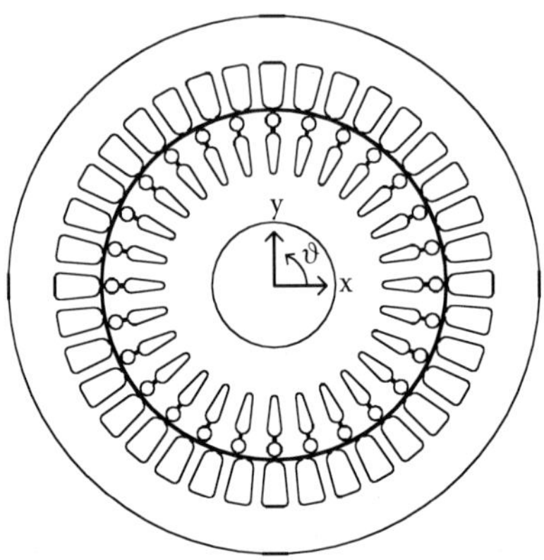

FIGURE 13.1
Frontal section of the three-phase induction motor.

With reference to Figure 13.1, the current density vector **J** and the magnetic vector potential **A** are normal to the considered (x, y) plane, i.e., they have only the component parallel to the z-axis. They are

$$\begin{aligned} \mathbf{J} &= (0, 0, \dot{\mathrm{J}}_z) \\ \mathbf{A} &= (0, 0, \dot{\mathrm{A}}_z) \end{aligned} \tag{13.2}$$

This implies that the magnetic field strength vector **H** and the flux density vector **B** have components on the (x, y) plane only, normal to the z-axis, i.e.,

$$\begin{aligned} \mathbf{H} &= (\dot{\mathrm{H}}_x, \dot{\mathrm{H}}_y, 0) \\ \mathbf{B} &= (\dot{\mathrm{B}}_x, \dot{\mathrm{B}}_y, 0) \end{aligned} \tag{13.3}$$

In this way, a two-dimensional analysis is carried out, yielding to shorter computation time and field solutions that are more easily interpreted. The 3D effects are particularly important in determining the three-phase induction motor performance and cannot be neglected. They are due to the finite axial length, i.e., due to stator end-winding and rotor rings, as well as the rotor slot skewing.

These 3D effects are considered including appropriate elements in the equivalent circuit, which are external to the field solution. The presence of these additional elements is sketched in Figure 13.2, which shows the motor section analyzed by means of the 2D finite element (2D-FE) method together with the electrical parameters that consider the 3D effects.

Concerning the stator winding, Figure 13.2 highlights the resistance R_s and the end-winding leakage inductance $L_{\sigma s,3D}$. Concerning the rotor winding,

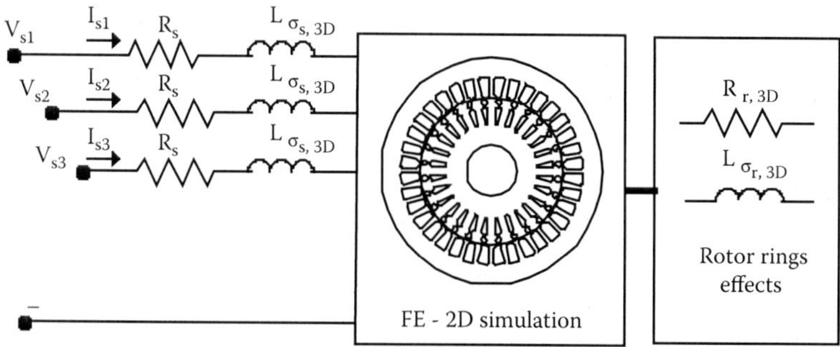

FIGURE 13.2
Sketch of the analysis of the three-phase induction motor.

Figure 13.2 highlights the resistive and leakage inductance of the rotor rings, i.e., $R_{r,3D}$ and $L_{\sigma r,3D}$. In addition, the parameters can be modified to take into account the possible rotor slot skewing.

Thanks to the geometrical and magnetic symmetry of the induction motor, the analysis domain is always reduced to a portion of the motor section. Such a portion is essentially established by the stator and rotor slot numbers and by the stator winding. Often the analysis domain is equal to one pole piece. In the case of the four-pole induction motor of Figure 13.1, the analysis domain is as shown in Figure 13.3, to which we will refer in the following sections.

As far as the geometry symmetry is concerned, usually there are no difficulties, especially if the slot number is a multiple of the pole number. On the contrary, particular care has to be paid to the magnetic symmetry, essentially due to the stator winding.

The analysis domain can be reduced, imposing suitable periodic boundary conditions. Moreover, the stator and the rotor can be analyzed separately. This kind of field analysis is based on the studies by Prof. S. Williamson, reported in the References, but is omitted hereafter.

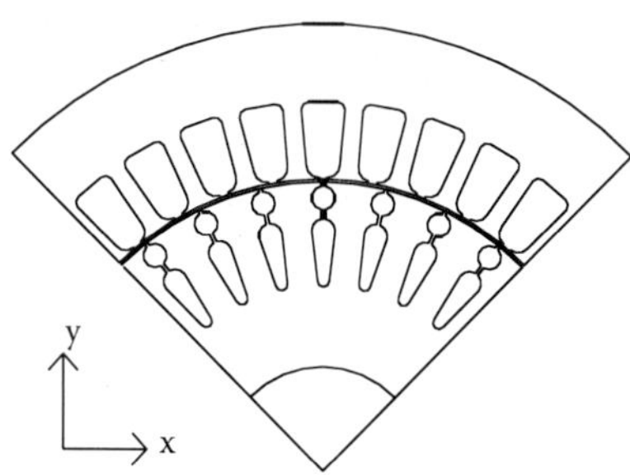

FIGURE 13.3
A quarter of the induction motor section of Figure 13.1.

13.2 Simulations of the Indirect Tests

In this first analysis approach, the induction motor is modeled by means of a simplified equivalent circuit and the field analysis is carried out with the aim of determining the parameters of the circuit. Once the latter is completely built, it is possible to assess the induction motor characteristics with a good accuracy, in a wide operating range.

The finite element analysis consists of the simulations of the indirect tests carried out on the induction motor: the no-load test and the locked rotor test. In the equivalent circuit, nonlinear parameters are adopted. Such parameters are easily found by means of the finite element analysis and allow some limitations of the equivalent circuit approach to be overcome. For instance, the nonuniform distribution of current in the rotor bar varies the rotor resistance, or the saturation of the magnetic materials varies the magnetizing inductance.

Once the equivalent circuit is built, the overall performance of the induction motor is computed, in particular, the torque, the stator current, the power factor, the Joule losses, each of which is a function of the rotor mechanical speed. Particular care is paid to the estimation of the motor efficiency, taking into account the mechanical and iron losses. The latter are not considered in the field solution; they have to be computed by means of the classical analytical methods.

13.2.1 Equivalent Circuit

The variable parameters equivalent circuit of Figure 13.4 is considered. In such a circuit, the parameters that are obtained by the 2D field analysis are

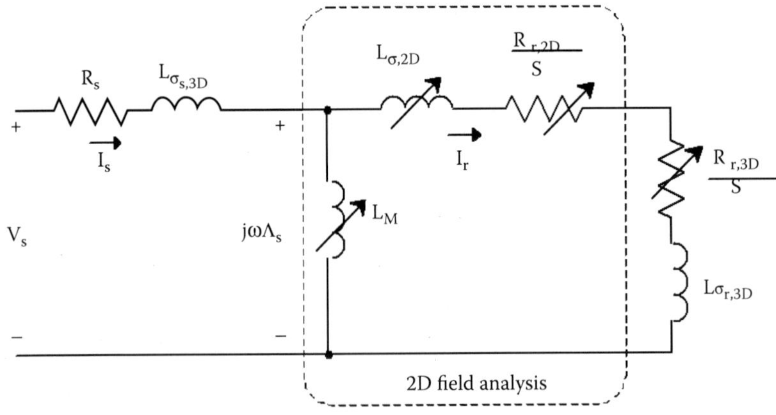

FIGURE 13.4
Equivalent circuit of the three-phase induction motor, highlighting the parameters obtained from the 2D finite element analysis.

placed inside the dashed box, while outside there are the parameters due to the 3D effects.

Figure 13.4 highlights all the parameters inside the box that are not constant. The magnetizing inductance L_M is a function of the stator flux linkage Λ_s, considering the saturation effects, even if attributed to the stator current only. The total leakage inductance $L_{\sigma,2D}$ and the rotor resistance $R_{r,2D}$ are functions of the rotor frequency f_r, which is of the rotor slip s, considering the effects of the current distribution in the rotor bars.

Conversely, the parameters corresponding to the 3D effects are considered constant. The stator winding resistance R_s and the end-winding leakage inductance $L_{\sigma s,3D}$ are assumed to be independent of the nonuniform distribution of the current. Similarly, the rotor parameters $R_{r,3D}$ and $L_{\sigma r,3D}$ are considered constant. They take into account the rotor rings and the bar skewing. They are computed analytically and are thus included in the equivalent circuit. The iron losses are considered separately; therefore the resistance R_o, that represents such losses is not considered.

At last, let us observe that a Γ-type representation is adopted for the part of circuit associated with the 2D parameter (inside the dashed box). A transformation from the classic T-type representation to the Γ-type representation is operated (see Chapter 6).

13.2.2 No-Load Test

At first the no-load test is simulated. The purpose of such simulations is the evaluation of the nonlinear magnetizing inductance L_M.

Assuming the field harmonics to be negligible and the rotor speed to be synchronous with the rotating magnetic field generated by the stator winding (i.e., a slip s = 0), then no current is induced in the rotor bars. The rotor becomes exclusively a part of the nonlinear magnetic path for the magnetic flux. A magnetostatic field analysis is then possible. The electric quantities are frozen in an useful time instant, say t*. This is chosen so as to satisfy the magnetic symmetry conditions. The field problem is then nonlinear and described by time-varying quantities. The simulation of the no-load test corresponds to the circuit of Figure 13.5.

Using Equation (13.2) and Equation (13.3) and letting J_{osz} be the z-axis component of the stator current density at no-load, the magnetic vector potential distribution can be obtained by integrating the following differential equation:

$$\frac{\partial}{\partial x}\left(\frac{1}{\mu}\frac{\partial A_z}{\partial x}\right)+\frac{\partial}{\partial y}\left(\frac{1}{\mu}\frac{\partial A_z}{\partial y}\right)=J_{osz} \tag{13.4}$$

where μ is the magnetic permeability; it is μ_o in the air-gap and in the conductor, while it is given by the B-H curve in the magnetic materials. If

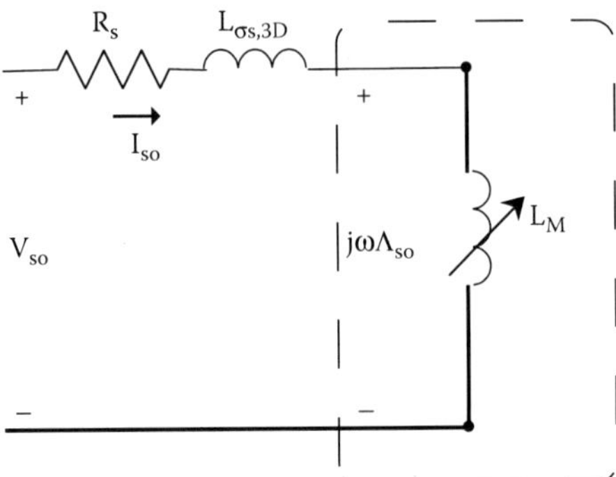

FIGURE 13.5
Reduction of the equivalent circuit during the no-load simulation.

the shaft is magnetic, it is considered as a part of the domain *D* and characterized by its B-H curve; if it is not magnetic, it can be omitted from the field analysis.

In a similar way to the computation of the synchronous inductances (see Chapter 8), the three-phase currents I_a, I_b, I_c are fixed and the corresponding flux linkages Λ_{sa}, Λ_{sb}, Λ_{sc} are computed. An equivalent coil side is considered in each slot. Such a coil side carries an equivalent current given by the algebraic sum of the currents flowing in the actual conductors within the slot itself. Let n_q be the number of conductors per slot, then $k_{jq}n_q$ indicates the number of conductors of the j-th phase within the q-th slot. The definition of k_{jq} is given in Chapter 8. Let $i_{oj}(t^*)$ be the current of the j-th phase at the time instant t*, corresponding to a given RMS phase current I_{so}. Such a current may be referred to as the line current, using the ratio 1 or $1/\sqrt{3}$ according to the stator winding connection, star of delta connection, respectively.

Then the equivalent current in the q-th slot is given by

$$I_q = \sum_{j=a,b,c} \frac{n_q}{n_{pp}} k_{jq} i_{oj}(t^*) \qquad q = 1, 2, \ldots, Q_s/2p \tag{13.5}$$

where n_{pp} is the number of parallel paths of the stator winding.

The currents I_q (q = 1, …, $Q_s/2p$) are forced within the slots and the field problem is solved. Hence, the stator flux linkages are obtained by means of the integral loop of the magnetic vector potential A_z. Using the equivalent coil sides, all of them with the same section $S = S_q$, the instantaneous value of the j-th phase flux linkage during the no-load test is given by

$$\Lambda_{oj} = 2pL_{Fe} \frac{n_q}{n_{pp}} \sum_{q=1}^{Q_s/2p} k_{jq} \frac{1}{S_q} \int_{S_q} A_z dS \qquad j = a, b, c \tag{13.6}$$

where L_{Fe} is the net axial length of the lamination stack. From the three-phase flux linkages, the RMS value of the no-load flux linkage Λ_{so} is obtained.

The same stator flux linkage may be also obtained starting from the air-gap flux density distribution. It is necessary to evaluate the radial component of the flux density $B_r(\vartheta,t^*) = \mathbf{B}(\vartheta,t^*)\cdot\mathbf{u}_r$ along a circumference in the middle of the air-gap. Using the Fourier series expansion, from the distribution $B_r(\vartheta,t^*)$, the amplitude of the fundamental harmonic B_{r1} is computed. Then, the average value of the fundamental harmonic distribution in a polar pitch is $(2/\pi)B_{r1}$. The corresponding average magnetic flux per pole is

$$\Phi_{o1} = \left(\frac{2}{\pi}B_{r1}\right)\frac{\pi D L_{Fe}}{2p} = B_{r1}\frac{D L_{Fe}}{p} \tag{13.7}$$

and then the corresponding stator flux linkage (the peak value) becomes

$$\hat{\Lambda}_{so} = \frac{k_w N}{p}\Phi_{o1} \tag{13.8}$$

where k_w is the winding factor of the fundamental harmonic and N is the number of series conductors per phase. The flux linkage is variable in a sinusoidal way. Then the RMS value of the induced EMF, due to the only fundamental component of the flux density, is given by

$$E_o = \omega\Lambda_{so} = \omega\frac{\hat{\Lambda}_{so}}{\sqrt{2}} = \frac{\pi}{\sqrt{2}}\, f\, k_w N \Phi_{o1} \tag{13.9}$$

At last, the nonlinear magnetizing inductance L_M of Figure 13.4 is

$$L_M\left(\Lambda_{so}\right) = \frac{\Lambda_{so}}{I_{so}} \tag{13.10}$$

It is assumed that the dependence of L_M on the saturation is the same at no-load and under load, that is, $L_M(\Lambda_s) = L_M(\Lambda_{so})$.

13.2.3 Locked Rotor Test

The locked rotor test is simulated. During the test, the rotor is locked in a fixed position and three-phase sinusoidal currents are forced in the stator windings. The dependence of the parameters $R_{r,2D}$ and $L_{\sigma,2D}$ of Figure 13.4 on the rotor frequency f_r can be assessed by changing the source frequency. The rotor frequency is the same as the source frequency f, because of the locked rotor.

In the locked rotor test, the rated currents correspond to relatively low source voltages, then to low magnetic fluxes. As a consequence, the magnetic materials are assumed to be linear. The field problem is then linear and characterized by field quantities varying sinusoidally, at the electrical frequency $\omega = \omega_r = 2\pi f_r$. The magnetic vector potential has only the z-axis component, as shown in Equation (13.2), that may be expressed by means of the phasor notation of Equation (13.1), which is

$$\dot{A}_z = \hat{A}_z \, e^{j(\omega t + \vartheta)} \tag{13.11}$$

with $\omega = \omega_r$.

Thus, with constant magnetic permeability μ, the field problem is described by the differential equation

$$\frac{1}{\mu}\left(\frac{\partial^2 \dot{A}_z}{\partial x^2} + \frac{\partial^2 \dot{A}_z}{\partial y^2}\right) = \dot{J}_{sz} - j\omega_r \sigma \dot{A}_z \tag{13.12}$$

where σ is the electrical conductivity and ω_r is the source electrical frequency.

The equivalent circuit corresponds to that reported in Figure 13.6. As shown in Figure 13.6, from the field analysis the rotor 3D effects are not taken into account. They will be analytically computed and added in the equivalent circuit later on. In other words, the parameters computed by the field analysis are only $R_{r,2D}$ and $L_{\sigma,2D}$. Considering the problem to be linear, the inductance L_M assumes a constant value, which was computed previously during the no-load test. The analysis is focused on the rotor parameters. From the locked rotor simulations, carried out at different frequencies f_r, it is possible to assess the functions $R_{r,2D}(f_r)$ and $L_{\sigma,2D}(f_r)$, which is to consider the effects of a nonuniform distribution of the current in the rotor bars at the different operating speeds.

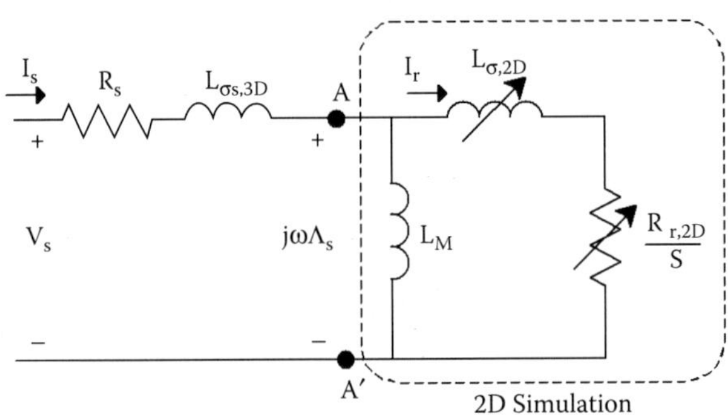

FIGURE 13.6
Equivalent circuit corresponding to the locked rotor test.

Also in this case, the field problem is solved referring to the equivalent coil sides. The shaft may be disregarded in the field analysis. In fact, the currents induced in the rotor bars shield the inner parts of the rotor significantly.

From the field solution, the Joule losses in the rotor bars are

$$P_{Jr} = \frac{1}{2\sigma_{Al}} 2pL \int_{S_{Al}} \dot{J}_z \cdot \tilde{J}_z dS \tag{13.13}$$

where L is the total length of the rotor bars, σ_{Al} is the Aluminium conductivity, and S_{Al} is the overall cross-section of the rotor bars in the simulated pole piece. The factor 2p considers that one pole piece only is simulated.

The average value of the magnetic energy in the magnetic material is

$$W_m = 2pL_{Fe} \frac{1}{2} \int_D \frac{\hat{B}}{\sqrt{2}} \frac{\hat{H}}{\sqrt{2}} dS \tag{13.14}$$

With the hypothesis of a rotating magnetic field, exactly sine-distributed, both P_{Jr} and W_m are constant in time. The real and imaginary components of the equivalent impedance at the terminals AA′ of the nonlinear circuit of Figure 13.6, indicated by $R_{eq} + j\omega_r L_{eq}$ in Figure 13.7, are obtained. Such parameters represent the connection elements between the field solution and their representation by means of the lumped-parameters circuit. They are computed as

$$R_{eq} = \frac{P_{Jr}}{3I_s^{\,2}} \qquad L_{eq} = \frac{2W_m}{3I_s^{\,2}} \tag{13.15}$$

where I_s is the RMS value of the current used in the test.

At this point, since L_M is known from the no-load simulation and equal to its unsaturated value, the resistance $R_{r,2D}(f_r)$ and the inductance $L_{\sigma,2D}(f_r)$ of the circuit in Figure 13.6 can be computed.

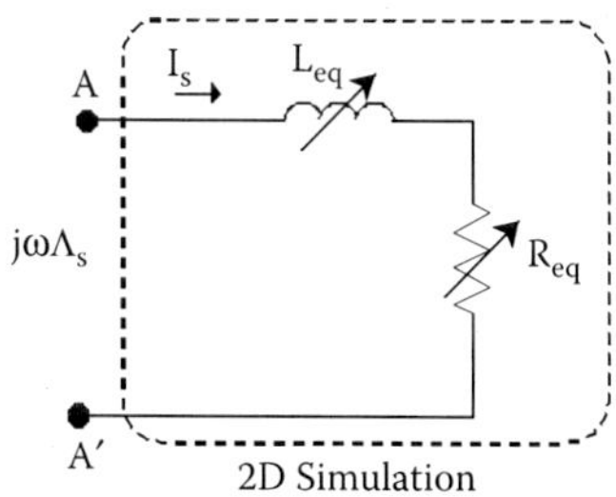

FIGURE 13.7
Electrical parameters corresponding to the locked rotor test.

13.2.4 Construction of the Equivalent Circuit

The parameter $L_M(\Lambda_s)$, obtained from the no-load test, together with $R_{r,2D}(f_r)$ and $L_{\sigma,2D}(f_r)$, obtained from the locked rotor test, constitute the electrical parameters within the dashed box of the nonlinear circuit of Figure 13.4. In order to complete the circuit, the parameters corresponding to the stator end-winding, the rotor bar skewing, and the rotor ring effects have to be included in the circuit.

The leakage inductance $L_{\sigma s,3D}$ includes the stator end-winding leakage inductance and the leakage inductance caused by skewing. An estimation of the leakage inductance of the stator end-winding is

$$L_{\sigma s,ew} = \mu_o \frac{N^2}{2p} L_{ew} \lambda_{ew} \tag{13.16}$$

where L_{ew} is the effective end-winding length, and λ_{ew} is the specific permeance coefficient. Its value ranges between 0.35 and 0.55, according to the winding type.

The leakage inductance caused by skewing is estimated by

$$L_{\sigma s,sk} = L_M \left(1 - k_{sk}^2\right) \tag{13.17}$$

where the magnetizing inductance L_M is given by Equation (13.10) and k_{sk} is given by

$$k_{sk} = \frac{\sin\left(\varepsilon_{sk}/2\right)}{\varepsilon_{sk}/2} \tag{13.18}$$

and ε_{sk} is the skewing angle in electrical radians.

The short-circuited rings resistance is given by

$$R_{r,3D} = \rho_{Al} \frac{3\left(k_w N\right)^2}{2\pi p^2 S_{ring}} \tag{13.19}$$

Once the model of the induction motor is obtained, i.e., all the parameters of the equivalent circuit of Figure 13.4 are found, it is possible to predict the motor performance at different operating conditions. The motor characteristics are computed, for instance the mechanical characteristic, the current-speed curve, and so on. As explained earlier, the iron and mechanical losses, needed for the computation of the efficiency, have to be considered separately.

TABLE 13.1

Main Data of the Three-Phase Induction Motor

Rated power	22000W
Rated voltage	280/485V
Line rated current	32.7 A
Rated frequency	60 Hz
Rated speed	1750 rpm
Number of poles	4
Stack length	300 mm
Air-gap length	0.5 mm
Rotor bar skewing	1 slot
Service	S1
Protection	IP55
Insulation	Class F
Cooling	IC 41

13.2.5 Example

The finite element method is applied to a commercial three-phase induction motor with a double bar rotor. The geometry is shown in Figure 13.1. The main data of the induction motor are reported in Table 13.1.

The number of stator slots is $Q_s = 36$ so that the number of slots per pole per phase is $q_s = 3$ and the electrical slot angle is $\alpha_c = 20$ degrees. The stator winding is a double-layer winding, with two coil sides per slot, each of them formed by 14 copper conductors, thus $n_q = 28$. The coil pitch is $y_q = 9$, corresponding to one slot chording. At last there are $n_{pp} = 2$ parallel paths. A delta connection is adopted. The number of rotor aluminum bars is $Q_r = 28$.

13.2.5.1 No-Load Simulation

In the no-load simulation, a large number of elements of the mesh is used in the air-gap and in the iron regions, where the highest field gradients are expected. With no or low saturation of the iron paths, it is observed that the magnetic energy in the air-gap equals almost 80% of the total energy. The magnetic material is a TERNI 2650.

Because of the delta connection, the phase current is $I_o = I_{Lo}/\sqrt{3}$ where I_{Lo} is the line no-load current. The RMS values are considered. The current flowing in each coil, since there are $n_{pp} = 2$ parallel paths, is given by $I_c = I_o/2 = I_{Lo}/(2\sqrt{3})$.

The no-load test may be carried out in different ways, choosing a suitable reference time t* and coherent boundary conditions. Even if a smaller part of the motor may be simulated for the no-load analysis, the simulation is carried out on one pole piece of the motor, so as to use the same domain for the no-load and the locked rotor analysis with fixed winding distribution,

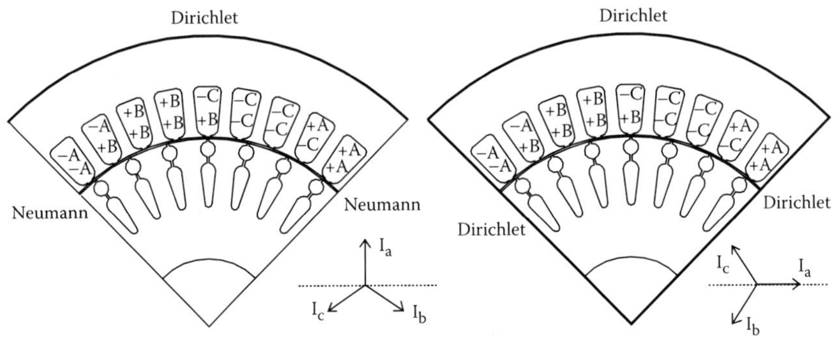

FIGURE 13.8
Two solutions of assignment of the stator currents and the boundary conditions for the no-load test: solution (a) and solution (b).

TABLE 13.2
Three-Phase Currents at the Time t* for the No-Load Simulation

Phase	No-load simulation	
	Solution (a)	**Solution (b)**
A	$+\sqrt{2}\, I_c$	0
B	$-\frac{\sqrt{2}}{2}\, I_c$	$+\sqrt{2}\frac{\sqrt{3}}{2}\, I_c$
C	$-\frac{\sqrt{2}}{2}\, I_c$	$-\sqrt{2}\frac{\sqrt{3}}{2}\, I_c$
–A	$-\sqrt{2}\, I_c$	0
–B	$+\frac{\sqrt{2}}{2}\, I_c$	$-\sqrt{2}\frac{\sqrt{3}}{2}\, I_c$
–C	$+\frac{\sqrt{2}}{2}\, I_c$	$+\sqrt{2}\frac{\sqrt{3}}{2}\, I_c$

the currents and the boundary conditions may be assigned as illustrated in Figure 13.8. They correspond to those used in Chapter 8. Equivalent coils are used: the corresponding current distribution is directly found once the instant t* is fixed, as reported in Table 13.2, where I_c is the RMS current flowing in each coil.

The flux lines and the air-gap flux density corresponding to the two no-load tests are shown in Figure 13.9 and Figure 13.10. It is easy to observe, in both simulations, the flux plot symmetry. This symmetry might be properly

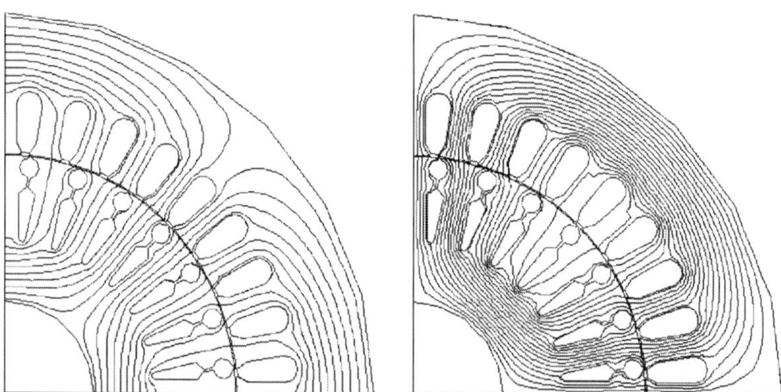

FIGURE 13.9
Flux plots in the two no-load simulations: referring to Table 13.2, solution (a) and solution (b).

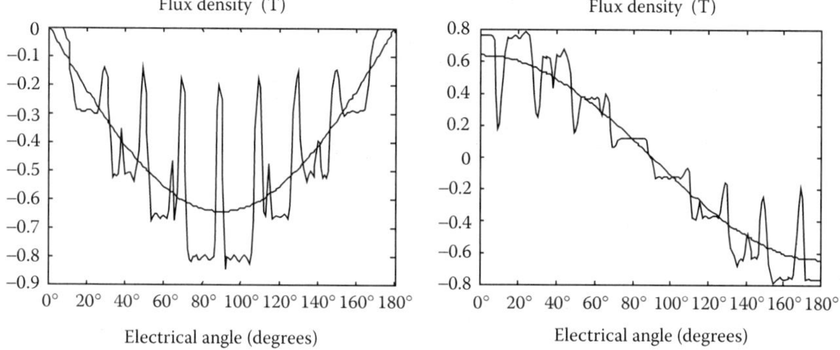

FIGURE 13.10
Air-gap flux density distributions in the two no-load simulations: referring to Table 13.2, solution (a) and solution (b).

used to halve the analysis domain. Finally, the effect of the slot openings is clear on the distribution of the air-gap flux density.

The magnetization curve is illustrated in Figure 13.11, showing the magnetizing current as a function of the source voltage. In order to stress the effect of the air-gap length, the no-load simulation is carried out on induction motor with different air-gaps.

At last, referring to the equivalent electric circuit of Figure 13.5, the magnetizing inductance L_M is computed as a function of the stator flux linkage $\Lambda_s = \Lambda_{so}$. It is shown in Figure 13.12.

It is worth noticing that using a current source allows the obtained result to be applied to motors of different stack length, but with the same laminations and the same winding distribution. In fact, the result is extended to motors with different stack length maintaining the same slot current. Of course, the values of line current and voltage have to be arranged according to the number of the actual conductors.

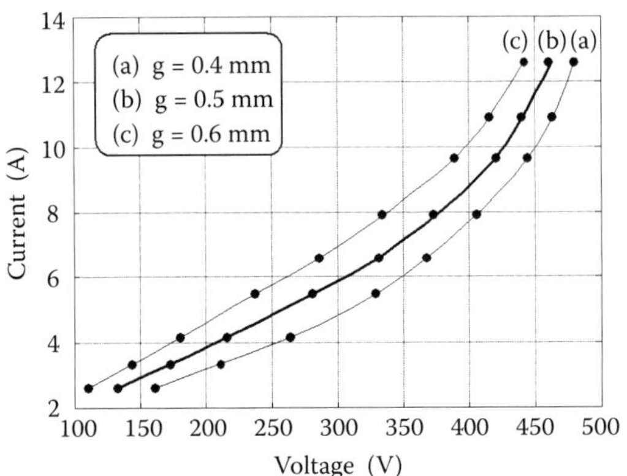

FIGURE 13.11
Magnetizing current versus source voltage with different air-gap lengths.

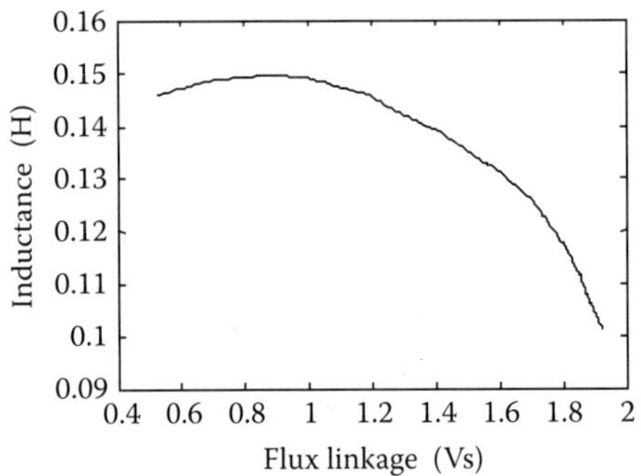

FIGURE 13.12
Magnetizing inductance L_M versus stator flux linkage Λ_s.

13.2.5.2 Locked Rotor Simulation

In the locked rotor simulation, a large number of finite elements is chosen in the region surrounding the air-gap and in the rotor bars. In fact, in the air-gap there are high field gradients, and in the rotor bars there is a non-uniform distribution of the current, caused by the operating frequency.

As depicted above, to the aim of freezing the value of the magnetizing inductance, the magnetic material is described by a constant relative magnetic permeability. The latter is fixed equal to the initial permeability of the used B-H curve, i.e., $\mu_r = 4000$. The conductivity of the aluminum is fixed to $\sigma_{Al} = 2.549 \cdot 10^7$ S/m, corresponding to a temperature of 120°C. An adequate value of the rotor cage conductivity is crucial for a correct analysis of the induction motor.

The induction motor is locked in a position. However, now the forced currents are sinusoidal, and are described by means of the complex notation. Let I_L be a fixed RMS value of the line current, then the RMS value of the

TABLE 13.3

Three-Phase Currents for the Locked Rotor Simulation

Phase	Locked Rotor Simulation
A	$+\sqrt{2}\,I_c + j0$
B	$-\frac{\sqrt{2}}{2} I_c - j\frac{\sqrt{6}}{2} I_c$
C	$-\frac{\sqrt{2}}{2} I_c + j\frac{\sqrt{6}}{2} I_c$
–a	$-\sqrt{2}\,I_c - j0$
–b	$+\frac{\sqrt{2}}{2} I_c + j\frac{\sqrt{6}}{2} I_c$
–c	$+\frac{\sqrt{2}}{2} I_c - j\frac{\sqrt{6}}{2} I_c$

current is each coil is given by $I_c = I_L/(2\sqrt{3})$, because of $n_{pp} = 2$ parallel paths. The three-phase currents are directly computed, as reported in Table 13.3.

As far as the boundary conditions are concerned, Dirichlet's boundary condition is assigned to the external circumference of the stator, forcing the flux lines to remain confined within the stator yoke. Similarly the same boundary condition is assigned between the shaft and the rotor lamination (the shaft is excluded from the analysis domain). Finally, along the lateral borders the periodic boundary conditions are assigned.

Figure 13.13 shows the flux plots relative to the locked rotor simulations. Figure 13.13(a) corresponds to a frequency $f = f_r = 60$ Hz, while Figure 13.13(b) corresponds to a frequency $f = f_r = 5$ Hz. The shielding effect of the rotor bars is as evident as the frequency increases.

From these simulations, the behavior of the resistance $R_{r,2D}(f_r)$ and the total leakage inductance $L_{\sigma,2D}(f_r)$, of the equivalent circuit of Figure 13.4 are obtained as a function of the rotor frequency f_r. As explained above, the latter corresponds to the supply frequency, since the rotor is locked. The two behaviors are shown in Figure 13.14.

13.2.5.3 Motor Performance

Once the variable lumped-parameter equivalent circuit is built, it is possible to assess the overall performance of the motor. In particular, the mechanical characteristic is computed at nominal voltage and nominal frequency, as reported in Figure 13.15 by the solid line.

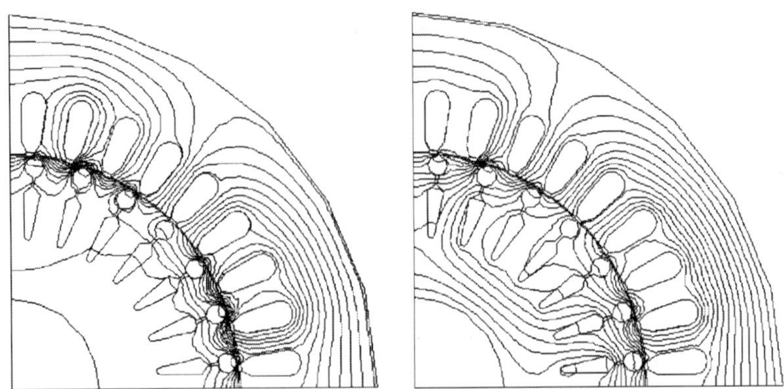

FIGURE 13.13
Flux plots of the locked rotor simulations, with source frequency (a) $f = f_r = 60$ Hz, and (b) $f = f_r = 5$ Hz.

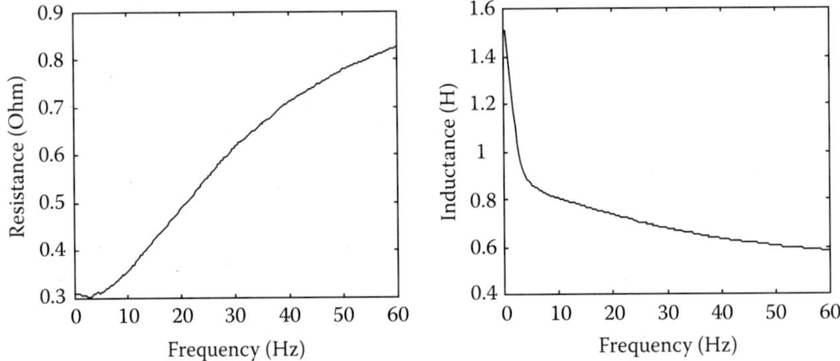

FIGURE 13.14
Variation of the resistance $R_{r,2D}$ and of the total leakage inductance $L_{\sigma,2D}$ with the rotor frequency f_r.

In the same Figure 13.15, some experimental results are reported, by circles. Comparing simulated and measured data, the agreement is satisfactory, during operations both at standstill and under load.

Analogously, Figure 13.16 shows the stator current as a function of the motor speed, while Figure 13.17 shows the power factor as a function of the phase current.

13.3 Motor Analysis Using Simulations Under Load

The second approach, alternative to that proposed in Section 13.2, is based on the simulations of the indirect tests. The motor operations under load are simulated directly. The aim of such an analysis approach is to assess each

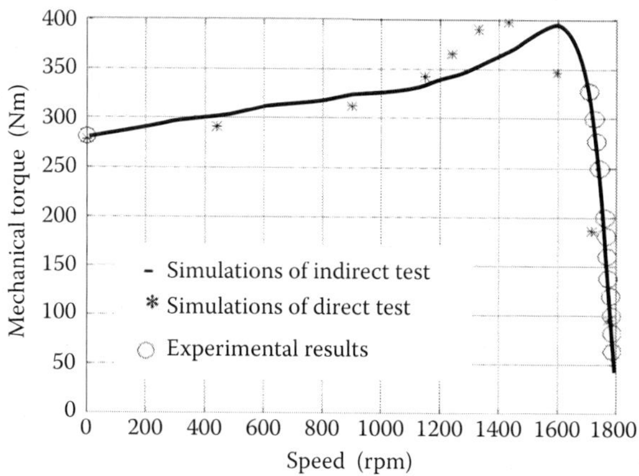

FIGURE 13.15
Torque-speed characteristic.

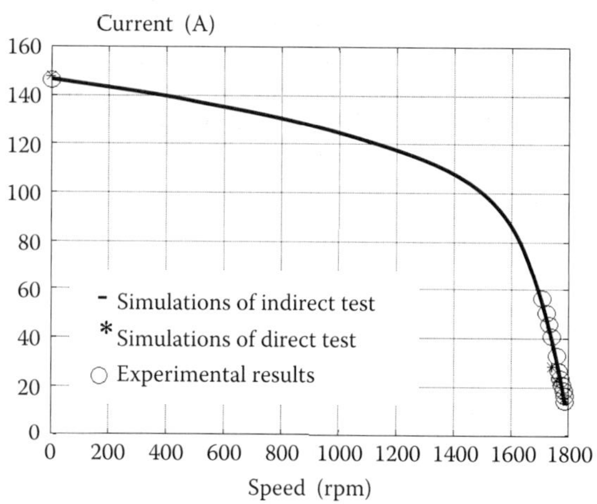

FIGURE 13.16
Stator current versus speed.

operating point, considering a sinusoidal voltage source. The motor characteristics are obtained by a set of simulations of the motor.

13.3.1 Magnetic Field Equations Under Load

Since a generic operating point is investigated, the simulations have to consider both the nonlinearity of the magnetic material and the nonuniform distribution of the current in the rotor bars, as well as the rotor movement.

The field problem is nonlinear and time-variable, and requires that we compute an induced current due to motional effects. The current density may be expressed by means of the constitutive law in σ given in Equation

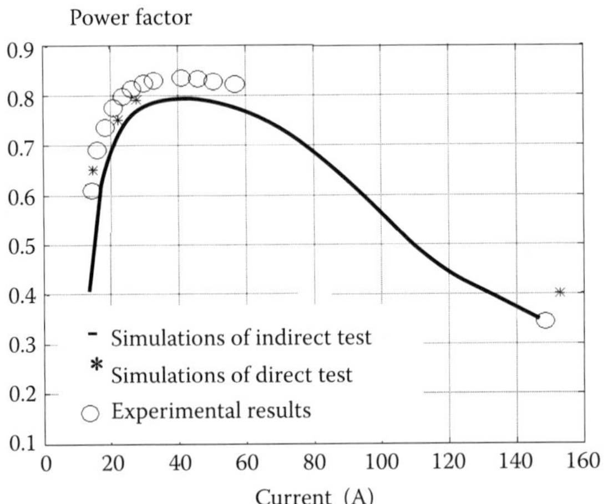

FIGURE 13.17
Power factor versus stator current.

(1.87), as a function of the specific electric force. Substituting for $\mathbf{E}_t$ in Equation (1.87) the expressions of the induced and motional specific electric forces, $\mathbf{E}_i$ of Equation (1.81) and $\mathbf{E}_m$ of Equation (1.85), respectively, and considering the forced current density, that can be expressed as $\mathbf{J}_s = \sigma(\mathbf{E}_c + \mathbf{E}_{ne})$, it results that

$$\mathbf{J} = \mathbf{J}_s - \sigma\frac{\partial \mathbf{A}}{\partial t} + \sigma\mathbf{v}\times\mathbf{B} \tag{13.20}$$

Because of the 2D problem, with only the z-axis components of the current density vector **J** and the magnetic vector potential **A**, as indicated in Equation (13.2), the field problem is described by the following differential equation:

$$\frac{\partial}{\partial x}\left(\frac{1}{\mu}\frac{\partial \dot{A}_z}{\partial x}\right) + \frac{\partial}{\partial y}\left(\frac{1}{\mu}\frac{\partial \dot{A}_z}{\partial y}\right) = \dot{J}_{sz} - \sigma\frac{\partial \dot{A}_z}{\partial t} + \sigma\,\mathbf{v}\times \mathrm{curl}(0,0,\dot{A}_z) \tag{13.21}$$

In the latter, please note that

$$\dot{J}_i = -\sigma\frac{\partial \dot{A}_z}{\partial t}$$

corresponds to the induced current density, while

$$\dot{J}_m = \sigma\,\mathbf{v}\times \mathrm{curl}(0,0,\dot{A}_z)$$

corresponds to the motional current density.

13.3.2 Alternative Formulation

The motional term, which appears essentially in the rotor, does not define a vector field. However, it depends on the particular reference system that is adopted. It is useful to change the reference system, so as to convert the motional term in a corresponding induced term.

Let us remember that, as in Equation (13.1), the z-axis component of the current density source is a sinusoidal function of the time, according to the relation

$$\dot{J}_{sz} = \hat{J}_{sz} e^{j\omega t} \tag{13.22}$$

In the same way, the z-axis component of the magnetic vector potential is

$$\dot{A}_z = \hat{A}_z e^{j(\omega t - \vartheta)} \tag{13.23}$$

The time derivative of the magnetic vector potential becomes

$$\frac{\partial \dot{A}_z}{\partial t} = j\omega \hat{A}_z e^{j(\omega t - \vartheta)} = j\omega \dot{A}_z \tag{13.24}$$

From the first Maxwell's equation (1.102), in which the displacement current density is negligible because of the low frequency, it is

$$\mathrm{curl}\mathbf{H} = \mathbf{J} \tag{13.25}$$

Substituting Equation (13.20) into Equation (13.25), it is

$$\mathrm{curl}\mathbf{H} + \sigma \frac{\partial \mathbf{A}}{\partial t} - \sigma \mathbf{v} \times \mathbf{B} = \mathbf{J}_s \tag{13.26}$$

Using the complex notation, and in particular Equation (13.24), it becomes

$$\mathrm{curl}\mathbf{H} + j\omega\sigma\mathbf{A} - \sigma \mathbf{v} \times \mathbf{B} = \mathbf{J}_s \tag{13.27}$$

In the cylindrical coordinate system of Figure 13.18, defined by the coordinate unity vectors ($\mathbf{u}_r$, $\mathbf{u}_\vartheta$, $\mathbf{u}_z$), the speed $\mathbf{v}$ of the generic point P of the rotor is given by

$$\mathbf{v} = r\omega_r \mathbf{u}_\vartheta \tag{13.28}$$

where ω_r is the rotor speed.

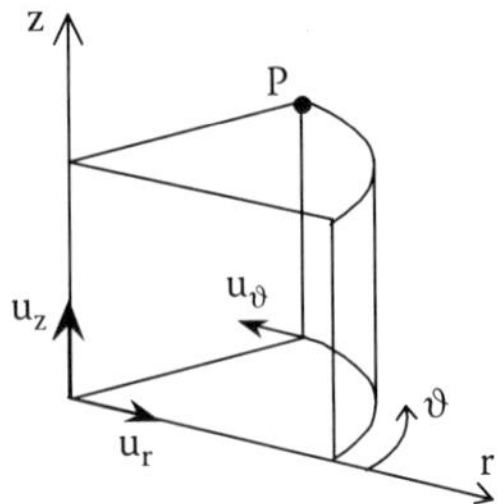

FIGURE 13.18
Cylindrical coordinate system.

Consequently, since $\mathbf{B}$ = curl$\mathbf{A}$, the motional specific electric force $\mathbf{E}_m$ can be expressed as

$$\mathbf{v} \times \mathbf{B} = r\omega_r \mathbf{u}_\vartheta \times \left(\text{curl}\mathbf{A} \cdot \mathbf{u}_r\right) \cdot \mathbf{u}_r \tag{13.29}$$

bearing in mind that the inner product $\mathbf{u}_r \cdot \mathbf{u}_r = 1$.

Taking into account that the magnetic vector potential $\mathbf{A}$ has only the z-axis component, the curl of $\mathbf{A}$ in cylindrical coordinates is

$$\text{curl}\mathbf{A} = \frac{1}{r}\frac{\partial \dot{A}_z}{\partial \vartheta}\mathbf{u}_r - \frac{\partial \dot{A}_z}{\partial r}\mathbf{u}_\vartheta \tag{13.30}$$

Substituting Equation (13.30) into Equation (13.29), results in

$$\mathbf{v} \times \mathbf{B} = r\omega_r \mathbf{u}_\vartheta \times \left[\left(\frac{1}{r}\frac{\partial \dot{A}_z}{\partial \vartheta}\mathbf{u}_r - \frac{\partial \dot{A}_z}{\partial r}\mathbf{u}_\vartheta\right) \cdot \mathbf{u}_r\right] \cdot \mathbf{u}_r \tag{13.31}$$

Since $\mathbf{u}_r \cdot \mathbf{u}_r = 1$ and $\mathbf{u}_\vartheta \times \mathbf{u}_r = -\mathbf{u}_z$, Equation (13.31) is simplified as

$$\mathbf{v} \times \mathbf{B} = -\omega_r \frac{\partial \dot{A}_z}{\partial \vartheta}\mathbf{u}_z \tag{13.32}$$

Now, the z-axis component of the magnetic vector potential $\mathbf{A}$ is assumed to be a sinusoidal function of the time and of the rotor coordinate, according to Equation (13.13), where ϑ is the azimuthal coordinate of Figure 13.18. Equation (13.13) can be interpreted as the description of a magnetic field sine-distributed along the air-gap and rotating at speed ω.

Substituting Equation (13.13) into Equation (13.27) results in

$$\text{curl}\mathbf{H} + j\omega\sigma\mathbf{A} - j\omega_r\sigma\mathbf{A} = \mathbf{J}_s \tag{13.33}$$

Finally, introducing the equivalent conductivity $\sigma_{eq} = \frac{\omega - \omega_r}{\omega}\sigma$, it is possible to write

$$\mathrm{curl}\mathbf{H} + j\omega\sigma_{eq}\mathbf{A} = \mathbf{J}_s \tag{13.34}$$

Expressing Equation (13.34) by means of the magnetic vector potential **A** only, it results that

$$\frac{\partial}{\partial x}\left(\frac{1}{\mu}\frac{\partial \dot{A}_z}{\partial x}\right) + \frac{\partial}{\partial y}\left(\frac{1}{\mu}\frac{\partial \dot{A}_z}{\partial y}\right) = \dot{J}_{sz} - j(\omega\sigma)_{eq}\dot{A}_z \tag{13.35}$$

In conclusion, the magneto-dynamic complex field problem expressed by Equation (13.21) is transformed in a magnetostatic complex field problem with induced currents: it would be just as simple to assign a suitable value to the term $(\omega\sigma)_{eq}$ at each operating condition.

The term $(\omega\sigma)_{eq}$ corresponds to $(s\omega\sigma)$, where s is the rotor slip. It can be obtained changing the conductivity of the rotor bars (assuming an equivalent conductivity $\sigma_{eq} = s\sigma$) or changing the source frequency (imposing in the simulation an equivalent frequency $f_{eq} = sf$), in any case depending on the particular operating condition of the motor (distinguished by the slip s).

In the first case, the analysis frequency is equated to the stator frequency f and the conductivity $\sigma_{eq} = s\sigma$ is assigned to the rotor bars. In the second case, the analysis frequency is equated to the rotor frequency $f_{eq} = sf$, keeping constant the conductivity of the materials. In other words, with the first approach the rotor is fixed in the stator reference frame, while with the second approach the stator is considered rotating, at the speed of the rotor reference frame.

13.3.3 Nonlinearity of the Magnetic Materials

The nonlinearity of the magnetic materials greatly affects the field solution. Since the field quantities are time-variable quantities, the problem (13.37) is sometimes solved using a step-to-step method. However, this method is very time consuming.

As an alternative, it is possible to take advantage of the fact that the variation of the field quantities is sinusoidal. To do that, the B-H curve is suitably adjusted so as to consider the variation of the permeability μ in a period. Starting from the dc magnetizing characteristic $B_m(H_m)$ of the material, an equivalent curve $B_{eq}(H)$ is computed, corresponding to a magnetic field $H = H_m\sin(\omega t)$, which is sinusoidal with the time. This is achieved imposing the average value in a time period T of the magnetic energy density computed on the dc B-H curve to be equal to the magnetic energy density computed on the equivalent curve $B_{eq}(H_m)$, as reported in Figure 13.19.

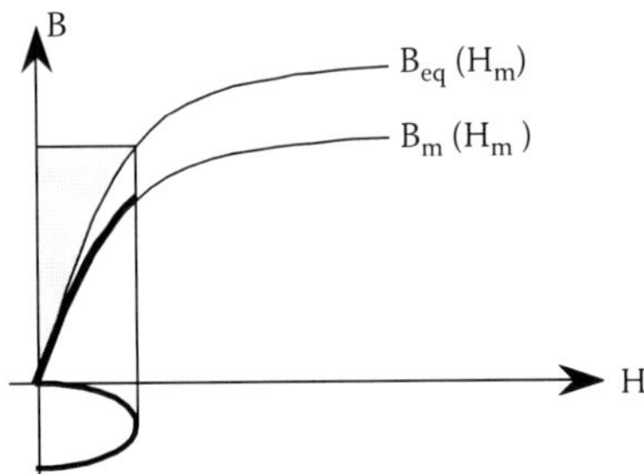

FIGURE 13.19
Equivalent B-H curve assuming a sinusoidal waveform of the magnetic field strength.

The corresponding magnetic energy densities are

$$\frac{1}{2}\int_0^{B_{eq}} H_m \, dB_{eq} = \frac{1}{T}\int_0^T \int_0^{B(t)} H \, dB \, dt \tag{13.36}$$

Applying Equation (13.36) with different values of H_m the curve $B_{eq}(H_m)$ is obtained, which is the curve $\mu_{eq}(B)$. The latter curve allows the time harmonics to be modeled.

13.3.4 Link to an External Circuit

The field problem described by Equation (13.35) has to be linked to an external circuit to study the exact operating point.

In other words, the field solution should consider at the same time the quantities of the field solution, corresponding to the central block of Figure 13.2, and the lumped-parameters of the external circuit, i.e., the elements external to the central block of Figure 13.2. It is necessary to fix a correspondence between the different region of the domain *D* and the lumped-parameters of the external circuit, which are not considered in the finite element analysis. Among them the 3D effects (end-windings and skewing) and the stator winding resistance.

It is possible to introduce directly in the matrix describing the field problem some integral conditions, which would otherwise be applicable only after the field solution is achieved.

The use of an external circuit allows the simulation of a voltage-fed induction motor to be carried out, describing the field problem by means of the differential equation

$$\frac{\partial}{\partial x}\left(\frac{1}{\mu}\frac{\partial \dot{A}_z}{\partial x}\right) + \frac{\partial}{\partial y}\left(\frac{1}{\mu}\frac{\partial \dot{A}_z}{\partial y}\right) = \sigma\frac{\partial \dot{V}}{\partial z} - j(\omega\sigma)_{eq}\dot{A}_z \tag{13.37}$$

13.3.5 Computation of the Mechanical Torque

Contrary to the previous approach in which the torque is assessed by the equivalent circuit, in this case the torque is obtained directly from the field solution. Either Maxwell's stress tensor or the virtual work principle may be used. Maxwell's stress tensor requires an integration on all the finite elements located between the fixed part (the stator) and the moving part (the rotor); the virtual work principle is based on a virtual angular displacement $d\vartheta$ of the rotor.

13.3.6 Example

The second analysis approach is applied on the same induction motor of the previous example. The simulation is carried out forcing three symmetric sinusoidal voltages with the RMS value $V_s = 385$ V at the motor terminals, which is in Figure 13.2. The various electrical lumped-parameters are added, to consider the 3D effects, and the magnetization characteristic is set up. Each simulation is carried out at a fixed rotor speed.

To have an easy comparison with the previous approach, the results are reported in Figure 13.14 to Figure 13.16, using stars. It is observed that the results obtained by means of this approach agree with the measured one. The highest error is observed on the mechanical characteristic comparing the maximum torque. The value of the maximum torque is the same with the two analysis approaches, while the speed at which this value is reached is different.

References

1. T. Trnhuvud, K. Reichert, "Accuracy problems of force and torque calculation in FE-systems," *IEEE Trans. Industry Appl.*, vol. 24, n. 1, pp. 443–446, 1988.
2. G.R. Slemon, "Modelling of induction machines for electric drives," *IEEE Trans. on Industry Applications*, vol. 25, n. 6, pp. 1126–1131, 1989.
3. E. Vassent, J.C. Sabonnadiere, "Simulation of induction machine operating using complex magnetodynamic finite elements," *IEEE Trans. on Magnetics*, vol. 25, n. 4, pp. 3064–3066, 1989.
4. R. Belmans, D. Verdyck, T.B. Johansson, W. Geysen, R.D. Findlay, "Calculation of the no-load and torque speed characteristic of induction motors using finite elements," *Int. Conf. on Electrical Machines*, Boston, 1990, pp. 2724–2729.
5. S. Williamson, M.J. Robinson, "Calculation of induction motor equivalent circuit parameters using finite elements," *IEE Proc.*, Pt. B, vol. 138, n. 5, pp. 264–276, 1991.
6. P.L. Alger, *Induction Machines. Their Behavior and Uses*, 2nd ed., Gordon and Breach, Basel, 1995.
7. S. Williamson, D.R. Gersh, "Finite element calculation of double-cage rotor equivalent circuit parameters," *IEEE Trans. on Energy Conversion*, vol. 11, n. 1, pp. 41–48, 1996.

8. N. Bianchi, S. Bolognani, "Design Procedure of a Vector-Controlled Induction Motor for Flux-Weakening Applications," in *IEEE Industrial Applications Conference Records, IAS '97*, New Orleans, Lousiana, 4–9 October 1997, vol. 1, pp. 104–111.
9. S.L. Ho and W.N. Fu, "Review and Future Application of Finite Eelment Methods in Induction Motors," in *Electrical Machine and Power System*, vol. 26, pp. 111–125, 1998.
10. N. Bianchi, S. Bolognani, G. Comelato, "Finite Element Analysis of Three-Phase Induction Motors: Comparison of Two Different Approaches," *IEEE Trans. on Energy Conversion*, vol. 14, no. 4, December 1999, pp. 1523–1528.

APPENDIX

Material Data

This appendix reports some characteristic data of conductive material, insulating materials and magnetic materials, useful for simulations.

Main Data of the More Common Conductive Materials

	Resistivity 10^{-8} Ωm	α °C^{-1}	Specific Weight kg/m^3	Specific Heat J/kg °C	Thermal Conduct. W/m °C	Expansion Coeff. 10^{-6} °C^{-1}	Breaking Load kg/mm^2
Copper	1.72	0.0043	8 900	383	389	16.5	21–24
Aluminum	2.78	0.0042	2 700	904	229	23.8	7–10
Iron	9.8	0.0065	7 800	452	66	11.7	30–45
Nickel	7.3	0.0065	8 900	444	88	13.3	45–50
Platinum	10.5	0.0039	21 400	134	71	8.9	34
Gold	2.4	0.0038	19 300	126	314	14.2	25
Silver	1.6	0.0040	10 500	234	420	19.3	30
Tungsten	5.5	0.0046	19 300	138	130	4.4	60–100
Molibden	4.8	0.0046	10 200	264	151	5.1	80
Tin	12.0	0.0044	7 300	226	65	23.0	1.6–3.8
Lead	21.0	0.0040	11 400	128	35	28.7	1.6
Zinc	5.9	0.0037	7 100	385	110	30.7	7–12
Aldrey	3.2	0.0036	2 700	920	188	23.0	30–35

Main Data of the More Common Insulating Materials

Material	Breaking Electric Field (kV/mm)	Relative Permittivity	Maximum Operating Temperature (°C)
Asbestos	2.5–3.5	—	200
Air	2–3	1	—
Distillate water	5–10	80	—
Asphalt	12–16	2.5	40
Bachelite	10–18	5–9	120
Paper	6–11	1.6–5	85
Paper bachelised	5–15	5	95
Wax paper	40–50	2.5–4	85
Presspan cardboard	8–10	2.5–4	90
Cellonite	8–10	4–5	95
Ebonyte	5–25	1.4	100
Fiber	2–10	2.1–2.3	100
Vulcanized rubber	8–20	2.5–3	60
Rubber (varnish)	10–35	1	90
Impregnated wood	8–30	3–3.5	90
Mica	60–160	5–5.5	700–800
Micalex	13–15	7–8	600
Micanite	20–40	3–3.5	90–120
Flax oil	8–18	3.5	—
Mineral oil	10–16	2–2.5	100
Apiroil	20–25	4.5–5	—
Solid paraffin	14–45	2	45–50
Porcelain	20–40	4.5–6	200
Steatite	14–16	5.6–6.5	600
Bachelised cloth	10–20	4.5–6	95
Sterlingata cloth	25–50	3.5–5.5	95
Glass fiber	45–50	4.5	150
Glass Pirex	30–150	5	400

Note: The maximum electric field varies with the material quality. In addition, it decreases when the dimensions of the sample are used in the test.

Magnetic Behaviors of Some Massive Magnetic Materials

Flux Density B (T)	Forged Steel and Melted Steel		Gray Cast Iron and Tempered Steel		Iron Steel		Nickel	
	H (A/m)	μ_r	H (A/m)	μ^r	H (A/m)	μ_r	H (A/m)	μ_r
0.1	70	1140	200	400	25	3200	215	372
0.2	90	1780	450	355	50	3200	420	380
0.3	100	2400	800	300	85	2820	1020	235
0.4	120	2660	1300	246	115	2780	2660	120
0.5	140	2860	2000	200	160	2500	8000	50
0.6	170	2820	2800	171	200	2400	24000	20
0.7	220	2500	4000	140	250	2240	88000	6.4
0.8	270	2370	5500	117	310	2060		
0.9	320	2260	8000	90	380	1900		
1.0	400	2000	11000	73	490	1630		
1.1	500	1750	15000	58	640	1380		
1.2	620	1550	20000	48	850	1130		
1.3	850	1230	—	—	1480	705		
1.4	1200	930	—	—	2800	400		
1.5	2000	600	—	—	5000	240		
1.6	3500	365	—	—	9300	138		
1.7	6000	226	—	—	17000	80		
1.8	10000	144	—	—	29000	50		
1.9	16000	95						
2.0	25000	64						

Magnetic Behaviors of Some Magnetic Laminations

Flux Density B (T)	Normal Iron Thickness 0,5 mm c_p = 3,6 W/kg		Silicon Iron Thickness 0,5 mm c_p = 2,5 W/kg		Silicon Iron Thickness 0,35 mm c_p = 1,1 W/kg		Grain-Oriented Thickness 0,35 mm c_p = 0,6 W/kg	
	H (A/m)	μ_r	H (A/m)	μ_r	H (A/m)	μ_r	H (A/m)	μ_r
0.8	230	2770	120	5305	140	4545	40	15920
0.9	330	2170	160	4475	200	3580	80	8950
1.0	470	1695	240	3315	290	2745	140	5685
1.1	630	1390	340	2575	430	2035	230	3805
1.2	800	1195	510	1870	680	1405	320	2985
1.3	1050	985	780	1325	1150	900	460	2250
1.4	1350	825	1320	845	1960	568	630	1770
1.5	1750	680	2360	505	3240	370	840	1420
1.6	3500	365	3720	340	5430	235	1110	1145
1.7	6200	220	5670	240	7160	190	1560	865
1.8	9000	160	8450	170	9000	160	2060	695

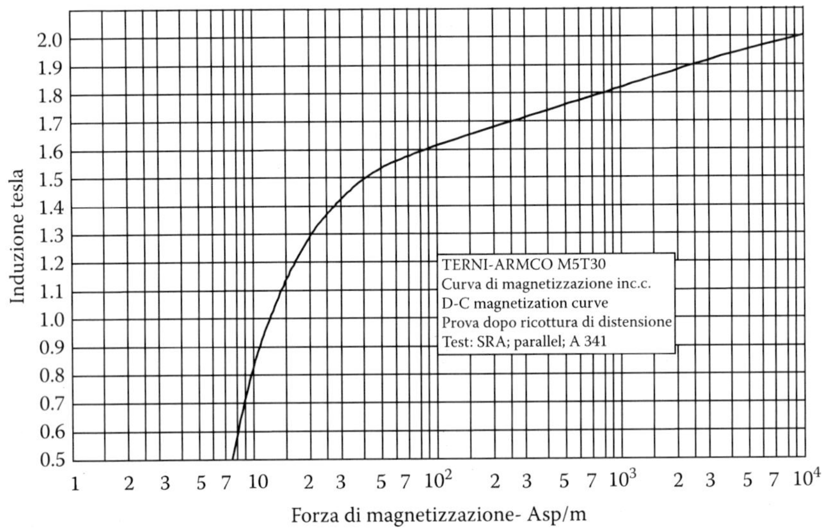

FIGURE A.1
Grain-oriented lamination.

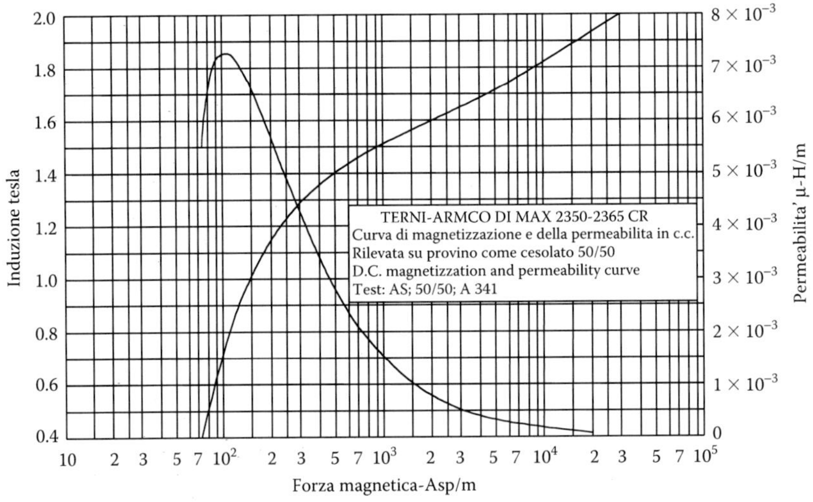

FIGURE A.2
Silicon iron laminations for rotating machines.

Main Data of the More Common Permanent Magnets

	Residual B_r (T)	Coercive H_c (kA/m)	Thermal Factor for B_r (%/°C)	Thermal Factor for H_c (%/°C)	Temperature ϑ_{max} (°C)	Specific Weight (kg/m³)
Ferrite	0.40	300	–0.2	0.4	200	4800
$AlNiCo_5$	1.30	60	–0.02	0.02	250	7300
$AlNiCo_8$	1.15	110	–0.02	0.02	250	7300
MnAlC	0.57	180	–0.12	0.25	300	5100
$SmCo_5$	0.90	700	–0.045	–0.250	250	8200
Sm_2Co_{17}	1.00	600	–0.035	–0.200	300	8300
NdFeB	1.20	900	–0.12	–0.70	150	7400

Note: The data refer to samples, obtained by sinterised magnetic material. Ferrite, SmCo, and NdFeB could be found bonded in plastic material. In this case, the values of B_r, H_c, and specific weight are practically halved. The maximum operating temperature ϑ_{max} is determined by the plastic material.

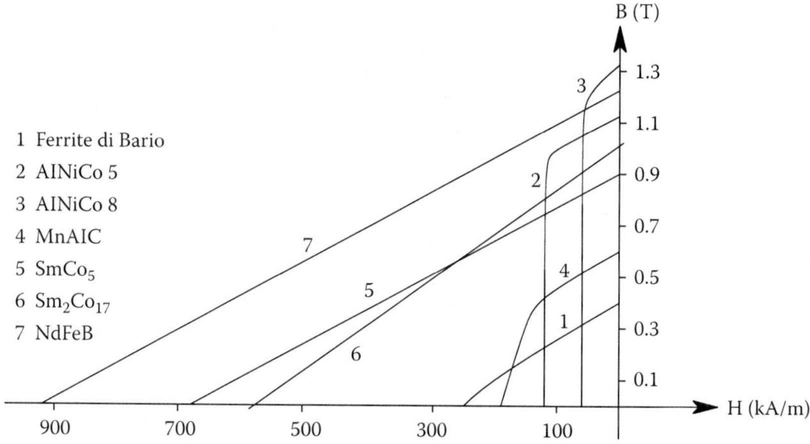

FIGURE A.3
B-H curves of some permanent magnet materials.

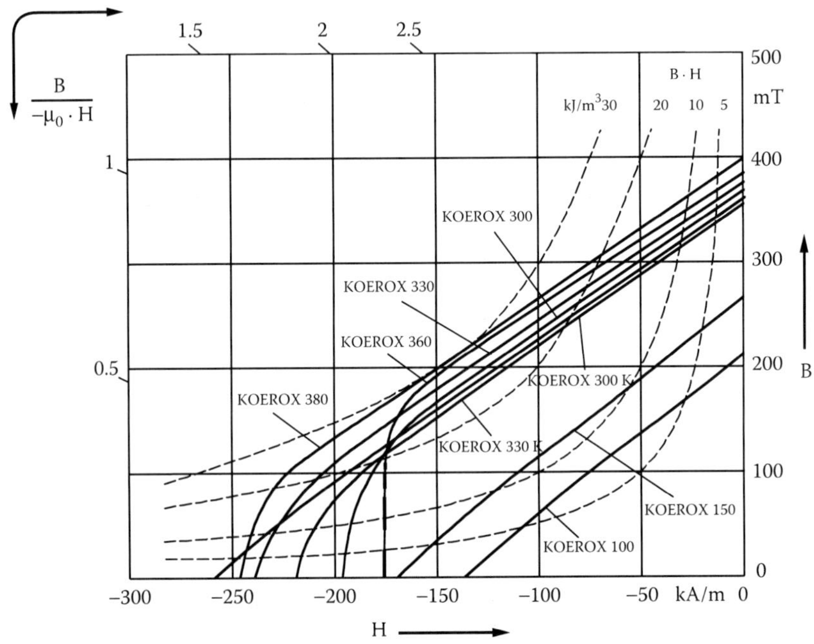

FIGURE A.4
Ferrite permanent magnet.

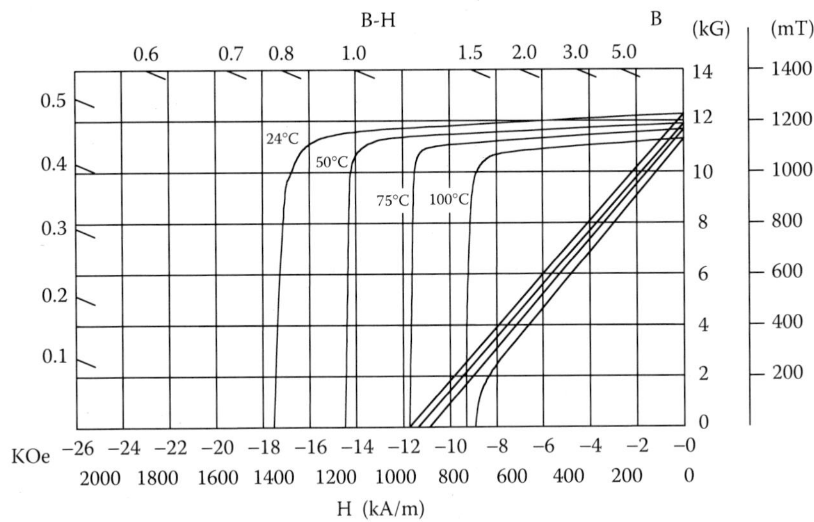

FIGURE A.5
Neodimium–iron–boron permanent magnet.

Index

G

H

I

L

M

N

P

R

S

T

V

W